Lucy's Legacy

Alison Jolly

Lucy's Legacy

SEX AND INTELLIGENCE IN
HUMAN EVOLUTION

HARVARD UNIVERSITY PRESS

Cambridge, Massachusetts & London, England   1999

Drawings by Stephen D. Nash
Title page painting by Gary Staab

Library of Congress Cataloging-in-Publication Data
Jolly, Allison.
Lucy's legacy : sex and intelligence in human evolution / Allison Jolly.
p.   cm.
Includes bibliographical references and index.
ISBN 0-674-00069-2 (alk. paper)
1. Human evolution.   2. Social evolution.   3. Women—Evolution.
4. Intelligence.   I. Title.
GN281.J6   1999
599.93'8—dc21   99-32252

FOR RICHARD B., ARTHUR, SUSAN, MARGARETTA,
AS EVER, RICHARD, AND IN MEMORY OF MY MOTHER,
ALISON MASON KINGSBURY BISHOP

I must put things in shape.
   I wonder why.
Surely there's shape enough
   In earth and sky?
I have to tidy rooms, fix flowers,
Paint, and phrase verses, number hours,
Change, meddle, analyze, reject, propound.
I have to bind things, even as
   I am bound.

                              A.M.K.B.

ACKNOWLEDGMENTS

I thank all who have inspired this book, especially my mother, who taught me about controlling and exuberant tastes in art; Evelyn Hutchinson, who taught me that everything connects; John Maynard Smith, for writing *The Major Transitions in Evolution*; and Richard Jolly, whose achievements in UNICEF and UNDP I observe and admire, though he comes to his own conclusions on globalization. John Bonner, Alison Carlson, Michael Fisher, James L. Gould, C. Stuart Hall, Donna Haraway, Arthur Jolly, Margaretta Jolly, Richard Jolly, Susan Jolly, Emily Martin, John Maynard Smith, and Frans de Waal offered helpful comments on part or all of the manuscript, for which I am grateful. Margaretta provided many poems, while R. B. Jolly tamed computer monsters. Noeleen Heyser, Executive Director of the United Nations Development Fund for Women (UNIFEM), and Maria José Martinez Patino, athlete, graciously supplied photographs and granted permission for their use. I also thank Susan Wallace Boehmer, who proves that great editors, not mere authors, create books, and Stephen Nash, whose artistic skill is exceeded only by his patience. Finally, at Princeton University, I am grateful to my colleagues in Ecology and Evolutionary Biology for support, colleagues in Women's Studies for putting up with me, Reality Check for provocative discussions, the superb Firestone and Biology libraries for research assistance, and all the students whose term papers have been an education for me. I end with appreciation for the copper beech tree, ca. 1900 to April 13, 1999.

CONTENTS

Contents

Prologue:
Beyond the Copper Beech Tree

Female ringtailed lemurs wholly dominate their males. I have seen females in Madagascar stride up to a male, cuff him over the nose, and take the tamarind pod he was eating right out of his hand. He retreats, squeaking, even if he is alpha among troop males. Lemur amazons guard territory. When two troops meet, females face off like American football teams, only with babies clinging to their fur. They escalate to lunges and feints and sometimes canine slashes: lemurs are knife-fighters. The males hang out well to the side, perfuming their beautiful black-and-white tails with their wrist glands, but both males and females scentmark branches to claim the site for themselves and their gang.

On the campus where I work, a copper beech tree stands at the frontier between science and humanities, nature and culture, the ecology building and the language lab. It is a noble dome, like a five-story cauliflower of beaten bronze. If Arts and Sciences professors were ringtailed lemurs, that tree would be smeared with genital scent marks.

I am about to trample right past the copper beech tree, a more daring foray than any travel around Madagascar. Biologists think they have a great deal to tell people. I am a biologist; I think we are right. What emerges is not just a heap of sexy or scary details. Biology offers an increasingly coherent view of human nature and humanity's place in nature. We claim to understand something about behavior: why she-lemurs scentmark, why the chimp wears a sexual swelling with labia like bright pink mangoes, perhaps even why humans evolved a sense of purpose. Of course humans differ from other apes, but we are subject to scrutiny like any odd species.

The theme of the tale is how we came together. You and I are not drifting chemicals in the primeval ooze. Chemicals organized themselves into primitive bacteria; bacteria joined forces to produce nucleated cells; cells cloned sister cells to build bodies. A few of the bodies evolved as social primates. One primate lineage, Lucy's kin, the African australopithecines, strode off into the savanna. Standing upright squeezed the australopithecines' birth canals awkwardly narrower from front to back. When Lucy's descendants grew larger brains, their children were born at an ever-earlier stage of development to traverse that tight passage into the world. We who inherit Lucy's legacy now have babies so helpless that they cannot even form their brains unless we tend them in a bath of language, culture, and love. Human interdependence grew with our species' history and now gallops forward to engulf the biosphere.

Global interdependence is not just hot money sloshing between nations. Nerve-nets of information potentially link all humanity—the first time in almost four billion years of life that this has ever happened to any species. It could be the fifth transition in evolution, on the same scale as the birth of bacteria, cells, bodies, and societies. This is a radical change even within the scope of life on earth, but it is a change resembling those that have gone before. If it is as big as I claim, it deserves to be seen in the context of our full history, beginning from those lumps in the ooze that turned, eventually, into us.[1]

I tell the story with my own slant, as a woman, teacher, lemur-watcher. I tend to take my examples from women's lives and women's biology. Female orgasm, menstruation, childbearing, and menopause are facts of life in the body I am familiar with—not to say that it is better or worse than bodies which ejaculate or play pro football. My heroine and hero are not usually in opposition. Much of the time they are in love. It still surprises a few people that their interests do not always coincide.

My other fascination is with intelligence. Not genius, but the intelligence of the lemur, the chimpanzee, the newborn baby. If human minds set us apart from the animals, how on earth did animals evolve a human mind? And what are the consequences for evolution when clever-monkey brains tinker with the biosphere? Our brains have now become the major force in evolution itself. Human purposes shape, or misshape, the world.

I do reflect a traditional feminine viewpoint in one major way. What interests me both personally and intellectually is cooperative organization, not competition. This is a fundamental dilemma for anyone trained in Darwinian evolution, with its emphasis on rampant individualism. Sociobiology offers the biologist the way out—a way to see how altruism evolved. For shock value, the principles of sociobiology can be reduced to the self-interest of selfish genes. But turn the phrase around. That means selfish genes, interacting with their environment, led to love between kin, trust between friends, the intricacies of the mind, and the emergent organizations of society.[2]

Ten years ago I started teaching about human evolution. I thought it was a one-off effort to tell bemused students in women's studies what any educated person should know about Darwin. A fair number of students hated the idea, to start with. They distrusted a biological straitjacket that might shackle them into traditional gender roles. We had good arguments. Ten years later, I have the opposite problem. A frightening number of people are ready to believe in genes for everything—not just gender but breast cancer, pesticide resistance,

3

homosexuality, extroversion. Is biology omnipotent, and its mysteries solvable? Dolly the cloned sheep baas in her pen; genetically engineered tomatoes redden in the fields; women long past menopause give birth. This book is not a polemic about the new power of biology, but I do attempt to sum up how we reached this point and to guess where we are going next.

In Part I, "Evolution," I build the idea that larger wholes evolved from the self-interest of their component parts. Darwinian evolution is all about competition, but the major transitions in evolution have arisen through integration—indeed, cooperation. Nature is amoral, rewarding short-term advantage. Out of this selfishness flowered our beautiful bodies and the minuet of sexual courtship and choice. Human intelligence swaps ideas for many of the same reasons that sexual bodies swap genes. We balk at imagining nature's indifference precisely because humans did in fact evolve intelligence, altruism, and a sense of individual purpose.

In Part II, "Wild Societies," we explore our primate cousins, especially our nearest relatives, the chimpanzees, bonobos (sometimes called pygmy chimps), and gorillas. Human nature resonates with the male bonding and warfare of chimpanzees and the exuberant sex play of bonobos. We humans have far more monogamous pairing off, indeed male ownership of females, than either chimp: like gorillas, we fall in love. It was in apish societies that we took our first steps toward human intelligence. The chief spur to our intelligence was outsmarting—and cooperating with—our apish colleagues.

Part III, "Developing a Mind," turns to people as individuals, with more explicit arguments on reduction and emergence. We are each a cathedral, not a pile of stones. Embryonic and adult hormones contribute to adult gender, inherited genes to adult behavior. Art, language, and our sense of self emerge from the whorls and vortexes of brains in league with their environment. The unfolding fetus and gurgling baby give clues to the construction of intelligence itself.

Part IV is "The Age of Humanity"—the past five million years and the near future. I touch gingerly on the transition to culture, though that term has iridescent, shimmering meanings as multicolored as an oil slick. I speculate on minds, mates, and families from the point four or five million years ago when Lucy's forebears walked away from other apes into the savanna, up until the birth of art: beaded necklaces and statues of women carved by hands dead now for twenty or thirty thousand years.

Modern times start with those artists' attempts to mentally shape the environment. Neolithic agriculture then inaugurated physical control, leading on to the growth of cities. The industrial revolution was a trivial precursor to the biological revolution that is restructuring our crops, our environment, and ourselves. We are all in on the act, from the doctor directing her sperm bank in Israel to the farmer who hacks down Madagascar's rainforest with his coupe-coupe. The extinction of wild species, the engineering of tame ones, and fiddling about with our own sweet selves will bring about changes far more drastic and intimate than anything due to iron or steel.

The closing chapter of this book can be dismissed as science fiction, if you choose. It is about the emergence of global organization—perhaps even a global organism—created according to the same evolutionary principles of competition and cooperation that spawned societies, bodies, cells, and life itself. Such an organism would be a new biological synthesis, as cells and bodies once were, directed in part through conscious human purposes.

The big ideas are not mine. Most are standard fare in biology as taught to college freshmen. Of course, though I claim this book represents a wide consensus, I do not expect there is a single other student of biology who will agree with it all, except perhaps an undergraduate in the moment of expedient gullibility before writing an exam. And some readers will have none of it. Nowadays many people react against science itself, from blank-eyed New Age mystics to high school students avoiding science courses to academic human-ists scrambling for some high ground from which to consider that scientists, too, are human. Critics on the "social construction" side of the science wars point to the hubris of scientists in claiming to actu-ally know anything.

Scientists are indeed fallible. We depend on our social context like a litter of puppies. We can't survive without the mother's milk of praise, preferably accompanied by the regurgitated meat of funding. To the critics, the fact that science confers power over chosen as-pects of nature and human life should not justify a claim to absolute truth. If anything it makes science and scientists more dangerous, the way people used to think of witches. But if the story of human evolution is worth telling, it is precisely because a great many biolo-gists would accept its general gist, even if not the details. And be-cause a great many nonbiologists haven't been told at all. No aca-demic territorial scent marks will keep us biologists from proclaiming our story beyond the copper beech tree.

Is it a tale of triumph? Having succeeded in building her body out of primeval slime, can our heroine embrace lovers, family, and friends, secure in her individuality? Will she reach out across the ether to network with her sisters in Beijing, and finally spread the light of human cooperation and achievement through the world? Or is this a tale of subjugation, of the once-free human being caught in the mesh of world economics, just as the once-free bacterium was half-eaten by another, to become a domestic slave within the preda-tor's cell? That I can't answer. Tell the story either way you please. What I can say is that it is all one tale, and worth telling.

I
Evolution

Figure 1. Two ringtailed lemur troops, each numbering about twenty males, females, and young, met daily in October 1998 by a guardian's house at Berenty Reserve, Madagascar. The dominant female of the near troop and her adult daughter glare and scentmark a sapling; females of the farther troop threaten back, while a male wristmarks. Male wrist spurs have left gouges on both troops' scent posts, which are only three meters apart. (After A. Jolly, photos)

1
The Evolution of Purpose

We evolved. We have only to look at the pouting face of a young chimpanzee to laugh at its reflection of ourselves. We know that more than 98 percent of our genes are shared with the chimpanzee, but we feel the kinship directly when the furry baby puts up its arms to be held.

Still, we are very reluctant to credit chimps with that most human quality, a sense of purpose. Surely no ape ever asked why it is here, what it evolved *for?* Somewhere in the progress from ape to human, the first philosopher appeared. Every human society we know of has asked, and offered some answer to, the question "What for?" Chimps do not puzzle over that question, but, strangely, they help us confront an even more abstruse one: *Why* do we ask what for? Why do we feel that the world and our own existence must have a purpose?

When we were evolving human minds, the environment posed many challenges. We had to find, recognize, and remember foods that were growing erratically about the savanna. We invented tools to smash nuts, dig for roots, and hurl at prey. We dodged saber-tooth cats and competed for food with troops of gigantic baboons. However, similar challenges confronted other primates. All animals have predators, competitors, and prey. What was different for us?

The difference was that we lived in social groups of creatures just about as smart as ourselves.[1] Our minds were honed in a social nexus generation by generation. We outsmarted rivals of our own sex by forming alliances with our friends against other networks of allies. Both males and females chose mates from a range of suitors. One way to be chosen was to be smart—smart at providing food, or distracting rivals, or wooing with primeval forms of music and rhetoric. Endearments—and insults—may have been among the earliest parts of speech. At some much disputed period we developed rights and commitments to particular mates, and affection and obligations toward both of our parents. Juggling relationships is the stuff of soap operas, but it made us smarter. Captive chimpanzees and bonobos love watching soap operas.[2]

The pressure of others' intelligence drove our minds in a forward spiral. We gained advantages for ourselves and our children whenever we could predict the actions of others, both friends and foes. The efficient way to predict a live creature's actions is not to do abstract, billiard-ball deductions about where it came from or which way it is headed but to ask what the chimp or dog or baby *wants*. In the long (Darwinian) run, we all want to survive and reproduce. All animals have developed a baroque collection of subroutines en route to this end. I want to eat a carrot or a chocolate chip cookie that is tempting me now, not because it builds vitamin-rich tissue or energy to bring up the kids. We enjoy sex for its own sake, now, not to make a baby nine months down the line.

A primate does not need to go beyond these immediate subroutines to sense that a rival "wants" to dig up the root I spotted, she "wants" to take the last cookie on the tray, she "wants" to cuddle up to the male I had my eye on. If a more subtle ape sees these things a little more clearly than the next ape, she may be a little more likely to turn the tables and gain her own ends. Then, somewhere along the trail of the generations, she becomes able to detach herself enough to think of her own behavior in such terms. At this point we drop the scare quotes; these are real goals, not just felt desires. Eventually the goals become Machiavellianly explicit: "He'll never see what I'm up to behind the bushes," or "I'll encourage her to grab that gooey cookie so he will see that I am pure but generous and she is a selfish slob," or "OK, I'll bite the cookie with gusto, and eye him over the top so he'll imagine enjoying other pleasures."

The chimp and its society tell us that our minds are shaped by the world of our ancestors. In that world, other ape-people were filled with purposes. From the purposefulness of our fellow humans we leap to the assumption that everything has a purpose. Richard Dawkins says, "It is almost as if the human mind has been designed to misunderstand Darwin."[3] Too right. We have evolved a social primate's mind that looks for purpose in everything.

The thought that nature itself might have no purpose was anguish in the nineteenth century, as it is to many today. The extinction of species raised the sacrilegious prospect of humanity itself as one more transient life form. Would we too be cast off by nature's indifference? Alfred, Lord Tennyson distilled that anguish in his elegy *In Memoriam*, which became so famous that even Queen Victoria kept it by her bedside. Man, last of nature's creations, so full of hope and splendid purpose,

> Who trusted God was love indeed
> And love Creation's final law—
> Tho' Nature, red in tooth and claw
> With ravine, shriek'd against his creed—

even Man might become shards of fossil bone, finally left to

> Be blown about the desert dust,
> Or seal'd within the iron hills.[4]

Darwin's Decision

Charles Darwin hesitated for twenty years before publishing his theory of evolution through natural selection in 1859. He was not idle, nor indifferent, nor unconscious. His early notebooks spelled out the whole theory in 1838. He occupied eight of the intervening years with prodigious labor on the anatomy of barnacles. Barnacles!—with the theory that would change humanity's view of itself already written, with instructions to publish if he should die. He hid his blasphemous idea in part because of the anguish it would cause to believers, including his own wife, if he could successfully explain the evolution of living forms without recourse to divine purpose. He also hid through fear of revenge on his social position and scientific respect.[5]

Then in 1858 Alfred Russel Wallace wrote a short letter from Indonesia outlining the principle of natural selection in a few pages. It finally catapulted Darwin into publishing a 500-page "preliminary sketch" of evidence for his theory.[6]

On the Origin of Species by Means of Natural Selection, or The Preservation of Favoured Races in the Struggle for Life was far from the first book to propose the evolution of one form from another. Charles Darwin's grandfather, Erasmus Darwin, had done so, as had the French naturalist Lamarck and the Scot Robert Chambers. Evolution was in the air, along with the innovations and extinctions it implied. Tennyson's poem was published nine years before the *Origin of Species*.[7] What Darwin did was give evolution a *mechanism*, a means of arriving at the diversity of living and fossil creatures without the intervention of conscious purpose. The philosopher Daniel Dennett calls this the difference between skyhooks and cranes. Evolution was no threat if a benevolent deity leaned over from on high with a skyhook to haul life upward for his own divine ends. When Darwin proposed instead a kind of crane, a physical means of building life's complexity from the ground below, it was horrifying.[8]

12

Darwin was a proper Victorian gentleman. He aspired to the respect of the illustrious scientists who ran the Geological Society of London, men who damned Lamarck for his "gross and filthy" views on evolution.[9] Biology in Darwin's time was almost exclusively natural history—gentleman's beetle collecting, the hobby that sent Darwin cycling round the countryside when he was supposed to be studying for the ministry at Cambridge University. Gentlemen belonged to the established Church of England. Their God personally designed every single species of beetle. As a theology student, Darwin read and reread Paley's *Natural Theology, or Evidences of the Existence and Attributes of the Deity, Collected from the Appearances of Nature.* Paley said that if, by chance, he kicked a stone on his path, he might assume it had lain there forever. If, however, he found a watch, he would know at once that someone had made it, and that its parts were put together for a purpose. Plants and animals, in turn, are far more complex and more beautiful than any watch. Therefore they must have arisen by divine design, not mere chance. Paley's God, the God of the scientific establishment, shared a major strand of Darwin's own character—delight in every fact and facet of living nature.[10]

Darwin was not born a grand old man with a beard. During the voyage of the *Beagle* he rejoiced with a young man's gusto in riding off through South America toward the Andes in pursuit of discoveries (though he never got over being seasick on shipboard). When he came home to London he was a bachelor-about-town, seeing much of his brother Erasmus and Erasmus's friends, especially the literary lioness Harriet Martineau. Martineau overcame the handicaps of being female, single, ugly, and deaf to write immensely popular political fiction. One of her soap-opera morality tales explained Thomas Malthus's principle that human population grows at compound interest unless checked by famine, war, disease, or moral restraint. Meanwhile, mobs rioted in the streets against the Malthus-inspired New

Figure 2. Charles Darwin, in his checkered vest, converses with Harriet Martineau and his brother, Erasmus, in 1838, when Darwin was formulating the theory of natural selection. (Charles Darwin after watercolor by George Richmond, 1839; Martineau after *Autobiography,* vol. 1, frontispiece, 1833; Erasmus Darwin after chalk portrait by George Richmond, 1850; 1840 "naturalistic style" teapot courtesy of J. Applegate)

Poor Law of 1834, which sent the destitute to workhouses made as harsh as possible to discourage paupers from seeking aid. Parliament justified the new Poor Law by quoting Malthus: the superfluous poor should be checked from reproducing.[11]

The idea of the theory of natural selection came to Darwin when he was reading Malthus "for amusement." He doubtless discussed Malthus's ideas with Martineau and others of their London circle. Their background, like his own family's, was mercantile, not aristocratic. These entrepreneurs dealt in the laws of gravity, steam power, and compound banking interest as well as compound population growth. Darwin became an increasingly explicit atheist. He turned from the God who sculpted each beetle carapace to the abstract grandeur of universal law.[12]

Evolution itself called to the hopes of the rioting mob. The idea that species could arise one from another appealed to a working-class audience—men who went to public lectures and night schools in the hope of self-improvement. If species themselves were mutable, so might a working-class Englishman hope to rise in station. If the history of life was one long march of progress, even laborers, or society itself, could expect to advance toward a better life. This doctrine was revolting to the wealthy classes, with their vested interest in the status quo—revolting, and revolutionary. Darwin's friends repeatedly warned him that espousing evolutionary progress would ally him with riffraff that the gentleman scientists despised.[13]

These three strands came together in the final theory: awe and delight at nature's diversity, explanation how that diversity of nature arose through a grand and simple law, and understanding that the law of natural selection at last explained evolution of new species and indeed the progress of life itself.

Natural selection is simple. Richard Dawkins remarks it is so simple that if we did not have a deep-seated distrust of mechanism we could hardly have waited millennia to come up with the idea. Indeed, Aristotle almost did, proposing that perhaps all possible forms had once existed, including monsters like "man-faced oxen." Then only those that were best fitted for life survived. Aristotle then veered away, seeing no way to produce the variety of such monsters in the first place. He concluded that all parts of living organisms arose both through mechanical "prior cause" *and* for a purpose "as leaves exist in order to shelter the fruit."[14]

After twenty years of labor, Darwin could summarize the "prior cause," the actual mechanism of natural selection, in one paragraph. He kept the last paragraph of his masterpiece unchanged through all successive editions of the *Origin*:

It is interesting to contemplate an entangled bank, clothed with many plants of many kinds, with birds singing on the bushes, with various insects flitting about, and with worms crawling through the damp earth, and to reflect that these elaborately constructed forms, so different from each other, and dependent on each other in so complex a manner, have all been produced by laws acting around us. These laws, taken in the largest sense, being Growth with Reproduction; Inheritance, which is almost implied by reproduction; Variability from the direct and indirect action of the external conditions of life, and from use and disuse; a Ratio of Increase so high as to lead to a Struggle for Life, and as a consequence to Natural Selection, entailing Divergence of Character and the Extinction of less-improved forms. Thus, from the war of nature, from famine and death, the most exalted object which we are capable of conceiving, namely, the production of the higher animals, directly follows. There is grandeur in this view of life, with its several powers, having been originally breathed into a few forms or into one; and that, whilst this planet has gone cycling on according to the fixed law of gravity, from so simple a beginning endless forms most beautiful and most wonderful have been, and are being, evolved.[15]

Darwin's "Laws Acting Around Us"

Self-maintenance, growth, and reproduction. The most fundamental property of life is active self-maintenance. Living things work to stay alive. They take in nourishment, repair themselves, and grow. But as soon as some accident, predator, or disease destroys a living thing, it is gone. If it has left no descendants, its genetic information is gone. If it has influenced no others, its ideas are gone. Working to stay alive can work only for a finite time. Leaving extra copies of oneself guards against such total annihilation. The second fundamental property of life, then, is reproduction. Life began, by definition, when the first chemical molecules that could make copies of themselves condensed from the primordial ooze.[16]

This fact that living things have the potential to grow and reproduce is what is wrong with the Gaia hypothesis that Earth itself is a living organism. Granted, Earth in some ways seems extraordinarily like a self-regulating system, like a living thing. James Lovelock, who proposed the idea in 1979, showed that marine algae produce a gas called dimethyl sulfate, which promotes cloud formation, partially regulating Earth's temperature. Most Darwinians thought this could hardly be selected for in evolution: what is the profit in it for the algal genes? However, if algae blooms are massive clones, then promoting clouds would allow their own genes to disperse—a reproductive advantage for the makers of the weather. Perhaps the self-interest of algae stabilizes temperature and climate in such a way that the rest of us can survive.[17]

Yet, so far as we know there is only one Earth, which neither grows nor reproduces. Even though the parts look suspiciously well coordinated, we still know of no mechanism that produces complex natural organization on the scale of the biosphere itself. It could be that the coordination is a case of observations in search of a theory, much like the data of continental drift, which were not believed until people recognized the mechanism of sea-floor spreading and plate tectonics. Or it could be that the notion of Earth as organism just is not viable.[18] Meantime, the simplest amoeba or bacterium has the first properties of life: self-maintenance, growth, and reproduction.

Inheritance. Darwin talked about reproduction not of genes but of parents' form and behavior. He did not know about genes, chromosomes, DNA. However, it was obvious to Darwin that natural selection could not work unless offspring inherited tendencies to develop the traits that made their parents successful.

Darwin worried about the problem of blending inheritance. Supposing, he wrote, that a white man were shipwrecked on an island with a black population. His offspring would be mixed, their offspring more so. If the genetic information were blended and dispersed, like a drop of water added to a bottle of ink, it is impossible that in some future generation a child would reassemble all the "white" genes and look like her shipwrecked ancestor. Suppose, however, that genetic information is not blended but rather the genes are shuffled and dealt like cards in a game of bridge. It is statistically no more unlikely for all the hearts to fall together than any other of the possible sequences of thirteen cards. With enough shuffling, children in a limited population could partially or wholly recreate some ancestral genetic pattern, as, at times, the offspring of a white South African couple show black background long since successfully denied in their forebears' past. This is called particulate inheritance. In the first edition of the *Origin* Darwin rejected blending inheritance, but eventually even he lost faith in particulate inheritance and watered down later editions. Not until the rediscovery of Gregor Mendel's work on genetics was blending inheritance finally put to rest.[19]

Mendel was an Austrian monk, a high school teacher, and a biologist who published just two papers in 1866 on the outcome of crossing peas. He annotated his copies of Darwin's works and saw how his own experiment fit with natural selection. He focused on a few characteristics: tall plants versus short plants, yellow versus green peas, wrinkled peas versus smooth—seven traits in all. He found that first-generation offspring all resembled one parent, the "dominant" one for each characteristic. In the second generation, they came out in the ratio of three to one, where the odd one inherited two recessive factors for the trait. Of the three with dominant characteristics, one had two dominant factors and the other two had a dominant and a recessive, but the dominant factor determined the look of the adult plant. The important point, to Mendel, was that the traits segregated independently from one another. He postulated that whatever factors produced the traits were sorting digitally and independently, like shuffled cards, not blending like a liquid.[20]

Very few characteristics in humans are as genetically simple as Mendel's pea plants. Our major blood types, A, B, AB, O, are based on single gene differences, but not much else is. Sex begins with a single gene difference, but sex is one of the very few traits where we have distinct castes: most people are either male or female. Most of the interesting traits, such as intelligence, involve an astronomically complex interaction of many genes with the environment. The basic principle, though, is that genes are discrete, particulate. That means that the information to rebuild parents' characteristics can pass intact from generation to generation.

Variability. For natural selection to work, individuals must differ within a population. If the line of descent consisted of perfect clones, no individual would be genetically better adapted to its environment than any other. Some would be lucky enough to survive while others died, but this produces no change in the genetic stock. Darwin pointed out that in fact puppies differ even within one litter; fancy show pigeons differ within an inbred strain. Evolution is not about the perpetuation of some static ideal type but about the differences and innovations that give some few individuals an edge in the struggle for life.

Darwin equivocated between what Ernst Mayr calls "hard" and "soft" inheritance.[21] Hard inheritance is what we now know to happen: Mendel's mutations occur randomly within the genome, and changes in the genome are transmitted to offspring. Environmental conditions that reach the genes are random mutagens, like ultraviolet light, mustard gas, atomic radiation, which hit the DNA like a scatter of shotgun pellets. Soft inheritance instead would allow external conditions to shape *particular* heritable factors, both directly, indirectly through increasing variability, and above all, through use and disuse of parts of the body—the so-called Lamarckian heresy. Lamarck's notion was that the giraffe, stretching its neck to reach tall trees, somehow stretched the neck of its unborn young. Hard inheritance says, instead, that taller giraffes swiped off leaves from tall trees out of reach of their shorter kin, and so survived to better sire and bear giraffe calves who inherited the parents' gangly genes.

It remained for August Weismann, at the end of the nineteenth century, to sort the story out and to place hard inheritance on a firm theoretical and empirical basis. He pointed out that many adaptations could not have arisen from habit or use during animals' lives—they are too complex, or their benefit happens after the animal is dead, or they produce highly adapted nonreproductives, such as the sterile castes of insects. Among a wealth of other observations and experiments, Weismann is remembered today for cutting off the tails of generations of mice. The distant offspring grew tails as long as ever, in spite of their parents' "disuse" of any tail. Only when Weismann interbred the naturally (genetically) short-tailed mice of his stock, generation after generation, did the descendants eventually inherit genes for puny tails. We now imagine only a mad farmer's wife would carry out this procedure, but Weismann recalled how he faced critics who believed that puppy dogs did indeed inherit their mothers' docked tails, and that "even students' fencing scars were said to have been occasionally transmitted to their sons, though happily not their daughters."[22]

Josef Stalin tested out Weismann's findings on a grand scale. Stalin had power to experiment on a whole nation's bread and borscht and vodka. The biologist Lysenko launched an attack on a flourishing Russian school of genetics in 1936, achieved full triumph in 1948, and did not fall from power until 1965. Lysenko argued that crops could acquire the habit of growing well for Communism. They should be fertilized and tilled, just as peasants should receive political education. It was immoral, according to Lysenko and Stalin, to save only elite seeds to sow. The crops' genetic stocks declined through lack of selection, playing a part in the collapse of Soviet agriculture, while Soviet evolutionary biology lost two decades of growth, as "hard" geneticists were fired, imprisoned, or killed.[23]

A ratio of increase so high as to lead to a struggle for life. This is Malthus's principle that every organism is capable of reproducing by exponential growth. This means every species can produce more off-spring than have room to live in a finite habitat.[24] Malthus's principle, acting on inherited variability, is the key to natural selection. If many more offspring are born than can possibly survive, those less well adapted to their environment will automatically be culled. In the same way, a purposeful human breeder produces new strains by saving only a few of each generation and culls those that do not measure up to the breeder's ideal. Malthus's principle is not so much about life as about the inevitability of death—the myriad shadowy might-have-beens.

Natural selection. The sum of all these elements gives the mechanism of natural selection, "entailing divergence of character and the extinction of less-improved forms." The actual divergence of character comes from tracking the environment. As a few individuals of a litter, or a species, are genetically better at coping with the environment, they drive their competitors to extinction. In this way the species as a whole is "improved" in relation to its environment, but through a wholly mechanical means, and *through the self-interest of each organism in its own survival.* There is no judgment about what constitutes "improvement." A parasitic barnacle which jettisons all adult organs except the gonads and feeds by sucking nutrients from a crab, its host, is improved if it survives to reproduce.

The War of Nature

Even in this triumphant final paragraph Darwin talks of the war of nature, of famine and death. Competition for him includes direct competition, as of two plants sucking water and nutrients from each other's roots, racing up toward the sun: the winner shades and shrivels the weaker. He also speaks of competition against the onslaught of the environment: to be the plant that can best stand the desiccation of the desert edge, or the frosts of winter. Competition, unthinking cruelty, the driving of others to death or extinction are the basis of natural selection of the few which survive.

Yes, it does sound a lot like nineteenth-century capitalism. Competitive individuality is the basis of natural selection, and of that brand of economics which believes in a free market, an idealization of individuality rarely or never found among actual human transactions. Humans recognize other individuals and cooperate, or at least cut deals, with those they know. It turns out that the major advances in evolution also involve cooperation.[25]

What do we mean by an advance? A very simple yardstick is size. Single cells are bigger than bacteria, multicellular bodies are bigger than single cells; societies are bigger than single bodies. Size usually implies complexity, as the number of different interacting parts grows from level to level. Each level involves a community of the simpler individuals of the level below.[26]

Cells originated as a community of originally free-living bacteria. Bodies are communities of mutually supportive cells. Societies, obviously, are a somewhat coordinated community of individuals. The great transitions in evolution have produced ever larger matrices of cooperation. In this sense there has been a genuine directionality in evolution, toward more and more complex interconnection. If we wish to think that complexity is indeed an improvement, we may call it progress.

How does such cooperation arise from the fierce competition of natural selection? Cooperation arises between biological entities much as it arises between people. It is advantageous to both partners, or else one imposes itself on another and forces the second to obey.

Natural selection is built on individual advantage. The agony or futility of the crab being sucked of its vital juices by the parasitic barnacle, the human being in spasmodic dehydration by cholera bacilli, or the rabbit skewered by an eagle's talons are unjudgmentally produced by the blind forces selecting for individual survival of barnacle, bacillus, eagle. But love, altruism, self-sacrifice are also real and have also evolved and survived, not just in ourselves. The lesson of Darwin's war of nature is not that unfettered selfishness is somehow good but that by understanding the war of nature we can also understand kindness and love in the natural world.

From So Simple a Beginning

Darwin surmised that present living things stem not from a few forms but from one. All of us construct our bodies using the same genetic code. The code has long been considered highly arbitrary, what Francis Crick called a "frozen accident." Recently, biologists are figuring out some of the chemical constraints that may have shaped the code's early evolution, and also finding that the code is extraordinarily efficient, more efficent than the artificial permutations that we can now jam into it, or simulate by computers. This means that the code itself evolved over time. Perhaps the ancestor who had the code we all use now was the fittest survivor among many early life forms, or perhaps those simple life forms exchanged components among themselves, and jointly converged on the pattern of our universal ancestor, the progenitor of you, me, the earthworm under your foot, the pigeon over your head, and the cold virus inside you.[27]

Some still consider natural selection only a hypothesis to be tested, although even the Pope now concedes that it is a "theory of vast explanatory power." Darwin claimed even more: that natural selection would prove to be a universal law of nature, an organizing principle like Isaac Newton's law of gravity. "There is grandeur in this view of life," he wrote, in awe of nature as majestically simple, whose universal laws explain the swarming embodiments of reality. In the end, Darwin was buried in Westminster Abbey, near the monument to Newton, England's hero of science.

Paley's test of the watch on the moor is trivial compared to the complexity of the merest beetle, even the merest cell. The DNA in each cell of a person contains as much information as an encyclopedia. And that is only the blueprint—the code to produce the interacting proteins that build the scaffolding that builds the structure. The final structure of a living thing is as much more complicated than its DNA as a cathedral is more complicated than the architect's blueprint. How can such things arise without a planned design?

The key is not just natural selection but *cumulative* selection.[28] Each step of survival does not start from a blank and random slate but from a previous parental blueprint already evolved over millions of generations. The embryo does not construct itself from scratch, like a tornado in a junkyard somehow producing a Boeing 747, to use Richard Dawkins's analogy. Instead, each change is a minor tweaking of a previous body plan. A few changes are larger—repeating a body segment, for instance—but it is extraordinarily rare that individuals carrying large single steps of modification survive. Dennett's "crane" building up from below is building itself, on itself. The mechanism of reproduction, inheritance, variation, struggle for life is enough, when building on what went before.[29]

Darwin's own test case was the evolution of the eye in all its precision. Generations of scientists have now traced the evolution of eyes. Each step was selected in turn, from the first light-sensitive pigment spot that might have registered the shadow of a swimming predator, up to the single-lens reflex of ourselves or, independently, the octopus, or the very different array of compound and simple eyes that a jumping spider uses to target its prey. It seems that eyes have evolved independently at least thirty times over. But even eyes are jettisoned when they are no longer useful. The mole rat's remnant eye is buried beneath skin and fur. Its brain circuits are rerouted to deal with the tingling of its vibrissae and to decode the sound of messages sent by other mole rats thumping their skulls against the roofs of underground tunnels.[30]

The same principle explains a great deal of the fine tuning in nature: enough of an enzyme, but not too much, fast enough legs to outrun the predator most of the time, but not legs like the cheetah, unless you are a cheetah. And cumulative selection does fine in explaining the presence, and even the absence, of the amazing eye.

Forms Most Beautiful and Most Wonderful

The entangled bank of "most beautiful and most wonderful" forms does not become simple and drab just because it has evolved through the fixed law of natural selection rather than divine intervention. Darwin certainly did not think so. Once he had found the mechanism of evolution, he did not feel there was no more to see or to say. He led a life of perpetual fascination with natural forms and behaviors: earthworms, barnacles, orchids, pigeons, peoples. His metaphors inspired the great literary minds of his time, and Darwin's image of the entangled bank remains with us today.[31]

The idea that science reduces the wonder of nature to sterile abstraction is the fallacy of reductionism. Nature is not "nothing but" a DNA blueprint, or atoms, or quarks. A Gothic cathedral is a pile of stones, pressing down upon one another according to the fixed law of gravity, with the addition of a design inspired by the architect. A Madagascar landscape of pinnacle karst—knife-tipped needles of limestone soaring 150 feet into the air—is as breathtaking as a Gothic cathedral. It too is a pile of stone, inspiring the beholder but shaped only by gravity and gravity's handmaid, erosion.

The idea that either cathedral or karst pinnacles are "nothing but" stone leaves out their form. A full description must include the architecture. The shape of a thing is no mystical essence, but it is not reducible to the materials which take that shape. A white sifaka, one of the glorious tribe of Madagascar lemurs, which dances over the ground and springs fifteen feet at a bound between branches and makes its home in hanging gardens among the pinnacles, is not "nothing but" molecules. Neither is it molecules imbued with some vital fluid of mystic life force. It is an architecture of molecules shaped by natural selection, just as the pinnacles are shaped by erosion. It is a sifaka.

Does wonder depend on ignorance? Is science merely Edgar Allan Poe's "vulture, whose wings are dull realities"? Or are we allowed to ask how the rain grooved the pinnacles, how the sifaka's wide eyes detect the predator, the nerves' electricity signals the leap, and the muscles contract as it springs in the air? Can we see how swooping hawks and the ravenous fossa (a giant, cat-clawed mongoose) provided the selection pressure that led to lemurs who leaped in time to survive? Does the sifaka become less wonderful if we imagine the layers of complexity in muscle and nerve and gene?

Our wonder itself is a product of natural selection. The emotions of wonder and of beauty evolved. They are an integral part of every human mind. Does that mean they are "nothing but" chemical products, like sugar and vitriol? The feelings, like the sifaka's leap, exist. Even if we succeed in understanding how they are produced, explanation does not explain them away.

This brings us back to the beginning. Evolutionary explanation gives us clues why we see the world in terms of purpose. Natural selection of bodies that survive refined the brains of social primates into minds that want consciously to survive, that think in terms of purposes. We see purpose in our comrades, in other living creatures, even in lifelike manifestations such as thunderstorms and cars and computers.

We have no proof whether there is a larger purpose in the universe or not—or perhaps it is fairer to say that some of us see no proof, while many others do. We may or may not have free will to choose between the courses of action offered up in the conscious part of our minds. All that is a larger metaphysics, an argument which has lasted three thousand years or so and will likely continue. Perhaps most can agree, though, that our own conscious purposes are real. A biological explanation of how the brain works, how the nerves offer up thoughts on our mental screen, does not deny the thoughts are there. That would be like saying if we understand the nerves and muscles and evolutionary history of the sifaka's leap, then suddenly the poor beast can't jump.

Our human sense of purpose is a product of our biological heritage; it now is changing the biology of the world around.

2
Life, Sex, and Cooperation

The great transitions of evolution have arisen from communities of individuals that take on an individuality of their own. The evolutionary biologist John Maynard Smith and his colleague the biochemist Eörs Szathmáry have pointed out four of the biggest transitions: the joining of molecules as living things, the joining of bacteria to make cells, the joining of cells to make multicelled bodies, and, finally, the organization of individual bodies as societies, whether a society of humans or of ants.[1] We seem now to be in the midst of a fifth major transition: the joining of human societies into a global network. Along the way, many smaller unions followed the same pattern, such as the chunking of genes into chromosomes or the retention of body segments instead of budding them off, which yielded both the rings of the earthworm and the parade of vertebrae down our spines.

The only really efficient way to grow more complex is to link subassemblies and subroutines. Herbert Simon, one of the founders of the information revolution, told the parable of Tempus and Hora, two overworked watchmakers. They were so popular their phones kept ringing with new orders. Tempus built his watches of a thousand parts each. Whenever his phone rang, he put down the partly made watch—in the worst case, with 999 parts assembled—but for want of the last, the whole lot fell apart. He rarely finished a watch and soon went bankrupt. The wiser Hora made subassemblies of ten parts each, which he grouped by tens and then again by ten to make the final thousand-part watch. When his phone rang, he lost no more than nine steps of any process. Simon pointed out that virtually any large organization is made of smaller subassemblies, from the living cell right up to the empire of Alexander the Great.[2]

Arthur Koestler, novelist, political activist, foreshadower of complexity theory, named these composite entities *holons*.[3] Assembled holons have three essential properties: *cooperation* among parts at the next lower level of the hierarchy; a *boundary* that compartmentalizes them from the rest of the world; and *differentiation* within and among the components.[4] If smaller entities cooperate to form some larger entity, the large entity must itself have some boundary. If it can be added to forever, like a growing crystal or a pile of pebbles, it is not an entity, just an agglomeration. Boundaries separate different holons from one another and from the world. Many separate holons can then exist side by side. Some of them undergo changes— damage or mutations. This differentiates their internal components and makes them potentially able to serve different functions within the holon. It also makes the holons themselves different from one another, potentially adapted to different environments. All life is built of holons, in successive layers of complexity.

The first step in building live holons is molecular cooperation. It may sound odd to describe molecules as "cooperating." I could pick a much less loaded term, like "interacting." But cooperation implies more: that they somehow help one another. One chemical reaction that promotes another that in turn promotes the first, or that promotes a second that promotes a third that in turn promotes the first, can become a simple circular chain reaction. The system feeds forward and grows, so long as raw material is available.[5]

The earliest known chemical traces of life go back to 3.8 billion years ago; the earliest bacterial microfossils are around 3.5 billion years old. The implication is that very shortly after the end of the rain of meteors that would have sterilized the earth—or, in wilder imagination, brought a few template chemicals to earth—organic molecules began to coalesce in the primordial soup.[6]

The basic building block of life starts with a molecule that replicates itself. There is still controversy whether that ancestor was a simple protein or RNA, and which RNA—a subject of intense debate among the clique who study the so-called RNA world. Whatever they were, early self-copying molecules absorbed energy. Their replication does not happen without energy input. Other chemical reactions give off energy, enough to power a chemical change in their immediate neighborhood. If the primitive replicator somehow managed to associate with those primitive energy sources, and if each increased the raw materials for the other, the first cooperative cycle could have begun to "grow."

The problem was that at incredibly dilute concentration in the ocean, the chances of the cooperating molecules' meeting were slim. Even though a few new molecules formed, they would drift apart and lose touch almost at once, stopping the chain reaction. One way to increase their concentration would be to stick to a two-dimensional surface.

In puddles near ancient seas, sea water alternately washed in and evaporated, concentrating the heavier molecules. It is a pleasing thought that tidepools may be the cradle of life. At first chemists argued that the clay at the bottom of the tidepool would be a suitable substrate. However, nitrogen bases, common components of life, do not form insoluble salts that would settle onto moist clay. It may be that they were attracted instead to the positive charge on crystals of iron pyrite. There they formed a film of interacting organic compounds on a mineral base, a "primitive pizza" rather than an ocean of primordial soup. It is perhaps an even nicer thought that the sparkling mineral we deride as "fool's gold" should be our original home.[7]

There are alternate versions. Perhaps the first pizza was on evanescent sea foam. Perhaps it was around vents of superheated water on the spreading rifts of the sea floor. Perhaps the early life forms took refuge within rocks, sheltering themselves from the sterilizing bombardment of meteors, as, today, bacteria hide from freezing and thawing deep in the rocks of Antarctica. Whatever their milieu, they were more likely to interact on a surface than in solution.

If a simple reproducing system made more of itself and colonized neighboring crystals of pyrite in the same puddle or along the sea floor by a hot vent, it faced a new problem. Some replicators might attach to others' energy sources and vice versa. If some were quicker to reproduce than others, they competed for others' raw materials. The competition within the puddle would favor those molecules that bound more tightly to their partners and those that were somehow sequestered, fending off their rivals.

Enter a third component: some form of boundary for the reproducing and energy-source molecules. A few molecules reproduce themselves to form a double layer of fatty protein, like the walls of cells today. If such a layer of membrane covered over the replicator and energy source, it shielded the other two molecules against the surrounding environment. This three-part globule now had replicator, energy source, and boundary membrane. The little system, a *chemotron*, could grow more safely. It was even free to float away intact, if storms or a high spring tide detached it from its substrate of fool's gold and swept it out to sea.[8]

Even at this stage, natural selection began. Some chemotrons were better constructed than others—a more active mix of amino acids or ribose sugars, a more durable membrane molecule. At first they probably were not competing much with one another directly in the wide ocean. Most of them just broke up under environmental stress. A few coped and kept growing. As they became more numerous, and more likely to meet, they influenced, stuck to, or degraded one another—direct competition between chemical complexes.[9]

Some met an accidental strike by ultraviolet light or encountered still other active molecules. An almost infinite number of the complexes stopped there, the chain reaction broken. A tiny few were "improved" by these accidents and proliferated faster. The improvements were not all the same. Different kinds of protolife were growing. Thus the first "species"—mere chemical species—began to differentiate.

If a fourth molecule catalyzed any part of the reactions, it allowed more molecules of the other kinds to form. This new molecule could then become part of the circular complex. Any further change that made or attracted more of the new molecule also helped that chemotron to outgrow its rivals. Above all, any change helped if it held all the essential molecules together when the complex grew so large it broke apart. This could not take place without already having boundaries. If the complexes were not separate but all their components sloshed about freely, they would settle into homogeneous chemical equilibrium in the sea—a clear soup, without the lumps that gave rise to life. Protolife sprang from lumps, each of which had individuality.

Already we are in the world of individuals, cooperation, and internal division of labor, even if you want to say the chemotrons were not really complicated enough to be called "alive." A half-billion years of this cumulative process produced the life forms that still dominate the earth: bacteria. Bacteria have specialized in chemical evolution, not growing much larger but able to survive in environments from boiling springs to the deep layers of Antarctic rocks, from dental cavities to oil spills. Five or six billion bacteria live on and in your body (and mine)—as many as the whole earth's human population.

Eventually bacteria achieved a startling new form of cooperation between individuals: sex.

Sex as Cooperation

"Sex" has many meanings in everyday speech. One is *mating*, intercourse, making love. Another is *gender* or *sex roles*, with a fuzzy confusion in most people's minds about the relation between X or Y chromosomes and adult gender. A third meaning is *reproduction*. We mammals have a parochial view here, because some form of sex (including artificial insemination) always precedes reproduction for us. The merest crabgrass or strawberry plant could challenge this viewpoint; intercourse is optional, if one can reproduce asexually by runners. Finally, there is the biologists' definition. For us, sex is the *recombination of genetic material between different individuals*, producing individuals with some new mix of two parents' genes. The new genetic mixes do not have to be newly created individuals. For bacteria and some protozoa, sex is just an exchange of genes without reproduction—it is the "parents" themselves who swim away transformed.[10]

Gender. Not every living thing has gender—not even an ambiguous gender. The so-called Gram-negative bacteria (those that do not take up the crystal violet stain developed in 1884 by Dr. Hans Christian Gram) lack the fatty coat that allows Gram-positive bacteria to stick to one another. If they can't stick, they can't mate, and therefore they can't recombine genes with one another at all—they have no possible gender. Other simple organisms have a single gene, a "mating type" that determines the class of others they mate with, but they do not have any other physical differences.[11]

Among many bacteria, the genes that determine donor and recipient are not on the main chromosome but on an extra little body called a plasmid, which looks suspiciously like a virus. A copy of the plasmid is transferred first when bacteria have sex. It transforms the recipient cell into a donor. Not all bacteria remain donors, because the plasmids can be lost in subsequent divisions. Still, it looks as though there is a takeover going on. One theory is that the origin of sex is parasitic. Perhaps it originally served the interest not of the host but of the viral plasmid.[12]

The whole idea of gender implies a physical difference in the creature's anatomy, not just its genes. Even if the body in question is just one cell, gender difference involves gene and environment and development. With us, that goes on to include mental environment and psychological development. Gender began to evolve when the gametes, the cells that share their genetic material, first became recognizably different from one another.

Females are the gender with the larger gametes, by definition. This implies no value judgments, no constellation of female characteristics that women must share with female sea urchins and female sea horses. In casual conversation and in classrooms, that kind of generalization rapidly emerges, and it sometimes takes a good deal of angry effort to sort out and demolish it. At the most general level, all that females have in common is bigger eggs. Males similarly have smaller sperm—that's all.

Reproduction. Reproduction may be sexual or asexual. It is not always linked to sex even in species that have sex. Some protozoa and bacteria mate and exchange genes without dividing. However, reproduction itself always happens, one way or another. As Darwin pointed out, it is a fundamental attribute of life. Without reproduction and growth we do not consider an object alive. (I will question this principle in the last chapter, on the fifth transition to global society, but it applies to all the rest of living things.)

Among the asexual reproducers are viruses, bacteria, protozoa, most plants including common dandelions and bracken ferns (which consist of a few clones spread throughout the world), many invertebrates, and some fish, frogs, and lizards. Most of these occasionally resort to sex for a generation or so when times get hard. A few species—the "ancient virgins"—seem never to have had sex. We will come back to ancient virgins in the next chapter, when considering why creatures bother with sex at all.

Recombination of genes. From a biologist's perspective, the fundamental definition of sex is recombining one's own genes with genes from another individual. When a human baby is born, parents are overwhelmed to meet a new person, one who is neither themselves nor grandparents nor aunt nor uncle, a face and a personality not like any who have gone before. (We haven't quite yet started cloning, though identical twins are clones of each other.) This recurrent miracle begins with the fact that each parent passes on only half of his or her genes, a quasi-random assortment, to mix with another unpredictable sample from the genes of the other parent.

Sexual recombination of genes is ultimately cooperation. I have talked of molecules combining into cells as being "cooperative." Cells, in turn, combine into a multicellular body. But each of the component parts in that kind of cooperation retains its own construction and nature, even though teamed up with other entities. Sex, in contrast, entails taking an individual organism apart and passing on only some of its components at a lower level of the hierarchy. If you are a paramecium, you take your whole self apart. If you are an oak tree or a human, you set aside special cells of your self—eggs or sperm—that encode half of your DNA, and this is what you pass on to your offspring. All your hope for the future rests in identities built up anew not from your own identity but from a sample of your genes freely mixed with the genes of another. How did we ever get into such a convoluted arrangement?

35

Sex for DNA Repair

Not all bacteria have sex. Even those that do have sex have a peculiar version, from our point of view. When they meet, the donor builds a long thin tube to the recipient, duplicates its genome, and slips one copy down into the other. The recipient takes the new strand (actually a ring) of DNA it has just received and lines it up beside its own ring. It snips out portions of its own and swaps them with similar parts from the donor's. This is useful if the new piece has working genes in a region where the old one was defective. One of the two hybrid chromosomes is then destroyed; the other is the new genome for the receiver. The donor, who has merely passed over a copy of its genotype, swims away unchanged.[13]

Bacteria were doing their own genetic engineering in this way three billion years ago; they did not wait for twentieth-century scientists to discover how to do it. The enzymes we now use in molecular biology laboratories are copied from those that evolved in bacteria and are among the most ancient mechanisms of life. They recognize which fragments of chromosomes to match up, they excise the strand that is to be replaced, and they splice in the new strand.[14]

For bacteria on the early earth, these recombinant DNA techniques constituted what Lynn Margulis calls a massive health care system. Those bacteria faced a particular hazard. They evolved in an atmosphere that contained very little free oxygen, and so the earth had a negligible ozone shield. This means that the earth's surface was drenched in ultraviolet light—about 10,000 times more UV rays than at present. Overall, though, the sun gave less light than it does today. Under such a weak sun all water might have frozen, except that there was 100–1,000 times more carbon dioxide in the atmosphere, creating a greenhouse effect that raised the temperature enough to melt water and allow for life. Early bacteria could live only in water, but they also had to stay low enough in the water column to avoid the lethal ultraviolet rays.[15]

The first land plants and animals did not crawl out under the hostile sky until about half a billion years ago, and then evolved compounds to absorb UV light and protect themselves. Those chemicals later evolved into the yellow flower pigments, the lignin that gives strength to wood, and the tannins that deter herbivores and flavor our tea, but their first function was sunblock.[16] We use UV light today to sterilize laboratories and industrial kitchens. There are signs that even the small increase in UV from our present assault on the ozone layer is sterilizing frog spawn and other vulnerable creatures in mountain lakes.

Blasting early life with UV increased the rate of mutations, both good and bad ones. As soon as the organisms grew complex enough to have a status quo worth protecting, they desperately needed to jettison harmful mutations. This was probably the original benefit of bacterial sex. It was sex for stability, not sex for change.

Doubling genes: the backup copy. A second way to protect the status quo, in addition to having sex, is to have two copies of each gene—indeed two sets of the whole string of genes. This redundancy is called diploidy, as opposed to haploidy, just one set. Diploidy ensures that if UV light or some other environmental insult inactivates one of your genes, there is always another, probably in working order, that can serve as a backup copy. A spare copy might also hold more silent mutations that could express themselves if the organism is suddenly required to adapt to a changing habitat. Under changing environmental pressure, diploid cells, with their reservoir of favorable as well as unfavorable mutations, can change faster to cope.[17]

There is, however, a catch to carrying two complete sets of instructions in each cell. It is correspondingly longer and more cumbersome to copy two sets of DNA during normal cell division.[18] A solution is to alternate generations. This gives speed of growth in the haploid, or one-copy, phase, alternating with insurance for the DNA in the diploid, or two-copy, phase. Even better if the organism can adjust form to circumstances: the fast, risky version when the environment is benign, say a pond in summer, and the two-copy version under stress, say the pond about to dry up or freeze. A few primitive protozoans and familiar yeast act just like this. They do not bother with sex. The doubling and halving of chromosomes between generations is entirely for their own (or rather their own clones') benefit.

Thus even at the bacterial or protocellular level there was sex, that is, recombination of genes with others, and there was doubling and halving of genetic material within some individuals and their descendants. Put those two together and we can picture the evolution of cellular sex, where one full copy of each parent's genes gets passed to the offspring. But to arrive at a real cell we must go back to the other kind of cooperation, the kind that is not sex but just sticking together.

A Community of One

In 1970 Lynn Margulis proposed that the cell was a community. This seemed an outrageous attack on orderly biology. The cell! The fundamental entity that sets apart protists, plants, and animals from mere microbes! How could the cell be anything but the unitary building block of higher forms of life? The idea of cells derived from bacteria had indeed been broached before, but biologists like E. B. Wilson pronounced the thought "too fantastic for present mention in polite biological society!"[19] Margulis triumphed. Nowadays, junior high school students learn that true cells derive from cooperation between life forms that were, long ago, free-living bacteria. Our cells are actually "a community of one."[20]

The host form was probably a large bacterium that lost the stiffening of the ordinary bacterial cell wall. The new, soft wall could be poked inward and then closed off as a separate sac to surround the genetic material. That meant the genetic material could move to a safe spot, well inside the cell, as a nucleus with its own membrane. Cellular forms are called *eukaryotes,* which means having a real nucleus, as opposed to bacterial *prokaryotes,* whose ring chromosome is unprotected by a nuclear membrane. I just call eukaryotes "true cells."

This new creature could now become a mighty predator. Bacteria do not eat bacteria. If not mating, they meet and just bounce off. But a hapless bacterium meeting one of these soft-walled things could stick to it and be slowly surrounded by a layer of the soft membrane. Once it was engulfed by the larger cell, the "host" could digest its prey at leisure.[21]

These early steps probably took place long ago, before there was free oxygen in the atmosphere. Many anaerobic bacteria and primitive cellular forms apparently survive as living fossils from those early days. Some live in hot springs, which mimic the conditions of the early earth. One candidate for the ancestor of true cells is *Giardia,* bane of tropical travelers and, for that matter, scuba divers working off sewage outlets in New York's East River. Giardia looks like a Halloween mask: two identical nuclei for staring eyes, flagellae like stringy hair. It apparently evolved in the ancient anaerobic ocean and lurked in warm, oxygen-free zones until mammals evolved a gut that was hospitable to just such a life form.[22]

As the new soft-walled creatures evolved, with their one or two nuclei and anaerobic habit, they confronted the greatest environmental change the world has ever known: the poisoning of the atmosphere with oxygen.

Cyanobacteria. The bacteria responsible for this environmental disaster were cyanobacteria, sometimes called blue-green algae. They invented the capacity to use light as an energy source—the process of photosynthesis. As a byproduct of photosynthesis, they produced oxygen, which combines with many organic compounds and degrades them. True, it produces the ozone shield, but it was also violently poisonous to anaerobic bacteria. By about 2.5 billion years ago, when life had existed already for more than a billion years, cyanobacteria had polluted the air with as much as 1 to 2 percent oxygen. The proof is the "red beds"—layers of two-billion-year-old sandstone streaked with oxidized iron, essentially rust.[23]

The large, soft-walled cells now had a major advantage over anaerobic bacteria. Their nuclei were shielded by that second membrane from the increasing oxygen pollution. Furthermore, they could turn the cyanobacteria into a food source, a new sort of prey. And third, if a photosynthetic bacterium was engulfed but not at once killed, it could go on producing organic compounds from the sun's energy, to the benefit of the engulfing cell. As Lynn Margulis says, there was some confusion in those days between sex, cannibalism, predation, and indigestion.[24]

Indigestion of cyanobacteria, long enough prolonged, gave rise to the green cellular organelles we call *chloroplasts*. Most chloroplasts are green and color our world green. A few are red or brown, as in red seaweed or copper beech trees. These structures within plant cells use light energy to produce organic compounds, and hence food for all the animals that feed on plants, including ourselves. There are even layers upon layers. Many reef corals harbor algae within them that use light to produce building blocks for the coral animal, their host. But within the algal cells, as within all other plants, are the still smaller chloroplasts, descendants of the free-living bacteria that transformed the atmosphere.

Humans today are having the impact on earth of a fairly large meteorite. We will drive many other species extinct—perhaps as many as half of the world's species—before we reach some new *modus vivendi*. Yet it is sobering to think that these changes are child's play compared to the effect of the early plants. The only comparison

that occurs to me is this: Suppose we were to set off so many nuclear bombs that we wiped out life as we know it. Suppose, then,
some bacteria survived on the radiating earth (they usually seem
to survive) and evolved the power to use radioactivity as an energy
source. That scenario is comparable to the effect of the oxygen-
producing bacteria, whose poisoning of the atmosphere allowed the
rise of extraordinary new creatures which profited from oxygen pollution as a whole new source of energy. These opportunists are now
called mitochondria.

Mitochondria and other livestock.   Again, the precursors of mitochondria were eaten. Once again, some turned out to be more
useful alive than dead, though trillions must have died for every one
that persisted inside the host. They scavenged free oxygen molecules,
which helped the hosts survive a move into oxygen-rich environments. They also produced energy—the basic energy transformation
we find in all oxygen-breathing life today.

Perhaps most of the benefits went to the most powerful partner,
at first. The host cells might have kept mitochondria as ants keep
aphids, or as traditional Masai keep cattle, milking or bleeding them
for energy but not actually eating them. The first stage could have
been more like slavery than symbiosis. All the same, there were advantages for mitochondria as well: they gained protection from other
predators and transportation to new pastures. Perhaps mitochondria,
aphids, and cows assume it is they who tamed cells, ants, and humans.[25]

Spirochetes and sperm tails.   Still another kind of bacterium to
join the growing communities were the spirochetes, the wriggling,
snake-shaped forms that cause various ills in humans, including
syphilis. Long, whiplike sperm tails, flagellae like Giardia's hairs, and
the little beating cilia that slide mucus round your nose and food
down your gut may all derive from spirochetes. If an ancient host cell
engulfed the head end of a spirochete but left the tail outside, then

the lashing tail would propel them both. The host would find more food to eat, grow, and reproduce, and the spirochete would at least be safer from predators that simply digested them.

Not only sperm tails but the mitotic spindle—the contractile threads that pull chromosomes apart during cell division—have the basic structure of spirochetes. It may be either unsettling or inspiring, depending on your point of view, that the questing of the sperm and the division of the egg both depend on structures related to a syphilis germ.[26]

Community Sex and the Origin of Gender

Over the course of a couple of billion years of trial and error, among creatures that might divide as often as once an hour under favorable conditions, there was ample scope for false starts, death, and cumulative selection on the few survivors. Sex probably got started from cases of incomplete cannibalism, when the nuclear material of two related cells survived together. Once this happened, nuclei started to carry double sets of genes. When attempting sex, the cell would divide into haploid gametes with a single set each and get an equivalent set from the mate, which made up two full copies.

Mating types may have evolved to solve a chemical problem. If a bacterium is actually attracting another for the purpose of sex, it could have a compound on its surface that would adhere to its partner and let them stick together long enough to exchange genes. But the last thing an organism wants to do is stick to itself, like a wadge of Scotch tape. The simple solution is to have an opposite compound on the partner—not the Scotch tape model of sex, but the Velcro model.

This system would also allow an early version of inbreeding avoidance. The point of genetic recombination is to obtain genes that are somewhat different from your own—not totally incompatible, of course, but different enough to make up for your own deficiencies. If the partner bacterium has a different mating type—perhaps one of a variety of different types, for there is no need to draw the line at just two—then it may also have usefully different genes.[27] Then, it could be a simple step to advertising your chemical nature. Just release a few of the surface molecules in the environment. Now the flavored water in your neighborhood announces, "I'm interesting, and I'm over here."

To react, of course, the receiving organism should be able to move. If it can swim up a chemical gradient, it can actually search out the interestingly scented partner. This is a plausible start for attractive and active partners. But why should one type become larger, the other smaller? Why shouldn't both be simultaneously active and attractive? And why settle for just two sexes?

The old answer is that the different mating types grew increasingly specialized. One expended its energy in active pursuit of a mate; the other stored energy to provide the fertilized egg with food for growth. At first these differences were random and slight, but they increased through competition within the population. The fastest swimmers of the active ones were most likely to reach a mate. This process continued until we reached the frenetic activity of modern sperm. They run the risk of exhausting their energy supply, but a wait-and-see sperm stands no chance if the others are rushing forward. In the ancient oceans, a mating type that was only languidly active might lose out in the competition, leaving just one identifiable type of whizzing male. Similarly, females would be driven to produce more and more competent eggs, with greater food supplies to ensure growth.

This argument begs the question, however, how some protists manage with "intermediate" differentiation. Also, with some population structures, a third or fourth sex, with smaller but highly diffusable eggs or larger sperm that did their bit for the food supply, might

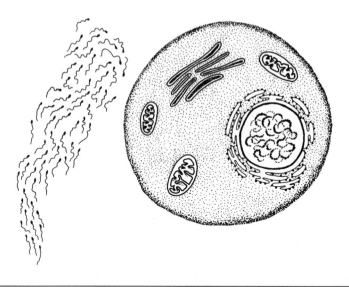

Figure 3.   A human egg's diameter is about fifty times the width of the sperm head. The egg is just visible to the naked eye, smaller than the period at the end of this sentence. One egg usually ripens per month; around 100 million to 1.5 billion sperm per ejaculation compete to reach it. In the egg, DNA strands are contained inside a nuclear membrane. Three oval mitochondria are enlarged to show their internal structure: they are actually the size of sperm heads.

theoretically exist in evolutionary equilibrium. These doubts led to a new proposal: mitochondrial wars.

Perhaps what really goes on is a conflict of interest between sexual members of the cellular community. Mitochondria, chloroplasts, and other organelles still reproduce asexually as a set of clones. If two mating cells each brought their own bridewealth of mitochondria, the two strains of organelles would compete for the same home castle. A cell in the throes of active interference between its own energy sources would not be likely to survive.[28]

There are two solutions. One is to pipe genetic material through from one nucleus to another without disturbing the mitochondria. Paramecia, for example, have many mating types, just like bacteria.

In general, forms that have many mating types do not fuse their cytoplasm and have no obvious differences between "sexes" except at the single gene level. The other solution is to actually fuse the cells, but to strip one of them of almost all its organelles. In other words, to have a sperm and an egg, or, at the one-celled level, to let one sex jettison its cellular body as it fuses with its partner. Just about everything that fuses cytoplasm has only two sexes.

In fact, the system is not perfect. Just behind the head of the sperm, a collar of mitochondria provides power for all that tail-lashing. Usually the mitochondria die when the sperm head enters the egg. Occasionally some survive and share genetic material with the resident egg mitochondria. This happens on the order of once in five hundred generations. That is enough to raise uncertainties for short-term calculations, like how long ago lived the human foremother "Mitochondrial Eve," whom we will meet in Chapter 16. For the really long term, the three billion years or so of sex, it doesn't matter. Sperm sacrifice their organelles for the privilege of achieving nuclear fusion. Egg mitochondria rule, OK?[29]

Macho Sperm and Motherly Eggs?

It is well to be aware (and beware) of a long, long history of describing sperm as active, thrusting, exploratory, penetrating, racing toward victory over one another and over the lumpen egg. In this scenario, the phalanx of warriors advances, led by one Achilles who will single-handedly scale the walls of Troy. From classical times on, philosophers believed that the life force, or indeed the immortal soul, passed only through sperm. The male force quickened the mother's womb. Even after human ova were discovered, they were still seen as inert bodies awakened to life by sperm.[30] This language persists in a few texts today. Textbooks speak of menstruation as "waste" of uterine lining and ova each month in which no child is conceived. Ova are present at birth; "far from being *produced,* as sperm are, they merely sit on the shelf, slowly degenerating and aging like overstocked inventory."[31] It is rarely considered wasteful, or even peculiar, that a man at a conservative estimate produces 100 million sperm a day over a reproductive life of sixty years.[32]

> A million million spermatozoa
> All of them alive;
> Out of their cataclysm but one poor Noah
> Dare hope to survive.
>
> And among that billion minus one
> Might have chanced to be
> Shakespeare, another Newton, a new Donne—
> But the one was Me.[33]

The sperm actually has a very weak propulsive force. Its head moves ten times more strongly from side to side than forward. When it nears the egg, a long thin filament emerges from the sperm head's tip. It may be twenty times longer than the sperm head itself. In the first scientific accounts, the sperm "harpoons" the egg. The anthropologist Emily Martin asks, since the filament just sticks, why not call it "throwing out a line"? When the sperm head touches the ovum's surface, a chemical bond forms between the two like a lock and key. There are many such bonds in nature between a depression on a protein, called a receptor, and a molecule with a protrusion that fits the receptor. Wouldn't you guess it? Just for this case, terminology reverses. The egg's binding molecules have protruding knobs, but they were called "receptors"; the sperm's proteins with their depressions became the "key that fits a lock."[34]

Martin warns, however, that reversing the image has its own dangers. The egg supplicated by a horde of wiggling pip-squeaks, its membrane that actively sucks in the sperm, and the egg's nucleus rushing over to meet the newly entered sperm can also be interpreted as the spiderlike engulfing female, the devouring witch-mother. Politically correct biologists now try very hard to describe sex as cooperation between two equal partners. Metaphors matter.[35]

3
Sex—Why Bother?

If you add up the costs of having sex, it seems amazing that any organism bothers with it at all. The fundamental cost is the loss of what makes you yourself, as dictated by the very definition of sex. Mix up your genes with someone else's? Then the resulting individual (your offspring if you are a mammal, yourself if you are a paramecium) has a new mix of genes.

But why fix what ain't broke? Whatever you are, you are extremely successful to have survived up to the life stage where sex is an option. The majority of human fertilized eggs are aborted, a high proportion of them with chromosomal abnormalities. Your set of genes did well to get you through that major bottleneck, and then the rigors of birth. Surviving the diseases of childhood is more likely but still not assured. If a genome has weathered such long odds even in humans, let alone if it is one of the hundreds of millions of eggs that stream out of a single oyster, it probably already has a successfully integrated set of genes, well adapted to work with one another. Why discard, when you're already holding a winning hand?

The Costs of Sex

But there are other costs, in addition to the risk of reshuffling the gene deck. As S. C. Stearns has pointed out, the costs of sex begin with its mechanical difficulties.[1] Sex is obviously more complicated mechanically than straightforward splitting. This is true even for bacteria. The common gut bacterium *Escherichia coli* can divide every 20 minutes under favorable circumstances. If it goes in for sex, that takes 90 minutes—3.5 ordinary generations. (Of course, a sexual act which takes 3.5 generation-lengths might sound tempting to some.)[2]

It's even longer for true cells, and cells cannot grow or feed or do anything else that involves the nucleus while they are dividing. A plant cell produces another body cell like itself in 15 minutes to 3 hours. Division to produce an egg or pollen cell takes 20–25 hours. First the chromosomes of the diploid (double) genome line up in two rows opposite each other like partners in a Virginia reel. The two lines draw apart, and the cell splits. The two new cells are haploid, that is, they have only one full copy of the genome, but the chromosomes have meanwhile each doubled, as though they were going to produce new cells asexually. This means the two cells must divide again. If the four resulting cells are going to be sperm, they each go off to try their luck. If destined to be an egg, only one of the four gets the bulk of cellular material, which endows it with resources to nourish the embryo, and the other three are thrown away.

There are many glitches in this complex process. Bacteria that are interrupted during sex pass over only part of the genome, and most die. Among true cells, very few anomalies survive to ever be noticed, such as the three similar chromosomes instead of two that lead to Down syndrome. The vast majority of mechanical failures simply do not become viable eggs or sperm.[3]

Having gone through the process of producing one's own sex cells, there is the drag of one's partner. Attending to mates and mating takes up time and energy that might be used to grow and maintain oneself or a batch of quickly grown clones. Sex multiplies the risks of predation. A lion can jump an unwary antelope while it is mating or is simply preoccupied with showing off to the opposite sex. Prey mammals minimize their time being so vulnerable: antelopes actually couple for only a few seconds at a time. Lions, on the other hand, can afford to court and mate intensively all day long.

Disease is another threat associated with sex, and AIDS is hardly the first. Syphilis drives insane and kills. Gonorrhea and chlamydia can make women barren. Various fungi and protozoans turn male plants and insects into females, which are then able to transmit the pathogen.

There are the obvious problems that go along with finding and gaining a mate. If the organism is rare—a Robinson Crusoe cast up alone in a new habitat—locating a suitable partner becomes very tedious. Organisms adapted to be long-distance colonists usually have an asexual option. Conversely, there may be too many others of the same mating type, leading to competition and choice.[4]

The twofold cost of males. A whole new cost appears in organisms that have differing males and females. It is easy to quantify: mating with a male leaves half as many copies of one's genome as either cloning or mating between organisms that can both be mothers.[5] Suppose there is an asexual lady named Agatha. Suppose also that for some reason she can have only two children, and her daughters are like her in this. Her two daughters are clones who carry all her genes. Her two daughters, between them, carry two sets of Agatha's genes. The four granddaughters total eight sets, and so on down the line.

Contrast her with Amorous Adelaide. Adelaide loves Bountiful Bill. They have two fine children, both girls. Each girl inherits half of her genes from each parent. The daughters total only the equivalent of one set of genes from Adelaide, along with one of Bill's. The

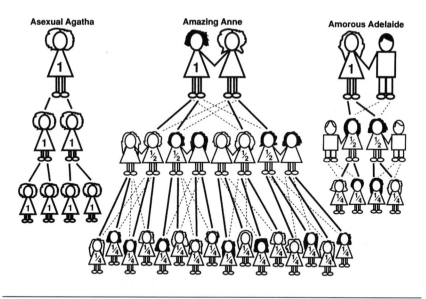

Figure 4. The two-fold cost of males. Having male partners produces half as many gene copies per generation as either cloning or sex between partners who are both capable of being female. (After A. Jolly)

girls marry Cautious Carl and Dashing Danny. Adelaide's four grand-daughters still carry only a total of one set of Adelaide's genes, diluted now not only by Bill's genes but by genes from the sons-in-law. Adelaide's genes have been halved in representation *in each generation,* compared with those of Agatha.

This is not actually a cost of sex per se. It is a cost of having sex with males. The real comparison is with Amazing Anne. Anne is a hermaphroditic snail or a lesbian beyond the cutting edge of current reproductive technology—the outcome is similar. She and her lover Bouncing Bertha each contribute nuclear material to fertilize each other's eggs. Each of them gives birth to two children. The four daughters have a half set of genes from each parent, so Anne has contributed two sets—as many genes as Asexual Agatha, though dispersed through more offspring. The girls approve of their mothers' orientation. They find partners: Carlotta, Denise, Elsie, and Fiona.

Among them all they present sixteen grandchildren to Amazing Anne, with the equivalent of four sets of genes from Anne and another four from Bertha. Again, this is as many gene copies as Asexual Agatha's.

You may object that males have their advantages, too. What if Dashing Danny's sons, Don Juan and Drifting Derek, father children by many women? Surely Amorous Adelaide will eventually see the few of her genes they inherit now multiplied many-fold? Fair enough, but the sons' mates are limited in how many offspring they can produce. By definition, having the larger egg makes females the limiting sex in most species. Derek and Don Juan's success prevents other males from fathering children with those women. As long as it takes two to reproduce, *on average* males leave no more offspring than females. Perhaps Cautious Carl's sons have no children at all. Later we will look at the extra considerations that arise if a parent can expect its offspring not to be average. But if one is looking at averages, and why a lineage should evolve two sexes, the twofold cost of males must be overcome in each generation. A high price for heterosexuality!

Advantages of Sex

We have already seen that DNA repair was probably the first benefit of sex in bacteria. Compensating for a defective parent also helps in the hazards of cellular life. Both sex and diploidy tend to appear when organisms are under pressure.

Many species that have both sexual and asexual reproduction switch to sex when times get hard and the food supply is scarce. The classic case is aphids. Aphids develop the next generation asexually inside themselves, visible within the mother's translucent abdomen, while granddaughters simultaneously develop inside the unborn daughters. (Eating for two in this case is eating for three generations!) When days shorten and grow cool in autumn, the aphids slow down, responding to changes in the food supply. They give birth to males and females, which mate to produce young that will live through winter. For these creatures, sex seems to protect the next generation. The offspring are more robust and more variable. At least a few of them will be well adapted to survive.

Sex is protective because damage accumulates. Most mutations are deleterious. A random change in a highly organized system is much more likely to disrupt it than to finetune it. The geneticist Herbert Müller pointed out that in an asexual clone, such mutations would "ratchet up," loading the genome with more and more damage in each generation and no way to get rid of it. Perhaps in very rapidly reproducing organisms, like bacteria and sperm, there are enough of each lineage so that some are superfit and potentially immortal, simply "outrunning" competition.[6] For a slower-lived creature, occasional sex would randomly assort some of its "bad" genes into different offspring. The bearers of many bad genes would die, while luckier sibs who inherit few or no deleterious mutations would thrive.[7] In humans, each person is conservatively estimated to have 1.6 new deleterious mutations; the real number is probably more like 3 per person per generation.[8] Taking out the bad mutations through the reassortment of sex thus can "cleanse" the germ line. Perhaps it comes as no surprise that Müller was a eugenicist as well as a great geneticist—we'll return to his dangerous ethics at the end of the chapter.

So far, we have been talking about the advantages of *stability*—the ability of sex to repair and return a genome to its ancestral state. This seems odd, because most of us think of sex as producing *variability*— many offspring, each individually different from the parents and able to cope with a variety of environments like the ones in Darwin's entangled bank.[9] In a diverse environment, is it better to clone identical offspring or to diversify your chances by having many different, sexually diverse seeds? If the sexually mixed seeds land on sand or rich loam or stony ground, by a dandelion or a blackberry bush, a few may have the various adaptations they need to profit and grow. Besides, differences help avoid sibling competition. If the seeds need different mixes of chemicals from the soil, they will interfere less with one another than if they all are seeking exactly the same things. Of course, one kind of environmental niche may be overwhelmingly present. If you live in the Sahara, you had best give all your offspring a genetic ticket for dealing with sand.

A fundamental problem is that the tangled bank argument predicts that organisms which would profit most by variability are the small, fast-breeding forms living in variable habitats, including the high Arctic. But these are precisely the species that seem to get along best as clones. Instead, sex is commonest among big, long-lived creatures. Some other explanation is needed.[10]

The Red Queen. What if the environment is not just diverse in space but changing in time? The Red Queen scornfully remarked to Alice in Wonderland, "It takes all the running *you* can do to stay in the same place!" If the environment is constantly changing, in feedback with the species' own evolution, there is constant pressure to keep up. The old genome, however "perfect" for the old conditions, just won't be good enough. Why are lemurs apparently not as intelligent as monkeys? Lemurs live on Madagascar. Their predators are a few kinds of hawk and that overgrown mongoose, the fossa. Monkeys on the big continents face leopards, lions, jaguars. The rate of predation on lemurs may be the same, but if fossae are not as smart as leopards, then lemurs need not be as smart as monkeys, and in turn the fossae only have to outsmart lemurs.[11]

Red Queen evolution need not be an arms race between species. It is also a feed-forward process within species. Competition among protohumans led our own brain to outstrip other primates'. It takes all the running we can do just to keep up with one another, never mind compete with other species.

One group of species that runs a great deal faster than we do is our parasites. In five months, under favorable conditions in our gut, *Escherichia coli* produces around 10,000 generations, about equivalent to the number of human generations since the appearance of *Homo sapiens*. Our lineage split from the chimpanzees 5–8 million years ago. It would take *E. coli* just nine years to go through a similar 250,000 generations. This is not to say that *E. coli* evolves at a similar rate per generation, but just to say that there is the potential for a great deal of evolution in a human parasite even during one human lifetime. It looks as though evolution is one secret of the AIDS virus. No matter how many billion cells the immune system throws at the virus, it just keeps mutating until one strain gets past the defenses.

Given this prodigious rate of parasite reproduction, a parent can be quite sure that it will bequeath its offspring a set of parasites exquisitely tuned to live on and in the parental generation. Parents have many ploys to shield their young from the family germs, ranging from feeding them antibodies in breastmilk to sending them far afield before they catch a disease. But the most effective defense of all may be sexual reproduction, which assures the offspring of its own, individual genome and its own unique immune system. In one leap it outdistances the adaptations of the parental parasites.[12] Long-lived species have the most need to foil short-lived parasites. Asexual reproduction is commoner in those with short life spans and hence less time for pathogens to finetune themselves. Rainforest mahoganies and Rocky Mountain bristlecone pines instead wait decades, or even centuries, to mature and reproduce. You might think they would send out asexual suckers in case they die before such late consummation, but it turns out that most long-lived species, including ourselves, reproduce only through sex.[13] This advantage of sex might be summed up:

> Your kids must keep up with the Joneses;
> Relaxation's forever denied you;
> For the reason you've kids and not clones is,
> The Joneses are living inside you.[14]

One lovely experiment reveals that variable immune systems do stymie parasites and tip the balance in favor of sex. A little fish called the topminnow lives in southwestern American streams. Topminnows have both sexual and asexual lineages. The fish get infected by worms, which give them black spot disease. In one pool inhabited by both sexual and asexual fish, the clones had more black spots. The worms adjusted their own genetic "keys" to fit the commonest immune system "lock," the one presented by the stereotyped clone, not the many presented by the diverse sexual fish. In a second pool where sexual fish lived with two different asexual forms, the sexual one was fairly safe, and so was the rarer asexual one. Again, the worms evolved to set their keys for the commonest population of locks.

But the last pool was the clincher. It had once dried up in a drought, and among the very few fish that managed to recolonize it were both sexuals and clones. The descendants of the sexual immigrants, forced by circumstances to inbreed, became almost as alike as clones themselves. And they too were horridly subject to black spots, because the worms adjusted to the fishes' "monoculture." The sexual fish almost died off. The secret to saving them was to drop in a few more mates. Sex with strangers restored their genetic variability, and the black spot infection dropped to its usual low level among sexuals. Soon sexual fish were once again out-competing the real clones.

It is genetic variability, not the act of sex itself, that protects the genome from parasites as well as from predators and other environmental threats, and overcomes the costs of sex in some species.[15]

Ancient Virgins

Many sexless strains are obviously recent offshoots of sexual species. Dandelions flaunt blossoms to attract insects they no longer need. (If dandelions were hard to grow, hobbyists would vie to produce the finest lawns of yellow flowers.) A third of the 45 species of whiptail lizards, fast-running denizens of the U.S. southwest, consist of nothing but females. Two lizards go through all the motions of mating. The one in the female role waddles off to lay her eggs. Once they are laid, she switches hormonal state and goes off to play "male" for other females, or perhaps even her first partner, now in the female role. The reptilian ability to lay fertile, fatherless eggs ensures, among other things, a potentially infinite number of sequels to *Jurassic Park*.[16]

A few creatures seem never to have had sex, or to have given it up a very long time ago. These are the ancient virgins. Take, for instance, the bdelloid rotifers. Rotifers, or wheel animalcules, swim around in water propelled by a circle of cilia at the front end. They need no genitalia at the back end, or anywhere else. They seem to have been perfectly asexual for tens of millions of years. Perhaps they may shed their parasites in the stage when they are wholly desiccated spores, which also lets them blow on the wind into all the world's ponds. Rotifers remain unexplained.

Artemesia, the brine shrimp (available in pet stores to feed goldfish), also are sexless. A brine shrimp divides its own genes into two haploid groups, like producing gametes, then rejoins two of the haploid cells with their chromosomes reassorted. This at least provides for some variability in the offspring, though not as much as sex with somebody else.[17]

One in five of the planet's fungal species was thought to be sexless, but now it seems even these species engage in "covert sex" underground. Their DNA shows unequivocal mixing between differing strains as the threadlike hyphae meet in the soil. In short, ancient virgins have a variety of special means of mixing genes or cleansing deleterious mutations. If you do not have sex, you may need to compensate for the lack of it.[18]

The most interesting personal question is how our own mitochondria and Y chromosomes manage to last as asexual entities within sexual organisms. They have actually transferred most of their genes to the sexually reproducing nucleus. Mitochondria mainly keep their crucial genes for using oxygen to produce energy. One possibility is that in the early stages of the evolution of sex, when both gametes brought mitochondria with them, the mitochondrial line that reproduced fastest won the race. Stripped-down versions could divide more quickly, so they became the ones to survive.[19] Likewise, the Y chromosome is stripped down to a few genes concerned with male sex and some "housekeeping" genes that act in concert with their counterparts on the X chromosome. Our own personal clones are not entrusted with more information than they need for their specialized functions.

Selection Arenas

Many parents contribute to their offspring long after the egg is fertilized. If a parent can detect which offspring are not growing well, it can shift resources to those with a better chance of survival. This is particularly important for the sex (usually the female) that puts the most effort into each embryo's growth.[20]

My family's backyard apple tree set hundreds of little apples every year. Every year most turned into windfalls, which I would try to eat, not wanting them to go to waste even though the tree still had scores of apples left to grow seeds and produce sweet ripe apple-sauce. Every year I got a tummyache from green apples. Now I realize that for the tree it would be a far greater "waste" to try to turn all its blossoms into ripe apples. Most could not grow effectively. If they did, the branches would break. A "selection arena" in which the healthiest gets the most resources produces the greatest number of viable fruit.[21]

Selection arenas intensify the advantages of sex. If organisms are to escape from Müller's ratchet, for instance, they must allow the offspring with the most deleterious mutations to die, while the others survive. And if they are to escape from parasites, they must produce young whose immune systems are unlike their own. If a tree has some mechanism for detecting which apples are defective or which ones are too much like itself, it can shed them early and favor those with a better chance.

Selection arenas are widespread in nature, and human reproduction is no exception. Something between 30 and 80 percent of our fertilized eggs abort spontaneously before birth. The vast majority of these do not even reach implantation; the woman never knows that she conceived. For pregnancies that last long enough to implant and cause a woman to skip a period, about 12–14 percent miscarry. This is still a very high rate of "waste." Sometimes it is only after the traumatic experience of miscarriage that women talk to one another and learn that many of their friends have suffered the same loss.[22]

At least half of recognized first trimester miscarriages turn out to have visible chromosomal anomalies, mostly having three of a particular chromosome instead of the standard two. It is not clear whether these conceptuses are themselves failing in some way or are rejected by the mother's womb. Most likely, it is both.

Older mothers may have a lower threshold for rejection. The higher rate of chromosome anomalies like Down syndrome in older mothers may be due not so much to anomalous eggs as to the mother's body's greater acceptance of differing embryos as her fertile years run out. There seem to be only two to six times as many chromosomally abnormal eggs in women over 40, as compared to women aged 20, but a 25–50-fold increase in genetically based birth defects among infants born. (This is still a very small number; the vast majority of infants are normal.) Scott Forbes argues that if older mothers reject fewer embryos overall, they also raise their chance of having a normal child. Whatever the chemical communication is between developing embryo and mother, it must be complex and therefore subject to errors. If a given pregnancy is the woman's last chance at procreation, from a strictly evolutionary point of view it is best to take a chance on what comes.[23]

Wild Types versus the Logic of Diversity

Herbert Müller, who discovered Müller's ratchet and an enormous amount of basic fruit fly genetics, became a eugenicist. He was convinced that random, undirected breeding would degrade human genetic stock as humans escaped from the winnowing of natural selection. He saw his "ratchet"—the accumulation of damaged genes—as a falling away from an original perfect type, a type with no deleterious mutations.

In the fruit fly, this is called the "wild type." The "wild type" is an ideal fly. It has functioning wings, beautiful deep crimson eyes, and every bristle in the right place. Treating its eggs with mustard gas produces mutants—wingless, bristleless, scarlet-eyed, or possessing some of a huge number of other diversions of development. A later group of people moved conceptually from a wild-type fly to the ideal *Übermensch,* a blond SS officer provided with a brothel of Aryan women who were idealistically willing to reproduce his "perfect" genes.

Nazi eugenics triggered revulsion in science, as elsewhere. Not coincidentally, geneticists began to look at the actual distribution of different genomes in wild fruit flies. It emerged that almost all wild species have an enormous degree of variation from individual to individual. There were no ideal fruit flies in nature. Indeed, the few wild species of animals where individuals are genetically similar, like cheetahs and Florida panthers, apparently got that way by going through a genetic bottleneck. A bottleneck is a crisis that kills all but a few individuals, who then interbreed, leaving only a few genomes to their descendants. Such populations are at risk from any disease that can tap into their common set of genes, in the same way that the black-spot worms tapped into the genomes of the inbred topminnows of the dried-up pool.

The logic of an advantage in sex goes with the logic of differences. Sex produces offspring that differ from the parent, either compensating for the parent's deficiencies or escaping the parent's parasites or colonizing diverse niches or evolving in the face of a changing environment. If sex has a rationale, then you clearly do not want your mate to be the same as yourself, so that your offspring won't be the same. The whole notion that there is one perfect ideal or wild-type form for the species is contradicted every time we have sex. Sex is a way of rejoicing in diversity.

Intelligence Is Like Sex

I should say right now that this book is short on definitions. I did not define purpose, and I am not going to define intelligence, consciousness, or the self. I did give some criteria for recognizing an organism: self-maintenance, reproduction, a boundary, internal differentiation, and internal cooperation. These will recur from chapter to chapter, but I don't think that's all there is to say about organisms—neither the balloon flower in the garden nor the deer (or snail?) who ate it last night.

61

I will assume that your conception of intelligence overlaps a good deal with mine—something about learning capacity, mental agility, an inclination to connect ideas. Something to do with language and logic, but embracing other skills like visual imagination, musical ability, social sensitivity, self-understanding, physical aptitudes. I definitely mean something humans do to be human. My husband remarked, "I sometimes wonder how anyone would recognize that I am more intelligent than the cat." Resisting all tempting cracks, I have to admit that the cat is much smarter than either he or I would be at being a cat. For that matter, the snail in the garden is better at being a snail. As for human intelligence, I hope we can get by on what purists deride as "folk psychology."

The reason definitions itch me is that I once helped edit a ponderous volume about play. Many articles we read for it tried to capture play on the wing and pop it into the killing bottle of a definition. None succeeded. It left me thinking that complicated ideas are sometimes better explored from as many viewpoints as possible, with all their nimbus of connotations. If definitions take all the fun out of play, perhaps a hard-edged definition of intelligence would, by definition, be rather stupid.[24]

That said, in what way is intelligence like sex? If sex evolved so that your children are not condemned to be just like you, intelligence evolved so that you are not condemned to be just like yourself.[25] Ideas, of course, unlike germ cells, come piecemeal and from many parents. There is no doubling up of one's whole mental apparatus and then passing a complete single set of ideas to one's offspring. We are much more like the bacteria. We take a string of linked ideas from one source and add in bits and pieces from many others, just as bacteria absorb single genes that have hitchhiked on viruses. The ideas themselves, however, reproduce, grow, mutate, spread. Ideas are the only system we know that shares those basic properties of genes.

Richard Dawkins has named these genelike elements of our mental system *memes*.[26] The strength of the analogy between genes and memes is that it lets us think of memes in terms of natural selection. They compete for space in our minds. They compete for credibility, that is, whether they fit with ideas already in our minds. They link up to form larger chunks, as genes link into chromosomes—Christianity, elementary particle theory, postmodernism, and so on. They mutate: the next generations of an idea will have subtle differences in different minds. Memes produce many more versions than will survive. In short, memes have every property that Darwin prescribed for natural selection and innovative evolution.

Where the analogy falls down is the same place that Darwin faltered through not knowing the work of Mendel. We can imagine no particulate, minimal idea. It is as though we were still doing genetics by growing peas, without any notion of DNA. Genetics and molecular biology are possible only because we can break down the complexity into a hierarchy of separable, discrete units. The word "meme" is currently as loose as the word "idea." It can mean anything, from the flicker of the thought "Must buy teabags" to a song that gets stuck in your head (a mind virus that blocks out other thoughts) to the structured edifice of Islam. Meme agglomerations are subject to blending—the same problem that led Darwin to doubt his mechanism of natural selection would work.

It may turn out that ideas are not particulate, that they contain nothing analogous to a gene. However, the moderately coherent chunks of thought that we call ideas clearly are identifiable. They do reproduce and are subject to differential survival, as well as recombinant engineering. Memes have their own deaths and immortality under many of the same rules of cost and benefit analysis that apply to genetic evolution.

Costs. There are costs to letting memes into your life, very like the costs of sex. Accepting other people's ideas about how to construct your mind (and then your children's minds) has a strong similarity to accepting someone else's genes to construct your children's bodies. In both cases, why not opt to be a nice safe clone? Why not cling to fixed, evolved instinct, which has had millions of years to get the answers right?

The fundamental question in each case is this: Why break up a winning hand? If you developed from one of the few fertilized eggs that succeeded in surviving to reproduce, your mind as well as your body must be exceedingly robust. Surely you should pass them both intact to the next generation. Suppose your parents endowed you with the instincts to manage life in exquisite detail, from how to earn your living to how to deal with the senior prom. It could be done. Spiders do it (though cannibalism would not be welcome at a human senior prom). Just suppose you could give your baby a mind full of all the knowledge you now have yourself and save it from all those painful mistakes. That would be like cloning your splendid body. You know it works.

The mechanical cost of meiosis, of preparing the egg or sperm for fertilization, also has its analogue in intelligence. Brains are incredibly complex to produce. Brain tissue is the most demanding tissue in the body. About 10 percent of an adult human's metabolic energy, and up to 40 percent of a child's, goes just to keep the brain running. Major deficits in the brain, such as autism or acute schizophrenia, are more crippling than being born without arms or legs, and harder to compensate for. An idea-organ is an even bigger investment than reproductive organs.

Then there are the problems inherent to fertilization. Learning from others is long and slow, compared to knowing what you need from scratch. It entails time, effort, and even danger. The predator snatches the antelope in a few seconds of courtship; our children must be guarded from danger through years of helpless childhood—extending in some cases right through graduate school! Disease is even worse. The wrong teacher can infect you with mental viruses that lead to destruction just as surely as sex with the wrong partner.

Benefits. Benefits also resonate between sex and intelligence. Learning from others, especially from many others, can provide a kind of stability. Just as a bacterium can take another's DNA as a template to correct a faulty genome, we can improve inadequate ideas by copying better versions wherever we find them. Maybe your offspring will cope with the prom even better than you did.

We do not have diploid brains, a double dose of ideas, but we arrange for redundancy when it really matters. One case is the emotional attachment between parents and children. Baby ducklings follow the first moving object they see, if it is less than three feet high and makes a repetitive noise like "quack, quack, quack." Hence the unforgettable story of Konrad Lorenz, founder of the ethological study of animal instinct, leading a train of little ducklings into long grass. Horrified passersby looked over the fence to see a bearded scientist, squatting and waddling and saying "Quack, quack," apparently all alone in the grass, and assuredly insane.

Ducks have only one shot at learning who their mother is, though, and then the critical period passes. We, instead, may bond with a baby in the hour after birth, three days later after a more difficult birth, or three years later in the case of an adopted child. Even if we do begin bonding in the first hour, our bonding goes on and on, as learning redundantly reinforces itself.

Sex sloughs off errors by shuffling genes. Learning far more quickly inserts corrections to behavior. Instead of the wastage of eggs and embryos that happen to inherit many bad mutations, our minds simply jettison bad ideas.

Selection arenas offer the strongest parallel of all between sex and intelligence. An apple tree can in a sense "choose" which apples to hold and ripen on the branch and which to shed as windfalls. Whatever else consciousness may be, it is clearly an arena where we actively choose between courses of action. We model the results of one or the other idea and pick the best. As our minds evolved to juggle conflicting ideas, we could more and more choose one idea over another: what to develop and what to disregard. This works only if the apples or the ideas are different. There is little point in selecting from among clones.

All these examples are ways that learning can promote stability from generation to generation, to buffer against change and error. But we are more used to thinking of learning as a way to be variable, not stable. Learning allows us to adapt to differing situations—the tangled bank of various environments that we meet. Our forebears may have been much wider-ranging than other primates, even sometimes nomadic. As individuals we did indeed meet many environments. Within the family, we also meet the immediate need not to compete directly with siblings. We do this in part through learning and even conscious choice: if my brother is a track star, should I challenge him on his own turf, or should I take up swimming instead, or become a bookworm? If my sister is my parent's docile doll, shall I be the tomboy rebel? Flexible learning lets me choose my niche, to some extent, with regard to my physical environment or even my family.

The Red Queen plays her role, too. Instead of running genetically in the competition with other individuals and species, we can expect to make mental leaps. What was good enough for my parents will have changed by the time I get there. This is all too familiar in the twentieth century, as older people ask younger ones to extract the file attached to an e-mail message or to program the VCR. Did the Red Queen torment prehumans? For over a million years *Homo erectus* made hand axes to essentially the same patterns. Back then, did they have much use for learned innovations?

Yes, in that the *social* situation of the troop changed from generation to generation. Throughout human prehistory, every generation coped with a different set of tribal politics. This would not necessarily drive intelligence forward beyond the ploys of chimpanzees, except that once the process began, we had ever-cleverer competitors and friends, so that we drove one another onward to be able to cope with shifts within our lifetimes.

The Red Queen role of parasites and disease does not have close parallels in the early evolution of human intelligence, though it is true that every tribe formulated its rules of health. We are told where to defecate, what medicinal plants to chew and what taboo foods to avoid, which saves having to vomit the nasty things ourselves. (Pork and shellfish are particularly lethal, carrying trichinosis, brainworms, cholera, and good old salmonella.) We also turn to the psychotherapy of trance and faith. However, there is no sign that these cultural remedies developed at anywhere near the rate that parasites evolved to counterattack, up till the advent of modern medicine. Until this century we relied on genetic defenses—an immune system within generations, sexual recombination between generations. Now the arms race is on between germs and medical research. At the moment, smallpox has lost, polio is losing, but malaria is winning the race against us, while the Red Queen impassively umpires gene against meme.

Mental genders? The costs of learning are like the costs of sex, except when it comes to the two-fold cost of males. We don't have mental males. Small boys and girls apparently differ in approaches to thinking, but that is another story (and we'll get to that in Chapter 14). If we had real mental mating types, each of us could only absorb ideas from people who differed from ourselves. A teacher could not be the same type as the student. There would be some distinct ideas only teacher-types could hold in their minds. If a teacher-type did transfer those ideas to a receptive student, the student would perforce become a teacher-type, like a bacterium receiving the donor-factor from another bacterium. Little to recognize here among humans.

Suppose we did not have just mating types but mental genders analogous to male and female? That would mean that some minds produced many light, agile ideas. This gender would then depend on other minds to develop the light ideas, transforming them into viable, heavyweight ideas while adding ideas of their own. In other words, "male" authors' memes would jockey to be chosen by the limiting memes of "female" programmers, publishers, and producers, who alone have the means to make ideas grow.

This notion leads us on to sex roles, and the asymmetry between those who court and those who choose.

4

Courtship and Choice

By definition, the female produces fewer gametes than the male for a given amount of energy and effort.[1] A woman starts life with about half a million oocytes capable of developing into eggs—clearly more than she will ever need. We are not limited by the number of eggs we produce but by our gestation and lactation, not to mention conscious decisions about how many children we wish to raise. So if the number of eggs is not a limiting factor, is there any good reason why females are usually the ones with greater investment in each child in other ways? Only the historical reason. The vertebrate ancestor, something like a sea squirt in the Cambrian ocean, actually had asymmetry directly related to the relative numbers of eggs and sperm: a few million more of one, a few million less of the other. Its descendants, the ancestral female fishes, could perhaps emit a mere million eggs. For many fish, the asymmetry in number led on to active male courtship or harem guarding.

Reptiles, which evolved much later than fish, were still more limited in their production of eggs. A reptile mother provides her eggs with waterproof coats, which allows her to nest on land. Each egg is endowed with enough food for the hatchling to face the physiological challenge of life in air, instead of slipping out into welcoming water. As female investment in each egg increased, so did the differ-

ence between the sexes. Some reptiles go even further in investment. Boa constrictors give birth to live young—twenty-odd
snakelings each wiggling out, an amazing process. Crocodile and
alligator mothers guard their nests and carry the infants to water in
their lethal jaws.

Through this historical progression we arrive at the biology of
human motherhood. There is no inevitable logic that says bigger eggs
mean you are stuck with being maternal. The logic just says you are
more likely to continue specializing in what you are good at. That was
the evolutionary path of our ancestors. Every lineage of organisms
boasts some species where the males put as much or more effort into
raising each offspring as the females—only these species are not so
common, especially not for those of us constrained to reproduce as
mammals.

As a historical footnote, our name, Mammalia, the creatures with
breasts, was not inevitable, either. Linnaeus gave it to us in the eighteenth century when he classified the kingdoms of plants and animals. He could have called us hairy creatures, or hollow-eared ones.
He opted for breasts because he thought natural motherhood mattered. Not coincidentally, he campaigned against the common practice of wet-nursing. Linnaeus announced that even aristocratic
women should be proud to suckle their own children.[2]

> Said a fervent young lady of Hammels,
> "I object to humanity's trammels!
>   I want to be free!
>   Like a bird! Like a bee!
> Oh, why am I classed with the mammals?"[3]

Parental investment.   The sex that has fewer potential offspring,
either because of producing fewer eggs or because of the demands
of gestation, lactation, and care of immature offspring, needs to find
high-quality co-parents for the few they do have. The other sex is

limited by access to mates rather than by gamete supply. They put their energy into competing for access to partners and are less selective over any one choice. Even in fruit flies, males who have more mates sire more larvae than other males; females with more mates, on the other hand, do not lay more eggs and larvae. Females with more mates may still do better in offspring survival than other females, but because they can test for mate quality, not quantity. Overall, among plants and animals, females are choosy and males compete to be chosen; males go for quantity, females for quality.[4]

The test of the argument is those species with "reversed roles," where the males raise offspring: the sea horse, the midwife toad, the human father who puts his kids through college. In all these cases, males also become exceedingly choosy about their mates—or at least about their official mate, whose offspring they recognize and invest resources in. Meanwhile, females compete to be accepted by such a male. Sea horse males grow a brood pouch in which they incubate the eggs laid by their mate. The female still donates all the nutrients. The two cooperate, reaffirming their pair-bond with daily courtship rituals. For any pair of animals who will cooperate to rear the brood, whether they are sea horses or people, each sex has a large stake in finding both Mr. and Ms. Right.[5]

Supply of mates. Number of available mates affects choosiness. If there is some imbalance, the rarer sex can afford to be discriminating, the commoner sex less so. This is foreseeable in China and India as this generation matures. There are significantly more men, so the bargaining position of brides (and their parents) should improve. The opposite, a dearth of men, comes out in personal testimonies of World War I—for instance, Doris Lessing's memories of her parents. Her mother, a nurse, eventually married a wounded patient with one leg and a gentle, loving personality, but she never quite stopped dreaming of her first love who was killed in the war. We can see the choices clearly in ourselves because we see ourselves generation by generation in a historic timescale. Over evolutionary time-scales, the relative numbers adjust—but animals, like ourselves, re-act to the numbers of suitors they meet personally, not to some many-year average.[6]

Exercising Choice

How does the choosier sex actually opt for a mate? In some cases, it isn't very choosy at all; it simply accepts the first mate who comes along. Natterjack toad females do not reject males who clasp them—any male who has hopped across the lethal English roads to reach the mating pond will do.[7] Some females have to wait so long, or have such a short window of opportunity, that they may best be very unse-lective.

The beautiful long-spurred comet orchid of Madagascar keeps each bloom fresh for a month. It may take that long for the only insect capable of pollinating it to come by: a pink-furred hawk-moth with a 30-centimeter tongue whose existence Darwin pre-dicted from just looking at the flower. When entomologists finally found the moth, they named it *Xanthopan praedicta*. (The moth may have evolved before the flower, hovering with a long tongue to avoid predators, but the flower's evolution tracked it.) Even orchids may be choosy: a related orchid that lives on rock outcrops in the Mada-gascar highlands has highly skewed fatherhood for its seeds. Bota-nists tagged individual pollen grains and found that insects brought pollen from many other plants to each flower. But the flowers them-selves only permitted favored pollen to grow tubes to reach their ova.[8]

Choice at the level of sperm or pollen that reaches the female reproductive tract allows the female to compare suitors simultane-ously. Much of the time, however, the alternative mates are not both present at once, either as individuals or as sperm. How does a female decide to reject one candidate and go on to look at the next?[9] If choosers cannot select from a simultaneous array of mates, they keep looking until one is better than some preset threshold—that is, until someone lives up to their expectations. As time goes on, or the end of the breeding season approaches, the threshold falls until almost anyone will do. A chooser may also adjust the threshold to an assessment of its own quality. If you know yourself subordinate, or think yourself ugly, you might start with a low threshold. Finally, if you are not sure of the quality of the information you are getting, you may go back and check out someone you already looked over, to see if he or she is better or worse than you first thought.[10]

Animals from pine engraver beetles to geese actually seem to use variants of these tactics. They visit several potential mates—typically two to six possibles—before accepting one. Even these may be only a subset of the possible possibles, chosen for a visit on the basis of some long-distance signaling. An animal's rules of thumb suggest how much information it gathers and retains to make its choice. What is the equivalent for goose or gander of thinking itself ugly? And do repeated visits to potential mates convince us that some animals do follow the optimal strategy of choosing "best of n" suitors, with updated information on each?

Courtship may not lead to mating, and certainly not to immediate mating. Indeed, male ring doves back away from females who are too eager. Female ring doves have a well-understood hormonal system that controls their readiness to mate. Early stages are turned on by male courtship. A female is not primed to copulate even with her eventual nesting partner without a decent length of time working through the hormonal preliminaries. Thus, in effect, she tests that a male will court her assiduously; when he invests so much time and effort himself, he may be more willing to help brood her eggs than to go off and start the whole process again with someone else. Male ring doves, in turn, are turned off by females who are too quick to mate. Such a female has already been courted by some other male, who has already primed her hormones. The male who accepts a too-ready female may be a cuckold.[11]

This account roused the ire of feminists, who saw females of yet another species being stereotyped as being either virgins or whores. And unfortunately, there is some truth in the human parallel—not in the hormonal level but in the social logic. If a woman grows up in a strict human culture and hopes to marry, it behooves her to be coy when courted, or else the male may draw the same conclusions as a ring dove—that someone has already beat him to it.

It is less often pointed out that playing hard-to-get is also a male trait wherever the male has high parental investment in the offspring of his official mate. Among monogamous bird species, the male gathers seeds or earthworms for much of the summer and stuffs them into hungry beaks. At this level of investment, he has a high incentive to find the best mate—a much higher incentive than in temporary liaisons with other birds' mates, which, we now know from DNA matching of eggs and fathers, is quite common. The triumph ceremonies of geese, dancing of cranes, and nest soliciting of doves are mutual tests and bonding, not just performances by the male before a critical female eye. In species like our own, a male may be quicker to arouse sexually than the female, but equally hard to tempt into responsible marriage.

Competitive Gambits

Scramble. One way to compete for mates is just to get there first. A male moth may scent a single molecule of a female's pheromone a mile downwind. He flies up the concentration gradient of the chemical until he finds her. There is enormous scope for selection: to have more feathery antennae to catch the molecule, to fly faster toward the seductive scent even burdened with antennae that branch like ferns. These traits involve no direct contact or conflict between the males. It is just a scramble for physical priority.[12]

Contest. Another way to impose is to physically chase, wound, or kill one's rivals. Konrad Lorenz believed there was no murder in the animal world, only among degenerate humans. Sadly, animals do wound and kill. Male ring-tailed lemurs stage elegant "jump-fights," during which the opponents leap nimbly in the air and come down with a canine-slash that can lay open an ear or skull or thigh. Polychaete worms seize one another in homosexual "rape," secreting a glue that cements the victim males' genitalia together.[13]

Our nearest relatives, the chimpanzees, are all too much like us. Males fight and bite in the rare battles for dominance. Alpha males in the Gombe Stream Reserve and in the Mahale Mountains of Tanzania typically hold their position for several years. Aspiring rivals form coalitions that threaten and chase, answered by coalitions of the political boss and his own friends. The final reversal usually involves a real fight. In the toppling of Goblin, lead male of the Gombe Stream, a gang of younger males targeted Goblin's scrotum; he would have certainly died of his wounds without the care of Jane Goodall and her team. In the Netherlands' Arnhem Zoo, crafty old Yeroen and his powerful henchman, Nikkie, attacked the lead male, Luit, who dominated each of them individually. The three males had lived together in relative peace for five years beforehand. Even after the final attack, Luit tried to remain with the others. It was only after zookeepers and scientists managed to remove him that they found his body covered in deep gashes and both of his testicles on the cage floor. In spite of heroic veterinary efforts, Luit died within twelve hours.[14]

The question whether chimps specifically attempt to castrate rivals is still open; few such violent fights have been seen. However, this kind of wound is very rare in other animals. It is hardly feasible when animals fight one on one: the loser can generally keep his teeth or

Figure 5. The killing of Godi. Six chimpanzee males of a neighboring troop met Godi alone in the Gombe Stream Reserve, Tanzania, in January 1974. One sat on Godi's head and held down his legs; five beat and bit him, all screaming loudly. As the attack ended, one threw a rock at him. Godi, badly wounded, was not seen again. (After H. Matama, eyewitness account, in Goodall, 1986)

head toward a single rival until he can break and run. But chimpanzees live in a male-bonded society, which means a male may have the misfortune to be caught by a gang of cooperating enemies. With two or six against one, the victim can be held and mutilated.

Perhaps even more horrifying, since apparently more systematic, are chimpanzee attacks on animals of other groups. In the Gombe Stream, chimpanzee gangs repeatedly attacked a neighboring group. One male held down an enemy while others beat and bit him. They never quite killed the victims but left them fatally wounded. One victim was seen some months later, dreadfully emaciated, with his scrotum shrunk to a fifth of normal size. The entire victim group ceased to exist. The males died, but some females transferred to the victors' group. In a similar case in the Mahale Mountains, all the males of one group disappeared and the females joined their rivals; the killings were not seen, but were inferred.[15]

Chimpanzees may be aware to some extent of the goal of their actions. We will return to consciousness much later in the book, but it is worth recording the opinions of two people who know these animals well. Jane Goodall feels that the violence, duration, and focus of the Gombe attacks, and the mental capacity of chimpanzees, implies that the attackers mean to kill. Frans de Waal writes, "I found myself fighting this moral judgment, but to this day I cannot look at Yeroen without seeing a murderer . . . But 'murderer' implies an intent to kill, something impossible to prove or disprove in this instance."[16]

Overwhelmingly in the animal kingdom, the chosen sex is more violent than the choosers. One measure is the difference in body size and weaponry between males and females. When there are tusks or antlers in only one sex, it is always the males; avoid male narwhals, male mountain goats, and male baboons. If both sexes are equally well armed, like forest duikers—tiny antelopes with horns like little poignards—or the elegant, furry gibbon, with its overgrown canines, you often find a monogamous mating system where both are choosy and both may fight rivals of his or her own sex. If you are in a position to get bitten by a gibbon, beware of both sexes. The size differences are subject to a complex interplay of natural and sexual selection, but the trends are clear.[17]

In polygamous species, males may have many offspring—one hundred to two hundred pups for a 2.5-ton elephant seal harem-master. Or they may have none. In contrast, female elephant seals who reach breeding age average four pups, but are not likely to die childless. No wonder that the male takes a high-risk strategy when he plays for so much higher stakes.[18]

Men in the United States commit about ten times as many violent crimes as women; in other cultures, the ratio is also high. Whatever the human cultural variables, it seems likely that this outcome rests on an initial biological bias toward aggressiveness in men, as in other male mammals.

Mate guarding.   Aggressive tactics can also turn on the opposite sex. Male baboons and chimpanzees sometimes beat and drag a female while she cowers and screams. Interestingly, chimp males very rarely use their canines; biting seems to be unchivalrous, whatever else happens. But punishment can force the female to go with the male, or, rather, not go with his rival. There is a continuum in such behavior between other primates and ourselves, but women become even more vulnerable than other female mammals because of explicit compacts among men regarding access to women. We can see the roots of this behavior in our closest primate relatives.[19]

Chimpanzees have four mating tactics. An alpha male has priority and generally takes estrous females first, because other males fear him and females seek him out. However, a whole group of males may also tolerate one another mating promiscuously, in plain view. Third, a male and female sometimes leave the group entirely, "on safari," out of sight of the rest of the group during her estrus. To sneak off successfully, the female must cooperate. If she makes noises and resists, a more dominant male may well come and prevent the couple from departing.[20]

The Ngogo community of wild chimpanzees in southern Uganda is exceptionally large, with 26 adult males and at least 40 adult females. There, males resort to mate guarding in coalition with one or two friends. A guarding male has to herd the female while chasing off all his rivals. She only needs 10 seconds out of his reach to mate with somebody else. When two or three males share the guarding, and also the mating, in a group of 14 or more males all vying for one female, the allies may all do better—even the alpha—than taking their chances alone.[21]

Now, if males could consciously trade favors, a male might let his subordinate leave on safari but expect similar treatment in return. They could agree that "she is yours, but this other female is mine." This would set the stage for our own human tendency to mate in private and to practice monogamy, or at least serial monogamy. From this point of view, the female's choice could be severely limited by the males. They might sort it out among themselves and enforce their agreement by physical strength and punishment of females. In societies around the world, between 17 and 75 percent of women say they have been physically assaulted by a partner. We must consider the degree to which mate guarding in humans depends on the collusion of other men.[22]

Before making anthropomorphic or even political judgments, one needs to look at each species' or even couple's real situation. The prancing horse rounds up his mares, but in a normal year a harem mare actually gets more to eat if her stallion protects her from other males' harassment. In a drought year, some mares may not play his game at all. Thirsty mothers with foals trek off to the waterhole to drink. Only a few nonlactating females stay put with the stallion, who is stuck out in the remaining pastures. The male has no power to keep his harem together without the mares' consent. He is reduced to staking out a territory and trying to intercept females as they wander by, just as zebra stallions do in the drier parts of Africa.[23]

Another case: it is all too tempting to see the female hornbill, imprisoned for three months in a tree hole behind a wall of clay and dung, being fed through a slit by the male, as in imposed purdah. The fact that she builds the wall herself and eventually dismantles it herself could just mean that she is instinctively brainwashed into accepting sexual captivity. A soberer look, however, suggests that she foils predators who might eat her precious chick. Meanwhile, the male flies back and forth for all the daylight hours gathering fruit to keep her provisioned. He brings up to 250 little figs in a single load, which he rolls delicately one by one to the tip of his massive beak and pops into the slit. If anything, *he* is the prisoner of an unbelievable innate work ethic.

Rape.   In the plant world, rape is nonexistent except by the most tortuous definition; and in the animal world, it is uncommon, since a female can usually run away. Among mallard ducks, drakes do sometimes jump females and copulate with them, despite the fact that the female is bonded to a different male and is struggling in the water. There has been violent controversy, however, whether such a loaded term as rape should be applied to mere ducks, because of the implication that rape may be "natural."[24]

In one primate species besides humans, rape is unequivocal. Orangutan males are slow to grow. They may not develop adult cheek flanges and throat sac until they are sixteen or eighteen. Perhaps some males never "grow up" physically at all. Big males have a long call that rolls round the forest like thunder. When a female is in estrus, she proceeds toward a big male, following his call. (If she lives in a zoo where a door just big enough for her but not for him connects the cages, she moves into the male's cage mainly when in estrus.) At other times she is not interested in company and travels alone. Sometimes, however, when an undeveloped male happens to meet her, he may grab her, holding her with three of his four hands, leaving one for the branch. She tries to escape, struggling and screaming. Given the small size of orang male genitalia, the male may not achieve intromission, let alone have much chance of fatherhood. Probably there is a small chance, or the male would not even try. It does seem that this should really be called rape. In fact, hand-raised orang males have attempted to rape women—their own species' behavior misdirected toward the species that reared them.

It shows, by contrast, that real rape is very rare in primates besides ourselves and orangutans. Perhaps other primate females submit to distasteful matings without much struggle, but that is not the usual meaning of rape.[25] Does this mean that rape is a "natural" behavior in humans? Yes, of course it is—not because orangs do it but because humans clearly do. The question whether what is "natural" is inevitable, let alone right, will be broached later on.

Bearing gifts.   Tangible gifts are another ploy to get mates. Male scorpionflies offer their females freshly killed insect prey. (Some male scorpionflies steal one another's gifts, quicker than catching their own.) Male katydids produce two-part sperm packets: the female curls up and eats some of it, rich in protein, while the rest is fertilizing her. Male and female African ticks bite the host right beside one another. The male secretes a blocker for the hosts' immune system defenses. This helps his mate gorge with a single enormous blood meal that she converts into thousands of eggs for him to fertilize.[26]

All these are trivial compared to the gifts offered by some spiders and mantises. The Australian redback spider, once he has achieved intromission, somersaults so the front part of his abdomen is just above the female's jaws. Males who allow themselves to be eaten fertilize more eggs than males who are not. The females permit longer copulation with the back end of the abdomen when they are busy chewing on the front. The males seem to take a calculated risk; some females are not hungry.[27]

Chimpanzee males share meat and even bunches of leaves with favored females.[28] We humans give presents, too, although cash downpayment implies that the partner is less, rather than more, likely to make a long-term commitment. The giving and accepting of gifts is always a delicate balance:

> A single flow'r he sent me, since we met.
> All tenderly his messenger he chose;

Deep-hearted, pure, with scented dew still wet—
One perfect rose.

I knew the language of the floweret;
"My fragile leaves," it said, "his heart enclose."
Love long has taken for his amulet
One perfect rose.

Why is it no one ever sent me yet
One perfect limousine, do you suppose?
Ah, no, it's always just my luck to get
One perfect rose.[29]

Honest promises.   The one rose is, of course, a token or display that promises later benefits. Here is a question raised by Darwin: who is the audience for such displays? As he pointed out, "A sprightly bearing with fine feathers and triumphant song are quite as well adapted for war propaganda as for courtship."[30] If the audience is other males, the male transmits a measure of aggressive capacity. His basso roar resonates to his body size, his erect hair and stiff-legged posture exaggerate his weight and musculature, the fact that he can jump up and down, up and down, up and down for hours on end shows his stamina, whether he is a Masai warrior or a bird of paradise. All these behaviors can be used to scare off opponents without actually fighting them. But the ultimate test of honesty is sometimes to escalate into real fighting. If a male is not prepared to follow through on his threats when actually challenged, they may do him no good. If he can get a little ahead of rivals by exaggeration without being found out, he will—so the others keep testing.

If the audience is the choosers—that is, the females—the same displays show what good genes the potential mate has and how likely he and his sons are to win if it comes down to a fight. In species where males are offering only genes, displays escalate. He must posture and strut to persuade the female to accept his genes.

The females' interests are different. She may be just as concerned with genetic stamina that will benefit her daughters as well as her sons. In many species the real question is the male's likelihood of guarding her and the nest, or feeding her and the young. In other words, indicators of kindness and reliability may count as much as aggression: hence the perfect rose. But females seeking such care still have a thorny problem in assessing displays and whether they are simply being seduced. Once they have mated and conceived, the male may not continue his investment in the young. Even in species where the male normally contributes only sperm, and the female is only attempting to find the best genetic father, her decision is nearly irrevocable once the eggs are fertilized, the child conceived.

Many displays accurately reflect the show-off's fitness. Hence the jumping up and down, which is actually costly in energy, or the roaring, whose intensity really does correlate with body size. Feather color is not so sensitive an indicator of bodily condition as are comb, wattle, and beak color, so hens prefer roosters with vivid wattles and combs.

Parasites probably play a role in the evolution of particularly telling displays, just as they do in the evolution of sex itself. More brightly colored birds live in regions of great parasite load, especially the tropics. One of the clearest signals that a potential mate is not suffering from an infection is the condition of his feathers or, in the case of mammals, fur. This idea is very difficult to prove, but one study suggests it works at least for guppies. In Trinidadian streams where the guppies have predators, they are dull colored, to escape predation. But in streams with no fish predators, or in the same stocks brought into the laboratory, females select the males to be brighter in each generation—and brightness increases with increasing parasite load in the water. It seems that the female guppies have studied sexual selection theory.[31]

Paradoxically, *exaggerating* honest displays—those that tell you something truthful about a mate's fitness—may go so far as to become an actual handicap. But handicaps have their advantages, too. They tell the choosers that the chosen have so much energy that they can, for instance, survive even when equipped with a peacock's tail. Peacocks are wild birds of India, which is full of leopards and other peacock eaters. (Having seen wild peacocks trying to fly, I suspect they also taste terrible—something must protect such overburdened prey.) Even so, the peacock who survives its predators must be superfit. He will sire not only beautiful sons but redoubtable daughters.

The handicap principle was resisted by biologists when it was first proposed, in part because we like to think of creatures as perfectly adapted unless proved otherwise. The very word "handicap" made us uneasy, and still does—could choosers actually prefer a disability? However, this concept is now widely accepted.[32] Jared Diamond, in one of those jokes that is not a joke, suggests that ads for smoking and drinking, with their macho cowboys and model babes, do not just say that cigarettes will blow-dry your hair and buy you a horse and a yacht. They imply that smokers and sots who stay handsome even while ruining their health must be the true supermen and superwomen. In short, the peacock is attractive to peahens not simply because he has a beautiful train but because he can get away with it.[33]

Running away with beauty. These exaggerated displays are said to be the result of "runaway sexual selection." This process can also drive the evolution of display. Suppose there is an initial bias of preference in one sex, and by chance some members of the opposite sex have the key that fits the lock. Perhaps females like red, and some males are a bit ruddy. Those males find more or better mates among the red-fancying females. The offspring would inherit genes for both traits, redness and red-appreciation, particularly if the two become linked on the same chromosome. This sets the stage for a feed-forward cycle. The females who act most decisively on their preference mate with the reddest males, who leave the reddest, sexiest sons, and the females leave daughters who choose the sexiest males. Before you know it, you wind up with scarlet tanagers.[34]

In a few cases—notably among sword-tailed platyfish and the Tungára frogs of Mexico—such a linked trait can be traced to its origin. For both of these groups there are more primitive species that lack the distinctive male display—the elongated lower tail fin in the case of the platyfish and the distinctive croak in the case of the frog. If you show males of the exaggerated species to females of the other, all the female fish prefer the dashing swordsmen and all the female frogs have ears tuned to the pitch of the Tungára male croak. In other words, it was a widespread, primitive female preference that originally drove the males to their antics.[35]

> The Tungára Frog is a classic case
>   Of Sexual Selection.
> "Tungára!" the male sings all night long,
> For the female's predilection.
> He tunes his guitar in the steamy swamp
>   To the female amphibian ear;
> For a wooing male is constrained to croak
> What the female is ready to hear.
> Oh, lecturers heed the Tungára Frog
>   As oratory you hone:
> Your hearers will hear what they're ready to hear,
> Or else they will leave you
> Alone.[36]

Females sometimes even prefer a "supernormal" mate who outstrips any real one. Kenyan widowbirds are black, crowlike birds that pop up out of the savanna grass and do a frisky dance, flirting tails longer than their bodies. Malte Andersson clipped off some male widowbird tail feathers and glued them onto the ends of other males' tails. (He set the appropriate controls—males with their own feathers clipped and glued right back on.) The double-tailed birds attracted more mates than normal, even though their tails were so cumbersome they actually caused the birds to eat less and probably made them more vulnerable to predators. (The birds whose own tails were cut and reglued did not attract more females, so the effect was tail length, not handling.) Female preference drives the system forward toward ever more gorgeous tails to ripple on the wind; only hunger and predation keeps the tail within bounds.[37]

There is a somewhat disapproving note to the term "runaway sexual selection." It implies that the females are frivolous, swayed by arbitrary aesthetics rather than soberly looking for good genes. Meanwhile, the males pander to them, wasting energy on plumes and sword tails rather than concentrating on the serious business of being fitter in the face of the environment. Then if males are pushed to develop ever more splendid tails and plumes and coloring, checked eventually by being less able to find food or to escape predators, the female can also judge the fittest of the suitors. There is just this sort of constraint on the Tungára frog: its call attracts not just females but also frog-eating bats. Those frogs who sing lustily but still manage to dive into the water just ahead of a swooping bat are worthy mates indeed.

Sexual selection for beautiful displays is the best candidate for the feelings and skills that evolved into our own sense of beauty. Characteristics that strike us as beautiful flame forth in many animal displays. Exaggeration—the peacock's tail. Brilliance—the rainbow trout and the fluorescent blue wrasse and the scarlet tanager. Rhythm—the dancing mannequin or bird of paradise, the croaking and chanting of birds and frogs. Theme and variations—the nightingale's song. Symmetry—now take symmetry.

More cutting and pasting has shown that barn swallows can judge the symmetry of swallow-tail feathers to a couple of millimeters; female swallows prefer symmetry. A symmetrical tail may actually help swallows' aerodynamics; it is also a sensitive indicator of the power of developmental control. If the genes can impose regularity of control in the face of the fluctuating influences of the environment, this means powerful genes, well worth choosing in a mate. Humans, by contrast, are one of the more asymmetric species around. (Fiddler crabs are another.) We keep different capacities on different sides of our brains, and most faces and many torsos are a bit lopsided. We also have strong feelings about how much asymmetry we can tolerate in a "beautiful" face.[38]

It may well be that our aesthetic senses were honed through sexual selection rather than natural selection. Obviously, aesthetic delight is not the same as feeling sexy. We appreciate the grandeur of a snowy mountaintop, the precise mysteries of a painting by Rembrandt or Van Gogh, even the preposterous peacock, none of which are appropriate human sexual partners. But the biological relation of preference in the wider sense to partners' sexual fitness is intriguing. The one thing that we do not usually think of as beautiful is the ordinary. When artist or lover discovers the beauty in an "ordinary" scene, it immediately is transformed into something distinctive, resonant with rhythm, symmetry or complexly balanced asymmetry, and brilliance.

More than one way to do it. In most species, mate choice and courtship gambits are not made along a single dimension. Choosers may well assess more than one aspect of the chosen. A few species codify alternative courting tactics. Bluegilled sunfish males, which are large and territorial, scoop out shallow mating territories in the sand of English ponds. Females come to the territory to size up the big bright male and his boudoir. Meanwhile, a second kind of male, small, drab, and remarkably like a female, hangs around the periphery, ready to dash in and fertilize some or all of the eggs if the female decides to spawn. The original technical term for the second class is "sneaky fuckers," a term used in lectures to see what one could get away with. In sober textbooks, it is now abridged to "sneaks."

The relative numbers of each class depend on the other. Given a surplus of sneaks, the better strategy would be to go set up a new territory, not hang out in the throng around present territories. If there are a great many territorial males but few sneaks, the more profitable course is to be a sneak. Far from some single type being ideal, evolution itself often produces alternatives.[39]

Nature is not limited to just two alternatives. A little California iguana, the side-blotched lizard, has three reproductive strategies. Orange-throated males hold territories, blue-throated males run around trying to guard females, and males with yellow-stripy throats are sneaks. In our own species, males will try anything, from aggression to mate guarding to resource-holding to just being beautiful, to differentiate themselves from the competition. Women have their own ploys as well. Most of these gambits are not genetically fixed but are adjusted to the immediate situation in one's own circle of friends and rivals. The principle applies, though, that polymorphism in courtship may get you further.[40]

Role Reversal

Although I have said "chooser" and "chosen" wherever possible, the terms map onto female and male in most species. Male choice does occur, however, in monogamous (or quasi-monogamous) species like sea horses and humans, where both mates have a similar interest in their brood and are willing to spend time assessing, and being assessed by, potential mates. A very few species thoroughly reverse the usual pattern, making females compete for access to males. Mormon cricket females in food-poor areas actually fight to obtain males' packages of edible sperm, while the coy males refuse about half the copulations they are offered. Among jacanas, a lily-trotting waterbird, males incubate the eggs while females fight with one another and guard their current mate. Moorhen females are larger and brighter than males and more likely to fight and court. Small moorhen males are actually in better condition than big ones, and spend more time incubating. Females compete most for small fat males![41]

A nice case is the stomatopod *Pseudosquilla ciliata*. These snapping shrimp have an armored forelimb that jerks open to stun and stab prey. One hazard of working with this group is that some species jab right through aquarium glass. The females are big, aggressive, and take the lead, both chasing and clutching males. As courtship progresses, the two begin a "dance" of repeated passes in opposite directions, rolling so their backs brush along one another. Even then, the male may break off the ritual, if he can get away, before the acrobatic finale when his forelimbs clutch her back while his abdomen curls round to engage in ventro-ventral mating.

Males are the choosy half of this duo: they want large females with many eggs. It is not clear what the females are after. Males do not incubate, but they may pass nutrients with the sperm. Females mate frequently. They store sperm, only some of which is used, so the male who mates could be giving up his effort (not to mention his personal risk) for nothing. This is, in short, an almost total reversal: the act of mating itself has a higher cost for males, so females mate more often, more widely, and go for broke.[42]

Sperm Competition, Sperm Choice, and Female Orgasm

Competition for fertilization at the level of sperm and egg takes us right back to the original competition of something like sea squirt gametes in ancient oceans. Millions of one sex compete for access to the other. Any choice made is chemical, on the basis of whether antigens on the sperm coat are appropriate to the surface of the particular egg. In any one mating, sibling sperm compete to be fast, energetic, and chosen, but at least they all come from one father. If the female mates with several partners, the competition is between different lineages.

Sperm competition has one great advantage for males. It can replace physical combat. This is particularly advantageous if a male wishes to stay bonded to his buddies. The muriqui, the largest of New World monkeys, take this much further than we do. Males almost never dispute, but lay a blond-furred arm round one another's shoulders in reassurance. Their testes, however, are spectacularly hypertrophied. The male with the biggest balls, literally, wins the competition.[43]

Males who produce more sperm flood out their opponents. Chimpanzees and bonobos, whose females mate with many males, have far larger testes for body size than humans, and much much larger than gorilla harem masters. Chimp testes are about 4 ounces for a 100-pound animal; gorillas are about 2.5 ounces for a 450-pound animal. Gorillas keep rivals away by teeth, muscles, and reputation, and so do not have to share their mates. Human testes weigh about 2.5 ounces for a 175-pound man—suggesting that our females are more promiscuous than gorillas, but less so than chimps. All these gonads are rather insignificant compared to right whales—not the largest of whales, but multiply mating ones, where females are surrounded by a flurry of males. Right whale testes each weigh a quarter of a ton![44]

The extraordinary work by Baker and Bellis suggests that men adjust the quantity of their ejaculate to the possibility of two-timing. Baker and Bellis persuaded 35 couples from their university biology department to take part in research on human sex by delivering up the contents of condoms for study. When couples had been physically apart for several days (common with academic commuter lifestyles), men produced more sperm than their normal baselines, even if the time since previous intercourse was the same. It could be a way to react physically to the possibility of sperm competition.[45]

One means of exerting choice in ourselves and other primates is female orgasm. Baker and Bellis actually persuaded 11 women to squat over a beaker after each intercourse, so they could measure the "flowback," the few cubic centimeters of liquid that re-emerge when the woman stands up. Flowback was much smaller after orgasm. Orgasm retains sperm.[46]

The variety of speculation about what, if any, is the function of female orgasm is rather odd. The three leading theories were, respectively, the pole-ax, upsuck, and by-product theories. (These revolting names reflect biologists' general unease at considering the subject.) The pole-ax theory was that orgasm leaves women so fulfilled or exhausted that they don't want to get up in a hurry, and lying down longer facilitates sperm retention. The by-product theory was that female pleasure in sex, from clitoral sensitivity to orgasm itself, is just an evolutionary by-product of being in the same species as men, who need sexual sensitivity and orgasm in order to reproduce at all; without penetration and ejaculation, there would be no procreation.

The upsuck theory, which seems conclusively supported by the Baker and Bellis work, is that orgasm actively transports sperm higher in the female reproductive tract. Baker and Bellis go on to say that this reinforces a female's power of choice, especially when she is sleeping with more than one man, which does happen. If she refrains from orgasm with one partner, she is less likely to conceive his child. In other words, good sex makes it more likely that a lover will become a father. This seems to surprise a few scientists.

Human Display and Mate Choice

We see aspects of our own behavior in a great many other animals: shyness, display, aggression. The most fundamental point is that we are capable of just about anything. Far from telling us what is "natural," the range of animal comparisons tell us that most things are natural. Humans do not divide into two mating classes, like territorial or sneaky sunfish. In any one community, and often in one person, we have multiple options. This recalls the point that a species does not prosper by having a single "ideal" type, best at everything, but a wide range of types from the genetic level upward. This said, of course we have our own peculiarities: permanent breasts and buttocks in the female and an enormous penis, compared to other apes, in the male.

Female fat and concealed estrus. Women do not have external signs of estrus; indeed, estrus is usually not even consciously marked by the woman herself. Why did we change from the ancestral ape condition, in which estrus is signaled by odor and, for some, bright sexual swellings? Perhaps the sexual signals among human females are a "deceptive" or, better, a "supernormal" stimulus. Breastmilk is secreted by milk glands, not the fat that surrounds them, which indicates nothing about milk production. Flat-chested and big-breasted women are equally able to feed babies, as are chimps, who have nothing visible but a nipple. Wide pelvic bones aid easy birth, but panier fat deposits alongside the pelvis do nothing of the sort. Caroline Pond, who has dissected fat deposits in everything from road-killed badgers to Inuit-killed polar bears (a job requiring dissection up to the elbow) points out that women's fat is distributed in the same zones as other mammals, but particular fat depots are exaggerated, like any other sexually selected signal.[47]

This does not mean that fat itself is arbitrary. In people and other primates, fat level correlates with menarche (age at which periods begin) and the interval between births (which is longer when fat level is low). Many cultures "feed up" a girl before marriage. Our own overstuffed culture is little aware of the constraints we faced throughout evolution in getting enough calories to reproduce.[48]

The latest support for the theory that fat is important to reproduction comes from the discovery of leptin. Leptin is a protein that tells the brain it is "fat enough" and that one can stop eating. Failure of leptin chemistry leads mice, and people, to eat themselves into obesity. Now it seems that for female mice, leptin also triggers puberty. A hormonal signal that says "I am fat enough" is also involved in saying "I am now ready to ovulate and to risk pregnancy."

The hourglass shape of nubile young women signals the same primitive message to young men. The contrast between a small waist and wide hips and breasts says that the woman is fertile (as evidenced by her breast and hip fat) but not pregnant (as evidenced by her small waist). It is the waist-to-hip ratio that attracts mates. Of course, even waist-to-hip ratio, real and preferred, is affected by culture. Machiginga Indians of the upper Amazon think thin-waisted women look starved, not attractive. The ideal ratio grew in Western culture from the bustle-bottomed 1880s up to the straight-hung fringe of the 1920s, and again in the anorexic 1990s. Fashionable or not, though, women still run to hips.

Why, then, does a woman carry these signals of general fertility but does not specifically signal estrus itself? Cycling chimpanzees and baboons, who have multiple mates, flaunt large pink swellings that indicate just which days of the cycle are most likely for conception. Bonobos, who mate through almost all their cycle, and with almost everyone regardless of age or sex, have sexual swellings that last for almost the whole of their cycle. Even in bonobos, though, casual sex is distinct from estrus sex. When the female is fully swollen, the adult males know it and compete for access. Humans do not usually know. A few women feel a little pang when the egg is released, called *mittleschmerz*, but most women, as well as most men, are unaware of the timing. In the modern era, many women monitor changes in the cervical fluid and use that knowledge for birth control. However, such conscious checking was not likely among the ancestors who evolved our human body. Instead they, and we, greatly extended the occasions when we are receptive and eager for sex.

We will return to concealed estrus and its social role in Chapter 8, on human ape society. However, female humans are not uniquely sexually active. We used to think humans the sexiest of primates, but we now know that bonobos leave us in the shade for both frequency of sex and variety of partners.

The enormous penis. Another peculiarity of humans is that the penis is twice the size for body weight as that of any other primate. Indeed, gorillas and orangutans are strikingly underendowed in this respect. Among bonobos, the penis is longer than our own, but thinner. Male bonobos display a pink and obvious erection and go in for "penis-fencing." The two apes with most frequent sex, ourselves and bonobos, are the two with huge penises.[49]

Jared Diamond concludes that the function of a large penis is to show off to other males. He may be influenced by his time in New Guinea, among people whose chief clothing is the exaggerated and erect penis sheath. It is true that men's genitalia seem to interest other men, in urinals and locker rooms, more so than women. But the reproductive advantages of giving pleasure to females, promoting mate choice, and stimulating female orgasm by tactile means should not be underestimated.[50]

Faces and antigens: how much like yourself? The first signals picked up from the opposite sex are hardly so crass as breast fat and penis size. Most people look at other people's faces, and when they do, they seem to be looking for signs of health: shiny hair, clear skin, alert eye, youth (in women), and (in men) age a little older than the woman. It is also typical to like symmetrical faces more than asymmetrical ones, though most faces are slightly asymmetrical. But beyond these generalities, people seem to choose mates who are almost, but not quite, like themselves. In fact, people like people who look a bit like their parents, right down to earlobe size.[51]

The animal parallel is Japanese quail. Presented with an array of birds whom they had not previously met, quail prefer to mate with second cousins—not so like them as siblings (that would be dangerous inbreeding) and not so different as totally unrelated birds. People's choices are apparently much the same: Charles Darwin himself married his first cousin, Emma Wedgwood, but worried desperately about the potential effects of inbreeding on their children. This preference perpetuates differences between races and even local groups. On average, of course—plenty of people are also fascinated by the new and exotic.[52]

At the molecular level, people apparently choose mates who differ in their immune system. Mice clearly pick mates who differ in major histocompatibility (MHC) genes that influence the immune system. Female mice avoid males with the same MHC genes. This goes back to sex and the Red Queen. If a major function of sex is to get ahead of one's parasites, the most crucial thing of all is to change the antigens of the immune system. Mice can smell out mates who are sufficiently different.

Can people? Forty-four male graduate students in Zurich slept in tee-shirts for two nights, while not using deodorants and only odorless soap. Then each tee-shirt was put in a plastic bag and solemnly (or gigglingly?) sniffed by 49 female graduate students. They rated the shirts for attractiveness, sexiness, and similarity or dissimilarity to their own current or previous partners. (Sexiness turned out to be the same trait as attractiveness.) The women reliably and significantly chose as sexy and as like their own partners the tee-shirts of men with differing MHC loci from their own. Repetition of the experiment, this time with 121 men and women, confirmed that they were not selecting for some especially good combination—just for someone different from themselves. Interestingly, the men—not just the women—graded men's shirts with dissimilar MHCs as sexy and/or attractive, and those with MHCs similar to their own as not attractive.[53]

However, here's a wrinkle: those women on birth control pills chose tee-shirts with MHC genes that *matched* their own. This would make a certain sense, biologically. Birth control pills mimic pregnancy; if a woman is pregnant, she should stay with her own family, with familiar essential smells, rather than chase after new males who might abuse her and her infant. This finding does raise troubling questions about the effect of the pill on an ongoing relationship. What happens if a woman falls in love with a man who has a sexily different MHC and then happily bears his child? I wonder whether by the time she is pregnant she has assimilated her mate's scent to her own, so that the scent of him in bed is comforting and safe?

Status and kindness. David Buss gave questionnaires to students round the world, asking what they look for in a mate. The students were not representative samples of their societies. They were often not even very large samples from any one society, sometimes only a class run by one professor. The total sample, however, was very large indeed, more than 10,000 people from 37 countries. Buss believes the answers indicate amazingly consistent choices across our species.[54]

Kindness is what most people of both sexes said they seek first, more than any other factor. From there on, males and females diverge. Males look for youth and beauty—the physical signs that a woman will bear many healthy children. Females are more likely to look for status and wealth—a promise of resource provision for the family. Women may not say that they want high status in a mate, but in one U.S. sample, the men women chose to marry earned 50 percent more than the men women did not choose to marry. Further, women choose to marry men taller than themselves, an indication of health and vigor as well as social status.

Is this just a dubious reinforcement of banal stereotypes? Or did the stereotypes become stereotypes because they reflect real average tendencies of our species? To turn the question around, would you trust a research result on human mate choice that actually *contradicted* common stereotypes and "common sense"?

Even if Buss's conclusions lack subtlety, he is looking for common denominators of humankind. By definition, a believable answer should lack subtlety. What may be more important to say, yet again, is that we excel in variety. However interesting it is to identify the average, it is even more interesting to realize that many people are not average, and with good evolutionary reasons not to be.

Fashions. Almost all changes in sexual displays among animals happen over evolutionary time, usually hundreds of generations. Fashions among humans, however, change from year to year. Human fashions may wildly distort or even negate our evolved species-typical looks. What about lip-plugs that distend the lips into circular duck beaks? Four-inch-long dangling earlobes? Even changes that may threaten reproduction, such as Australian aborigines' subincision that opens the base of the penis, or anorexically starved, heroin-high American models of the 1990s? Fashions are self-imposed "handicaps" in the sense of "look what I can get away with." They have an extra dimension, because in humans they indicate social status as well as physical prowess. As we will see in Part IV, the earliest human clothing was probably beads and body paint.

We are the only mammal that I know of whose head hair is not self-maintaining. Several of our primate relatives sport hairdos that would look quite elaborate on a human, like the cotton-topped tamarin with its long white plumes, or the bonobo with its central parting and horizontal head-tufts. Their hair just grows that way. Human hair instead mats, cakes, and attracts lice: its very physical form could not have evolved without the culture of hairdressing. I suspect that we segued from primate mutual grooming and parasite-picking straight on to a sexually selected advantage of showing off that we had the manual skill to do our hair, and that we had friends who would help with the bits round the back. Local custom suggests whether tresses are washed or oiled or mudded or dyed, straightened or curled, cut or extended. However, we all judge hair that is totally unfussed with as a sign that the wearer is desperate or insane. Anthropologist Judith Berman points out in her article "A bad hair day in the Paleolithic" that we also assign unkempt hair to our notions of "cave men." Yet the cave people themselves, when they made the first human representations over 30,000 years ago, fashioned statues of women naked with elaborately braided coiffures.[55]

Keeping up with rapidly changing fashions takes physical stamina, cleverness, time, and often money—all desirable in a mate. The only animal parallel I know is the song of the humpback whale. All male humpback whales sing almost the same song at the same time, but it grows variants and new themes throughout the mating season. Females approach the massive singers to mate. We do not know how a female makes her choice, but we presume that some males just sing better than others. Someone must be ever so slightly in the lead toward next week's version. As mere visual, sea-surface creatures, we humans must match depth recordings of a singer to a glimpse of his tail flukes above the waves to guess who is who. Female humpbacks instead apparently sense half an ocean away who is at the top of the ratings.[56]

The Red Queen's Race between the Sexes

We have skipped throughout this chapter between confrontational and cooperative sexual behavior. Each sex needs the other, but each may also be imposing on the other, duping it, two-timing, or, for the male elephant seal, twenty-timing. If an unfortunate pup wanders too near a big bull, he can simply roll over and squash it, without apparently noticing or caring, in his single-minded concentration on defending his harem. Females, too, may be wounded, though elephant seals mate lying on their sides, which keeps a mating female from being crushed by the weight of the male behemoth.

The antagonistic game of opposite roles plays out, like any other Red Queen race, between evolving competitors. William Rice bred fruit flies for forty generations, allowing the males to evolve but holding the female type constant. (Forty generations is not very long: 800–1,000 years for humans, less than two years in fruit flies.) In each generation, he gave the males new females from a separate population of their original stock. Females who had once mated, and the daughters of those females, never returned to the experimental population. Only their sons carried on the experimental line. That meant that the males were selected to be better and better at being males, but the females never had a chance to adapt to dealing with those supermales.[57] As the males evolved, they developed seminal fluid that was toxic to other males' sperm. This aided in sperm competition, but it was also toxic to the unevolved females. The supermales also evolved a tendency to mate more frequently. The more of these toxic matings a female received, the more likely she was to die.

In a normal population, female adaptations would counter the effect. Perhaps females would evolve to neutralize the poison. Perhaps females would offer counter-benefits to cooperative males who increased the females' fecundity and their own life span instead of merely competing with the sperm of other males. Perhaps they would choose more agreeable mates. In Rice's experiment that removed half of the Red Queen race between the genders, there were no checks and balances. The outcome was not pretty.

Does this fruit fly experiment mean that gender war is inevitable? No. It means that the sexes adjust to each other, generation by generation. In species where each sex gains most by cooperating, the winners of the reproductive race are those who are best at cooperating and at persuading their partners to do the same. The toxic sperm of the fruit flies is no more inevitable than the edible sperm of the katydid, the male's gift that raises the female katydid's nutrition and woos her compliance.

Now we can turn to a much wider view of the costs and benefits of cooperation, played out not just between mates but among parents, children, aunts, and the soldiers' platoon.

5

Calculating Love

$\mathcal{M}y$ father and his brother were orphans. Their mother died in childbirth. The little boys remembered a tall, thin, dark, and lovely stepmother, who gave their father tuberculosis and died with him, a fate common enough in the 1890s. The boys were packed off to be raised by Aunty Marion and Aunty May.

Neither aunt married, though family legend tells that Aunty May refused the hand of the richest man in Brantford, Ontario. Aunty May became a triumphantly sarcastic and independent spinster, a nurse by profession. She toured Europe alone at the age of 80 because she had no intention of being bothered by a traveling companion.

If we view an organism as the transmitter of a group of genes, we can make a simple calculation of Aunty May's choice. Suppose she'd married and had even one child. That child would carry half her genes. Her dead brother also shared half her genes. The two little nephews had only a quarter each. Genetically, the two of them counted the same as one son or daughter of her own. As a practical matter they did count a bit more, because when Aunty May took on their care they were past the dangerous stage of early infancy when so many children succumbed. In biological terms, this gave them

Figure 6. Aunty May and her nephews, about 1896. (After Bishop family photos)

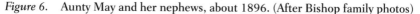

higher "reproductive value" than unborn babes. Aunty May's aged parents, whom she also nursed, had no "reproductive value" at all, except as they helped their grandsons with resources, baby-sitting, or useful advice.

My point is that all of this was perfectly obvious to Aunty May. She did not need biologists to tell her about degrees of kinship and the risks of infancy. She was quite able to make up her own mind, thank you, and to include the other factors that mean human love cannot be so simply quantified. It is also possible, of course, that she did not like her ever-so-suitable suitor.

Calculating Love

Kin Selection

One might not expect Aunty May's decision from a worker ant, who is also denied her own hope of reproduction when raising her nephews and nieces. Darwin saw ant castes as a real challenge—not so much their existence, because he thought sterile workers could evolve if it profited the community, but the fact that they had their own peculiar forms. "One special difficulty . . . at first appeared to me insuperable, and actually fatal to my whole theory. I allude to the neuters or sterile females in insect-communities: for these neuters often differ widely in instinct and structure from both the males and fertile females, and yet, from being sterile, they cannot propagate their kind . . . This difficulty, though appearing insuperable, is lessened or as I believe disappears, when it is remembered that selection may be applied to the family, as well as to the individual."[1] Note that he does not say the group but the family—the descendants of the sisters of the worker ant.

Following Darwin, there was a long period when zoologists talked of adaptations being "for the good of the species." From that point of view, the self-sacrifice of a worker ant did not seem such a problem. This kind of thinking pulled up short when it tried to explain sex. If sex, or indeed any other traits, evolve and are maintained, it is for the good of the individual's genes, not for the good of the species.[2] Whose genes does the worker ant benefit?

The answer is her full-and-a-half sister's. Male ants are haploid. They have just one full set of genes, so they pass the whole lot on to every offspring. That means every worker in the colony has identical genes from her father. The queen is diploid. She gives half of her genes to each daughter, just as a human mother does. If a worker mated and laid eggs, her offspring would each get just half the worker's genes. But when the worker helps send out new young queens to mate instead, she launches her winged younger sisters, who share a full three-quarters of her genes. Her sisters have half, on average, of the ones she got from her mother but an identical set of the ones she got from her father.

In contrast, the ant's brothers develop from unfertilized eggs. They share only a quarter of their sister's gene complement. The workers invest minimum effort in these half-brothers, and ant males don't lift an antenna to help with the housework.

In 1964 William Hamilton published an article which changed everything back to Darwin's view that the key to understanding evolution and behavior is selection on genes shared within the family, not just those of the individual alone nor the species as a whole. He worked out the general principle, as it would apply to everything from clonal protozoa to people. Then he realized that the haplodiploid relationship between ants, which had been known for a long time, was the perfect test case. He concluded that their odd means of sex determination would explain the workers' self-sacrifice and the drone males' lack of contribution toward the colony. Kin selection, acting over generations, would "calculate" the relative genetic advantages of selfish and selfless behavior. Among ants, bees, and wasps, social life based on the labor of sterile females should develop over and over in distant lines. (It did.) Among other insects and other animals, that kind of society should appear very rarely. It has done so just twice, in termites and naked mole rats, which turn out to form such inbred colonies that they are also super-related, like female ants.[3]

Let me quote E. O. Wilson, not for what was in Hamilton's article but for the most powerful description I know of the agony of confronting a truly new idea:

I first read Hamilton's article during a train trip from Boston to Miami in the spring of 1965. This mode of travel was habitual for me during these years . . . I picked Hamilton's paper out of my briefcase somewhere north of New Haven and riffled through it impatiently. I was anxious to get to the gist of the argument and move on to something else, something more familiar and congenial. The prose was convoluted and the full-dress mathematical treatment difficult, but I understood his main point about haplodiploidy and colonial life quickly enough. My first response was negative. Impossible, I thought. This can't be right. Too simple. He must not know much about social insects. But the idea kept gnawing at me early that afternoon, as I changed over to the Silver Meteor in New York's Pennsylvania Station. As we departed southward across the New Jersey marshes, I went through the article again, more carefully this time, looking for that fatal flaw I believed must be there. At intervals I closed my eyes and tried to conceive of alternative, more convincing explanations of the prevalence of hymenopteran social life and the all-female worker force. Surely I knew enough to come up with something. I had done this kind of critique before and succeeded. But nothing presented itself now. By dinnertime, as the train rumbled on into Virginia, I was growing frustrated and angry. Hamilton, whoever he was, could not have cut the Gordian knot. Anyway, there was no Gordian knot in the first place, was there? I had thought there was probably just a lot of accidental evolution and wonderful natural history. And because I modestly thought of myself as the world authority on social insects, I also thought it unlikely that anyone else could explain their origin, certainly not in one clean stroke. The next morning, as we rolled on past Waycross and Jacksonville, I thrashed about some more. By the time we reached Miami, in the early afternoon, I gave up. I was a convert, and put myself in Hamilton's hands. I had undergone what historians of science call a paradigm shift.[4]

Reciprocal altruism.   Soon Wilson was also arguing with a brilliant graduate student named Robert Trivers. Trivers claimed that not only family love but mutual support between friends evolved through natural selection. In the right circumstances, helpers do not even have to belong to the same species. According to Wilson, "We switched from concept to gossip to joke and back to concept. Our science was advanced by hilarity. My own pleasure in these exchanges was tinged with a sense of psychological risk, as though testing a mind-altering and possibly dangerous drug."[5]

Forty-five species of fish and six species of shrimp have evolved to clean parasites off other fish. A fish never, ever eats its cleaner. Even when startled, the big fish gives a distinctive wiggle to warn the cleaners to leave, or in the case of the moray eel, sharply jerking its head before the jaws snap shut.[6] In a large and randomly mixed population, there is no gain in altruism at cost to oneself. However, many animals actually live in fixed sites, like the cleaner fish, or in small social groups. If you are likely to meet the same companions day after day, and if you can be selectively nice to those likely to return the favor, altruism evolves so long as the cost to the giver is small compared to the benefit to the receiver. Trivers staked the claim that simple cooperation between individuals laid the basis for "friendship, dislike, moralistic aggression, gratitude, sympathy, trust, suspicion, trustworthiness, aspects of guilt, and [punishment for] forms of dishonesty and hypocrisy."[7]

It is obvious that friendship and dislike could be psychological correlates of such a system. But moralistic aggression? Well, that is a way to punish cheaters. Injustice and unfairness make us angry out of all proportion to the offense. Trust or deception are not isolated incidents, however trivial, but clues to a stance likely to accumulate over a lifetime. Friends may even kill each other over some trivial betrayal. The fury of the poet Archilochos still flares after two and a half millennia.

. . . slammed by the surf on the beach
naked at Salmydéssos, where the screw-haired men
of Thrace, taking him in
will entertain him (he will have much to undergo,
chewing on slavery's bread)
stiffened with cold, and loops of seaweed from the slime
tangling his body about,
teeth chattering as he lies in abject helplessness
flat on his face like a dog
beside the beach-break where the waves come shattering in.
And let me be there to watch;
for he did me wrong and set his heel on our good faith,
he who had once been my friend.[8]

Multilevel Selection

What if the population is not large and randomly mixed but so geographically inert that neighbors are likely to keep meeting over again? What if the population is not just viscous but subdivided into discrete groups? This is in many ways the core question of sociobiology. If genes are grouped for a generation within one individual, and individuals are grouped within societies or tribes, there are times when their survival and reproduction will depend on the success of the group as a whole.

Competition among peers leads genes, or individuals, or groups, to be "selfish." One out-competes the next, in simplistic Darwinian fashion, to leave more offspring. However, if selection at the next higher level is strong enough, a group containing altruists—at least some altruists—may out-compete other groups. By definition the altruists will be at a disadvantage within the group due to their self-sacrifice. Still, if their altruism helps the group as a whole to grow, prosper, and survive to reproduce, then the altruists' genes may increase in absolute numbers in the next generation. Their own selfish group-mates' genes increase even more. So long as they are cooped up in the same group, altruists help the selfish as well as one another. Then, when a new "generation" of groups reassort members, some may have a very high number of altruists, who in turn help their groups out-compete their neighbors in between-group competition. Altruism has still greater advantage if altruists tend to gravitate toward other altruists. Then they may form groups that work together to have very high overall success, in competition with groups riven by selfishness.[9]

Like kin selection and reciprocal altruism, multilevel selection is a numbers game. How intense is the selection between individuals, as compared with selection between groups? How tightly bound are individuals within their groups? What is the generation time of an individual, as compared with its group—that is, how often do the groups break up and reassort? What is the likelihood that altruists selectively join other altruists, either because they recognize and like one another or just because they stay in the same neighborhood?

Suppose the group is a species. Species live for millions, even hundreds of millions, of years. All that time, each individual competes against others under the laws of natural selection. Selfishness between individuals pays off. There is no way that an individual evolves to be altruistic toward its whole species—except, just perhaps, under the wholly novel conditions of humankind, as I will suggest in the final chapter.

Suppose, though, that the "group" is a set of genes cooped up together in an individual. There are in fact some truly selfish mutations at gene level, which drive meiosis in such a way that they wind up in the one egg of four that survives division, or which skew the sex ratio to pass on those particular genes. However, selection at the level of the whole individual is so strong that other genes are quickly selected to counteract them, increasing the chance of the whole individual's survival and reproduction. The surprise is not that some selfish genes exist but that there are so few. The fate of genes in a cell or body is largely determined by the success of the group.[10]

The interesting question, then, is where is the breakpoint? What actually tips the balance in favor of a powerful influence of group-level selection as compared to selection between group members? This does not mean either-or: both processes happen at once. It means under what, if any, circumstances, is selection noticeable at many levels, not just one?

"Group selection" as a concept has a tormented recent history. Vero Wynne-Edwards proposed in 1962 that animals and plants actually evolve to curb their own reproduction, to save their herd or species exploiting the environment below subsistence level. George Williams, John Maynard Smith, and William Hamilton demolished that argument. They showed that selection for the personal advantage of the individual must always outweigh selection on a large and nebulous herd, let alone a whole species. However, for more tightly bound groups—genes in an individual or idealized mice nesting in idealized winter-long haystacks—the question comes down to the force of selection at each level.[11]

For humans, and for a more qualitative answer, the problem seems to me to be simpler. We are all aware of challenges to our group (or various groups) as a whole which impinge on individual survival. Minimal acquaintance with human history tells that intergroup warfare has been ubiquitous in the past, and, I will argue, stretches back at least to our common ancestor with chimpanzees. Only in favored times have we merely battled disease and poverty, not the prospect of being killed by Romans or barbarians, Mongols or Han, Apache or Cavalry. We have had ample opportunity to evolve an altruism that stretches out to our group at large in crises of life or death.

In 1975 Wilson summed up the costs and benefits to genes, individuals, and society in *Sociobiology*, the book that has named the field. The following year, Richard Dawkins published the same ideas in *The Selfish Gene*. The wider public suddenly realized that biologists were quantifying the ways that kindness and love could grow from genetic competition.[12] Before that, most people had vaguely assumed that our behavioral traits have some evolved basis, just as people before Darwin and Wallace vaguely assumed that species evolved. As with Darwin's explanation of natural selection, what raised the furor of sociobiology's opponents was having their noses rubbed in a quantitative, mindless mechanism. It seemed *immoral* that morality should increase genetic fitness.

Of course, that is just the reaction one should expect. We evolved to be superbly sensitive to distinguishing an open-handed giver or an impulsive hero from a calculating Scrooge who never moves until he figures his own percent on the deal. The whole point of reciprocal altruism is the bedrock of trust. Calculating the pay-off for kindness is the ultimate ungenerosity. We are programmed to deeply distrust— or even detest—the theory of sociobiology.

Parent-Offspring Conflict

Calculating love at first seems much more threatening than considering conflict. But what about conflict within a loving relationship?

A child wants milk. Its mother wants to save some of her energy for taking care of her own health, or raising the next child, or caring for older ones already born. How much milk and attention go to the new baby instead of the toddler with her nose out of joint at the new arrival? And how can a baby manipulate the parent to get more resources than the parent wants to give? By honest signals? Or by giving the impression of great need, and thus persuading the parent that a great deal of benefit will accrue if he doles out a spoonful of honey, a bit of quality time, or (in a nest of robins) an earthworm?

One way to falsify signals is to "regress" to more infantile manners, which would mean the benefit of a given gift is proportionately greater, as it would be to a younger child. This strategy may explain that appalling babyish whine that sets in when a school-aged child is tired, or the ultimate blackmail of public tantrums.[13]

A child shares all its genes with itself, but on average only half of its genes with each sibling. Therefore, a child will favor its own survival twice as much as that of any one sibling. The parents are equally related to all their offspring—they want the whole brood to survive equally. Parents try to inculcate all kinds of mutual aid among siblings, but at the point where the cost to a child is more than half the benefit to its sibling, the child should be biologically inclined to balk.

Sometimes the parents give even *more* resources than the offspring wants to accept, manipulating the child into the parental home as a "helper at the nest," while the child struggles to leave and seek independence. Rereading Trivers, I suddenly wonder about my father's other aunt, Aunty Marion. She was not the sister of redoubtable Aunty May but the half-sister of the boys' dead mother. In addition to changing off with May the burden of caring for her two little half-nephews, she devoted much of her life to looking after her own widowed mother, an imperious brother, and a half-brother. They all told her how to run her life, and Aunty Marion was always so eager to please.[14]

War in the womb. Parent–offspring conflict starts before birth. The fetus wants nutrients for itself, with discounted altruism toward its future siblings, and with an extra measure of consideration for the fact that if its mother does not survive, or is weak, she won't take care of it. But genes from the father are not related to the mother—and are not even sure of being related to the mother's future offspring. We now know that a few genes are active, or not, depending on which parent has contributed the gene. The *Igf2* gene (which produces insulinlike growth factor 2), if derived from the father, makes mice 40 percent larger at birth than the corresponding gene derived only from the mother. It seems to "demand" nourishment for just this fetus. The mother has a counter-strategy, however: an active gene of her own that clears out and destroys the product of *Igf2*. They more or less cancel one another, countervailing evolution so the mother does not explode. Mice (and humans) are now poised like teams in a tug of war: if either side were to let go, the other would fall flat.[15]

There may even be psychoactive genes that differ depending on their parental derivation. An imprinted gene "knows" whether it comes from mother or father and thus whether it is related to the mother's or father's kin. Trivers expects that active maternal genes would promote behavioral altruism toward mother, siblings, and further kin in the female line, while active paternal genes encourage selfishness except to paternal kin.[16]

American deermice proliferate into many species and subspecies. Most are polygamous, a few monogamous. The genes of monogamous fathers are relatively "unselfish," since a monogamous male's interests coincide with his monogamous mate's: he wants her to thrive and bear his future litters. Polygamous fathers' genes, by contrast, push their offspring to grow as big as they can at the expense of someone else's future kids. The females, of course, are prepared to counter these genes, so it normally comes out all right. If, however, you cross the strains in the laboratory, polygamous mothers clamp down on the relatively unassertive embryos of a monogamous father: the young are born as half-sized runts. In the opposite cross, the fetuses grow and grow, and the mothers die. Monogamous mothers cannot counter the extra demands of polygamous fathers' young. The mechanism is still a mystery, but the outcome is clear.[17]

Stepfather infanticide. In the early 1970s, when Sarah Hrdy was a young graduate student in anthropology at Harvard, she was surrounded by sociobiological discussion and looked for an extreme case to test the emerging theory. She heard that male hanuman langurs sometimes killed infants of their own species when they took over a new troop. This contrasted oddly with another field study of langurs, which depicted them as a peaceful society. According to this latter scenario, a rising male was once able to reverse dominance in his troop simply by having the courage to shake branches at the far end of the grove.[18]

Hanuman langurs are the sacred monkeys of India, tolerated and fed at temples and shrines. They are silver-gray, long-legged leapers, who carom off trees and temple walls with the precision of a hurdler and the aplomb of a ballerina. Hrdy reported that "in the industrial city of Ahmedabad, I was astonished to see a band of langur males zigzag through traffic, ricochet off moving three-wheeler scooter-taxis, and then disappear out of sight."[19] No wonder Hanuman the God called up an army of these leaping monkeys to save the heroine of the *Rāmāyana,* in ancient Hindu legend.

Could male infanticide be an evolved, adaptive behavior? Hrdy spent three years observing the troops at Mount Abu, a pilgrimage site of forests and temples rising above the dry plains of Rajasthan. Pilgrims and tourists fed the monkeys with chickpeas and chapatis, so a watcher could stand right in the midst of the troops. Hrdy saw eight invasions of her five troops by new males, during which a total of twenty infants disappeared. She saw one male make nine attacks on the new troop's infants, wounding one severely, which eventually disappeared; another male made more than fifty attacks. Hrdy saw only one actual killing, but the summary of attacks, wounds, and disappearances made an overwhelming case for infanticide as a recurrent part of langur life.[20]

She could even explain why. Male tenure at Mount Abu was only about 2.5 years. If the females came promptly into estrus, a new male had just time to father his own offspring, guard them through infancy, and bring them up to the point where his little sons could survive being exiled from the troop by his successor and his daughters would be reasonably independent of their mother. If he waited until the females finished their current pregnancy and lactation, he would leave only a group of vulnerable infants to be killed by the next usurper.

What counter-strategies were open to females? When a new male entered the troop, the females often "simulated" estrus. They invited copulation with the peculiar juddering of the head that is the langur's come-hither sign. If they were already pregnant, this sometimes "fooled" the male into treating the infant as his own. Mothers who already had infants attempted to stay clear of new males. Old females, grandmothers with little to lose, actually sometimes attacked the usurpers, but the male was twice their size and usually prevailed. The evolutionarily effective strategy for wiping out infanticide would have to be for females to refuse to mate with killers, but for any given female that would have meant no infant for the next three years or more—far too high a price to pay.

An enormous controversy erupted when Hrdy published *The Langurs of Abu* in 1977. From a feminist point of view, the controversy is bizarre. Hrdy spelled out clearly, publicly, that male and female agendas can sometimes be in opposition, rather than cooperative. In the climate of the seventies, surely that was an important message.[21] Instead, many people found the whole idea of evolved infanticidal tendencies so shocking that it could not be true. If the females resisted only up to the point where they escaped injury themselves and in the end mated and prolonged the killer's genes, that was even worse. Meanwhile, more infanticides occurred before observers' eyes. George Schaller had reported the same thing in lions even before Hrdy's study, with the same suggested explanation. Some 48 cases have been actually seen now in 13 primate species, without counting the hundreds of suspicious infant disappearances when a new male takes over.[22]

Then there are people. Martin Daly and Margot Wilson surveyed 408 cases of child homicide recorded in Canada, where records list biological as well as official relationship. Of course, child murder is very rare, and probably even more rarely detected. The rate of murder by birth parents was under 10 per million children, but by step-parents it was about 700 per million, most of which were committed by the stepfather. Stepfather homicides were overwhelmingly concentrated among children under two—the age of infancy that might significantly delay conception of the next child.[23]

The Daly and Wilson study produced even more furor, because it seemed to imply that extreme cruelty among stepfathers is in some way "natural," and if an act is "natural" it must be right, or at least excusable. So we should never admit that infanticide is "natural."[24]

The human infanticide controversy is relatively dormant for the moment. The primate controversy has flared up again. Critics claim that most infant killings appear to be spin-offs of other aggressive episodes, rare and evolutionarily trivial. This matters, because infanticide is now suggested as a driving force in the evolution of monogamy. Perhaps the monogamous male lemur or marmoset or gibbon— or even dung beetle—sticks to one female because he has to guard his own offspring against lethal neighbors. Even multimale groups may have evolved in part because it is an advantage to the females to confuse paternity, and an advantage to the males to have allies against infanticidal takeovers, even if it means sharing matings within the group. The critics see all this as pushing a theory far too far, on slim evidence.[25] They want to see proof that the trait of infanticide is inherited and is reproductively advantageous before accepting the argument.

The genetic basis of infanticide is now clear in mice, and the reproductive advantages are well worked out in lions. (Incidentally, wild animals can be as lethal as they choose, but in controlled experiments on laboratory mice, infants are put in a kind of mouse-sized metal shark cage so they are not hurt.) Year by year, evidence accumulates for the frequency and adaptive advantage of stepfather infanticide in a wide range of mammals.[26]

Adoption.   Infanticide as evolved behavior is gruesomely fascinating, but it is, fortunately, very rare in humans. Adoption is far more common—and also demands evolutionary explanation. Why would a person invest so much effort and love in a child with no genetic ties?

It might, of course, be an evolutionary byproduct, just a spillover of emotions that usually go to biological children. But a more likely explanation recognizes that in a social group, having children means having allies. Children take your side in conflicts with other families. If you can raise an extra child to be allied with you, without sacrificing your own reproduction, that increases your support group. If, as a dominant, you kidnap the infant from a subordinate, it not only enlarges your family of allies but diminishes hers. From lemurs to baboons, when there is a kidnaping, it is a dominant who does it.

Fortunately, there is a kinder side as well. If a mother primate dies, the outlook is bleak for her young unless other adults help. In the Gombe Stream, there was a remarkable barren female named Gigi. She had magnificent sexual swellings and was greatly popular among the males, but she never conceived. Over the years she adopted four different orphans and brought them up. Probably they were not her close kin, since females migrate among chimpanzees, but they survived under her care. Among Japanese macaques, it is more likely to be males who take over yearlings if the mother is killed. But throughout the primates, even among lemurs (a fairly primitive group), older sibs, aunts, and uncles adopt abandoned young, like Aunty Marion and Aunty May. This is common behavior in a crisis. As Frans de Waal points out, a great deal of our own evolved behavior is similarly "good natured."[27]

Choosing Children

Most human cultures practice, and approve, infanticide under strictly determined conditions. This is quite different from murder by a step-parent, because it is approved or carried out by the parents themselves. Parental infanticide is rare in other mammals, though almost any mammal mother that is disturbed during or just after birth will abandon her young. Rodents eat their litters if the nest is opened. Zoo bears, being adapted to giving birth in a cave in total solitude, may eat their young if they are merely exposed to strange noises or odors. Zoo gorilla mothers are prone to abandon or batter their infants if the mothers are separated from the father. This makes sense. Male gorillas in the wild are highly infanticidal of others' infants. If the group silverback dies or is deposed, there is not much hope of raising his infant offspring.[28]

Culturally sanctioned human infanticide targets infants whose demands would put the family at risk: a physically handicapped child, a twin, an infant born too close to an older sib. Nisa, a !Kung bushwoman, remembered how she still wanted to nurse at the age of four and loudly clamored for the milk being denied during her mother's next pregnancy. When her mother gave birth, she asked Nisa to fetch a digging stick so she could bury the newborn and continue feeding Nisa. Nisa cried out, "Mommy, he's my *brother!* Pick him up and carry him back to the village! I don't want to nurse!" Her mother relented at the four-year-old's pleas and saved the child. However, if Nisa had been only two, a !Kung mother who walked some ten miles a day to gather wild food would know she could not carry and feed both a toddler and a new baby.[29]

Infanticide or abandonment of children is a last resort—the last step in a whole series of decisions whether or not to raise a child. The decision can be to abstain from sex, to use birth control, to abort. The decisions may often be unconscious—distaste for the partner offering sex or, later, spontaneous abortion of a fetus that somehow is not in tune with the mother's body.

If birth control fails or is not available, women in all cultures, through all times, have made the next difficult choice. The risk of maternal death from "safe" abortion, as in developed countries, is 1 in 3,700 cases; from "unsafe" abortion it is 1 in 250. The World Health Organization estimates that globally 20 million unsafe abortions are performed each year, resulting in the death of 70,000 women. In other words, women often choose to abort, after taking into account a terrifying risk.[30]

This is not to say that deliberate abortion in any one case is right or wrong, only that, like both spontaneous abortion and infanticide, it is natural. It sometimes kills the mother as well as the child; it sometimes allows a woman to lead a healthier life and become a mother to other, healthier children.

How many children? This one is obvious. An organism should have as many offspring as will maximize its long-term fitness. If you are human, you may leave ideas and influence instead of children, or in addition to them, but that is a later story. Long-term fitness means generations down the line. Therefore, it may pay off better in the end to have just two children and be prepared to put them both through medical school (if their talents lie that way) than eighteen kids destined for a hard-scrabble life. I mean pay off genetically, not in mere cash or prestige, though wealth and status may help the two doctors to ensure that their own children and grandchildren survive and reproduce.

Every living thing makes such choices. Dandelions do. The dandelions in a field behind the Harvard biology department were regularly mowed. They rushed through to flowering and set as many seed as possible to outpace their mechanical predator. Dandelions in a farther, taller field had other problems: they sprouted among goldenrod

and ragweed and all the other sneeze weeds. That strain of dandelions needed strong vegetative growth to elbow out their competitors.
They grew broad-leaved and tall, to get their heads above the ruck
where the wind could catch their parasol of seeds. Each seed needed
more food to give it a start in life in this environment, so this race of
dandelions set fewer, larger seeds than the mown-field lot, whose
only competitor was a lot of brow-beaten grass.[31]

This is *r selection* versus *K selection*. The strategy of setting as
many seeds as you can is limited, at an extreme, by the maximum
reproductive rate of the species, called *r*. The opposite strategy—producing fewer seeds but giving each a higher investment—is appropriate in environments already near carrying capacity, called *K*. Of
course, many individuals and populations have some intermediate
level, or a variety of levels, or, best of all, the ability to switch strategies according to environmental cues.[32]

A complex of traits describes r-selected and K-selected creatures.
K selection, with fewer, better-prepared offspring, is appropriate in a
full, predictable, competitive setting; r-selection, with many but possibly less-well-endowed offspring, is appropriate in a wildly fluctuating setting. If times are bad, no amount of parental investment protects them against the lawnmower. If the lawnmower has just passed,
times will be very good, with space for most seeds to survive and
flourish.

Certain kinds of life histories do the selecting. K-selected organisms have relatively secure adults. Once they have established their
place in the community, they will probably survive a slow-paced life's
challenges. In contrast, r-selected populations are vulnerable even as
adults. They have little certainty of surviving if they put off reproduction. They rush through their own growth, reproducing early and
massively. Chance may favor the next generation, even if chance cuts
short this one.[33]

Familiar? Of course. Not because we inherit genes to do this or that, but because we inherit the ability to make a relatively wide range of choices about how many offspring to have and to assess our local situation when making choices. The logic of what pays off is similar. In relatively stable or relatively dangerous environments, we follow the logic of the dandelions.

Sons or daughters? What sex children should one have? In a sexually reproducing species, the sum of all males and the sum of all females contribute equally to the next generation. On average, parents put the same amount of parental effort into sons and daughters, which is not the same as having the same number of sons and daughters.[34] Supposing, like Scottish red deer, you give more milk to a boy. That means fewer sons and more daughters for the same cost in parental energy.[35]

Suppose custom says that a daughter needs a dowry. Dowries and daughters' weddings in India are a main cause of families being precipitated into incurable, life-threatening poverty. Without the dowry, the daughter has no hope of marriage; in some social castes that means she can only be a spinster, a wage-slave, a whore. A daughter costs more than a son in cold cash rupees. Furthermore, a son pays back investment by staying with his parents and lighting the funeral pyre when they die. No wonder Indians do what they can to bias the sex ratio toward boys. In Bombay, of 8,000 abortions after the introduction of ultrasound testing for sex, 7,999 were of girls. This is generally known; it is not so widely known that the parents of the boy fetus sued the tester for misinformation. But given the relative costs and benefits, the merest fig-wasp would make a similar choice.[36]

Consider the Seychelles warbler, a tiny, orange-rust-colored ball of feathers. It is an endangered species, whose world population in 1988 was 320 birds all living together on Cousin Island, a 29-hectare chip of land in the Indian Ocean. Then Jan Komdeur and his colleagues translocated a few breeding pairs to the unoccupied islands of Aride and Cousine, giving the species two new chances of survival.

Seychelles warblers have nest helpers. These are young adult females who stay with their parents. On crowded, poor-quality territories in the original home, helpers actually reduced fledgling success by eating the food supply themselves. Parents produced 77 percent sons, who then dispersed. The same parents in the new, rich islands switched to producing only 13 percent sons. Broods of daughters matured and helped out. Fledgling success shot up. On a rich territory, helpers fed the chicks. Then, when the parents had two female helpers, all they needed, they went back to laying sons. In short, sex ratio for the warblers, as for humans, depends both on the investment the parents have to make and on which children pay back investment to the family.[37]

Sons or daughters that aren't average.   The principle that a population will invest equally in sons and daughters, because it takes both males and females to reproduce, is a calculation based on population averages. What if you (or your body) have some means of knowing you are not average?

If you are a dominant female baboon on an exceedingly sparse diet in Amboseli Reserve by Mount Kilimanjaro, go for daughters. Baboon daughters stay in the troop and inherit their mother's rank. If you are a put-upon female baboon in the same troop, have sons. The sons go to new troops, where they fend for themselves based on their own size, strength, and social skills. If a low-ranked mother succeeds in raising a smart, healthy son, he has about as good a chance as anyone else.[38]

If you are a baboon luxuriating in a corral in Texas, it doesn't much matter whether your daughter is dominant, because there is food for all. However, a superbly healthy and confident son may mate with more females than the next male. Dominant mothers under these circumstances should, and do, have more sons than daughters. For a long time primatologists argued over the conflicting data, since some populations were biased one way, some the other, many apparently random, until they realized it all depends on the food supply. The primates are ahead of the scientists. Toque macaques in Sri Lanka factored in both conditions: rank and bodily condition. High-ranking and robust mothers bore more sons, high-ranking mothers of lower body weight more daughters, while the lowest ranked mothers, in poorest condition, again bore sons. Of course this is a statistical trend, not a prediction for any one birth, but clear enough when a series of 589 births, for 26 troops, were totaled together.[39]

The whole idea started (again) from Robert Trivers and his student Dan Willard. Picture a polygamous species—say elk, or red deer on the Scottish Isle of Rhum—where the males' harems vary in size, so the males sire vastly different numbers of calves. Females vary, but not so much. Even a scrawny doe bears some calves, but a defeated male leaves none. The scrawny doe can rear a scrawny daughter, who has some prospect of reproduction. The glossy dominant doe raises a son, feeds him lots of milk, and may launch a future harem-master. For red deer, the system works much as Trivers and Willard predicted.[40]

No one yet knows how animals choose the sex of their offspring. Possibly the baboons make the choice by selective abortion, conceiving before the rains but more likely to keep a fetus of the costlier sex if the rains are good and grass is plentiful when they are in early pregnancy. Seychelles warblers actually lay eggs of the right sex. A lot of people and medical laboratories would love to know how.[41]

And humans? Almost all populations show a bias of 105 boys to 100 girls at birth, which soon evens out, as boys are more vulnerable to early disease and accidents. Only a few data sets reveal any bias— most do not. In medieval Portugal, sex ratio apparently tracked the weather. More boys were conceived in years of good rains and good harvest. The Mormons of Utah kept superb records during the nineteenth century. Their religious ranks reflected all aspects of wealth and social status, and they firmly eschewed birth control. The wives of higher ranked men, especially the youngest, most privileged wives, tended to have more boys.

A provocative study in the 1950s compared the reproduction of fighter pilots and bomber pilots. The fighter pilots tended to have daughters, while less prestigious bomber pilots had the usual equal ratio. This was the era of the "right stuff"—I remember a certain snickering at the time about the supermacho cadres fathering mere girls. A possible explanation is that fighter pilots' wives were under extraordinary stress, not being sure whether their husbands would survive the next flight, and that even the "right stuff" men were a bit stressed themselves. On the other hand, there is a mysterious tendency for women to have boys just after wars—perhaps in relief at the approach of better times? Whatever the explanation, it is not to provide soldiers for the next war, though this has been claimed by various politicians. Finally, there is the odd fact that Chelsea Clinton is the sixty-first U.S. presidential daughter, but there have been ninety presidential sons.[42] Nobody knows how it happens in humans, either.

What data there are indicate a possible Trivers-Willard effect, of dominance, security, and good years yielding a few more boys. However, there are so few such data in spite of the huge number of human records that you cannot argue much prenatal, biological bias of offspring sex. It is worth remembering that a few studies will seem statistically significant just by chance, if you look at enough different data sets. Until the spread of new reproductive technology, prenatal choice of offspring sex has been trivial compared to our massive, cultural choices once the child is born.

Tribalism

Both the good and bad sides of human nature imply that we spent an influential part of our evolution as members of small tribes. Kin selection implies treating others as family. Actual genetic similarity falls off rapidly with each step of relationship, as we saw with Aunty May and her nephews. Only in a tribe small enough to be an extended family does helping everyone also help your own genes. Reciprocal altruism extends cooperation beyond close kin, but only to individuals you are likely to meet again. Members of small communities see each other again, over and over, for life. The generous impulse to extend trust and help to others, beyond the family, evolved in a world where very few people you met were real outsiders. Outsiders, in fact, probably belonged to rival tribes. In the face of their competition, it paid even more to extend altruism to your own support group.

The evil side of tribalism is that people outside the tribe can be defined as subhuman. That gives grounds to fight "just wars," leading on to genocide. Anyone who doubts genocide is a fundamental human tendency can look at the sample of 17 genocides seen between 1950 and 1990, mapped out in Jared Diamond's book, *The Third Chimpanzee*—a map which precedes the 1990s horrors of Hutu and Tutsi, Serbs, Croats, Albanians, Kurds, Iraqis, Turks, Chechens, north and south Sudanese.[43]

We are all capable of genocide. The United States was founded on the genocide of Native Americans and the enslavement of African Americans. Americans of European descent can attempt to console ourselves that we are not much worse than anyone else—not even worse than Native Americans among themselves, or Africans who sold captives from other tribes into slavery, who were, by their captors' own definition, subhuman.

Part II, "Wild Societies," explores the tribes of our primate kin. For now, let us just say that sociobiological explanations work well to help us understand societies of 50 or 150, where kinship, reciprocal altruism, and intergroup competition forged group communion, much like soldiers on the eve of Agincourt: "We few, we happy few, we band of brothers!"[44]

Darwinian Fundamentalism

One of my colleagues in the humanities recently told me that sociobiology had been "worsted," by which she means that the argument is twenty years old and not a current problem for feminists. It's not a problem in biology, either. It's normal science that we accept and use. As Thomas Henry Huxley said in an earlier round of the argument, "Is the philanthropist or the saint to give up his endeavors to lead a noble life, because the simplest study of man's nature reveals, at its foundations, all the selfish passions and fierce appetites of the merest quadruped? Is mother-love vile because a hen shows it, or fidelity base because dogs possess it?"[45]

T. H. Huxley won his round, but fury broke out again when E. O. Wilson published *Sociobiology* in 1975. His colleagues Richard Lewontin, a geneticist, Stephen Jay Gould, a paleontologist, and Ruth Hubbard, an embryologist, as well as other Harvard faculty members, largely succeeded in ostracizing him. (They all still coexist on the Harvard campus, though no longer in the same building.) Students dumped ice water on his head at the American Association for the Advancement of Science. Sociobiology was denounced as attempting to provide a biological basis of human behavior. Gould and Lewontin claimed this would "tend to provide a genetic justification of the status quo and of existing privileges for certain groups according to class, race, or sex . . . [Such] theories provided an important basis for the enactment of sterilization laws and restrictive immigration laws by the United States between 1910 and 1930 and also for the eugenics policies which led to the establishment of gas chambers in Nazi Germany."[46] Wilson was attacked and vilified for about ten years for daring to write Chapter 27 of his book, which extended sociobiology to human beings. Richard Dawkins, who published *The Selfish Gene* around the same time in England, largely escaped such personal vendetta, though he was if anything more inflammatory.[47]

That controversy has long since cooled off. Wilson went on to champion the conservation of biodiversity, though recently he has stirred things up again by proclaiming that all knowledge can be unified by rooting the humanities in biology. Wilson's antagonists have moved on to other concerns. It no longer seems obvious that proposing biological bases for human behavior leads straight to justifying the gas chambers.[48]

The arguments are more technical now. They revolve around just how much you can turn to Darwinian adaptation for explanations, and whether there are other forces at work that seriously modify Darwinian arguments. As I write, there is a bitter public debate between Gould, the philosopher Daniel Dennett, and the evolutionary theorist John Maynard Smith. "Darwinian fundamentalism" is Gould's phrase for the worldview of the adaptationists, who, in Gould's opinion, have gone far too far. This particular argument will simmer down eventually, but there is sure to be another in progress, because a theory that potentially explains everything should always be suspect.[49] Let me try to ignore the egos involved (a bit like ignoring battling bull elephants) and give the gist of the disagreement.

Spandrels, stasis, and chance. First comes the major question: Is the world perfect? No? Then how can one assume that each wing and feather and tentacle and tendril of living things has been exquisitely shaped by evolution? The theory itself says there is random variation, wastage of "less improved forms," and eventual change. So why expect to find an adaptive reason for everything we see? Some species are lagging in the course of tracking the environment. Others are marooned by chance on suboptimal adaptive peaks. If you assume all is for the best in the best of all possible worlds, you wind up telling

Just So Stories, giving made-up reasons for how the leopard got her spots. Objection one to Darwinian fundamentalism: in a shifting scene, don't assume optimality.

That example indicates another problem. It is obvious that the leopard's spots hide her from prey. A plain yellow leopard would probably go hungrier than a spotted one. But why spots, not spirals? Perhaps spots match the sun-flecks of the jungle. In fact, spots are easy patterns to program genetically. Many animals are spotted or striped; few sport fun-fur coats with op-art spirals. (*Growth* in spirals seems to be easy, at least for plants and snails. Not so many creatures grow in spots.) In other words, there are genetic constraints on the "easy" directions to evolve. Take this further. If you start with a monkey's genes and lifestyle, you make a hand of your hand and maybe your tail. Start as an elephant's ancestor and you make a hand of your nose. Objection two: not everything is freely possible, given genetic history.[50]

Objection three: many of the most radical innovations were originally byproducts, or what Gould most elegantly calls "spandrels." When Gould visited the Catedrale di San Marco in Venice, he was overwhelmed by the four golden mosaics of the Gospel writers, who stare down from the curved spaces between the four great arches that support the central dome. The mosaics loom up and lean over you with their piercing Byzantine eyes. But they are essentially decoration. If you build a dome upon four opposed arches, it inevitably leaves spaces (spandrels) at the corners. The spaces must be filled up with masonry to focus the dome's weight on the arch pillars. The spandrels' existence and concave shape are fixed from the moment the architect decides on arch and dome construction. In San Marco they are so gloriously decorated as to seem like the chief purpose of the building, but this is an elaboration of what was originally an architectural byproduct.[51]

Suppose, in evolutionary terms, there is selection for a central feature, such as a back supported on four legs, with a body slung from the backbone. Then suppose the sides of the animal are decorated with spots or stripes or sexual iridescence. You would not say that the animals' sides evolved merely to show off its fine pattern. The billboard for display was given by the architecture. The interesting point is that new patterns can arise precisely where there is some play in the system, because coat color is not quite so fundamental as, say, legs and back and sides. A major mutation that produced a cripple would likely not survive in the wild, but a mutation in the spots makes the leopard a black panther that can lurk even deeper in jungle shadows. The point about spandrels is that play in the system may offer more creative potential than the "important" adaptations.

Objection four to Darwinian fundamentalism is that many of the changes in the fossil record happen relatively suddenly, interspersed by long periods of apparent stasis. This is of course a paleontologist's version of fast and slow: "sudden" means over a million years or so; a "long time" is more like a hundred million years. Perhaps, Gould suggests, we are looking at two different evolutionary processes, not Darwin's continuous slow increments of change.[52]

It is not too difficult to explain the fast periods. On the largest scale, when meteorites or perhaps chaotic collapse wiped out up to three quarters of the ocean's species at once, the survivors found a wealth of empty niches to colonize. On a small scale, a windblown pregnant female fruit fly might land on a new Hawaiian island, there to launch species that adapt to new environments at breakneck speed. This is familiar natural selection, with the lid taken off of the many variants. Because the founding organisms interbreed and have comparatively little genetic variation, recessive traits that would never be expressed under normal circumstances get a chance to be expressed and to be tested against a new environment. Most will be deleterious, but a few will be adaptive and become the basis for a new species. This is one way the "hidden" variation gets out—interbreeding in remote populations lets recessive genes be expressed. Such new opportunities probably explain many great bursts of speciation, though not all new species. The geneticist J. S. Jones remarks that it is difficult to imagine a windblown pregnant female brontosaurus.[53]

Hard to explain are the periods of stasis. If ammonites or horseshoe crabs persist over eons, does this mean that their genomes have somehow frozen into a single pattern? So far as we know, genomes do not freeze up or grow old: mutations keep happening to us all. It is true that apparently no new phyla, that is, creatures with wholly different body plans, have arisen since the early multicelled organisms of the Cambrian. However, within the broad outlines, there is a huge amount of play. Our own phylum, the vertebrates, holds creatures as different as hagfish and tree pythons, peahens and people, not to mention the pygmy mouse lemur of Madagascar, which counts as our close cousin on this scale but could sit on the ball of your thumb.

Peter and Rosemary Grant have uncovered at least one secret of stasis. The Grants study Darwin's finches on Daphne Major, an uninhabited island in the Galápagos (uninhabited except for the Grants, their students, and an occasional film crew, who leap ashore from a small boat to the one possible landing rock and trek up to the one flat campsite carrying all their gear, including water). The Grants have mist-netted and banded the entire finch population of the island every year, for 25 years.

It turns out that evolution happens all the time. In wet years, finches with slightly smaller beaks do well at opening the abundance of small seeds. In dry years, finches with larger beaks open the remaining large, hard seeds. Brood by brood, natural selection nudges average beak size first one way, then the other. If you look at museum collections assembled decades apart, it seems that nothing has changed at all. At the finest scale the apparent inertia turns out to be an average of averages—all the wet and dry year results added to-gether.[54]

Such oscillating selection explains why Darwin's finches appar-ently stay the same for twenty years, or for a century and a half since Darwin's own collections. Is that good enough to explain why living fossils like horseshoe crabs can hang on for eons? At Cape May in New Jersey, waves of horseshoe crabs scramble ashore to spawn, while the air is black and deafening with thousands of seabirds div-ing down to devour their eggs. It is not quite the wildebeest migration of the Serengeti, but you cannot doubt that it is one of the impressive animal hordes—and that natural selection by beak and claw and exhaustion climbing the beach is happening all around you. Genetic inertia is simply not enough to survive Cape May. In the face of the slaughter, the only conclusion is that horseshoe crabs—legs and carapace and all—are selected year by year to go on being very good at being horseshoe crabs. The objection about fast and slow change is probably resolvable into selection for change or for continuity.

A fifth and final factor is chance. Even a perfectly adapted horse-shoe crab can be squashed by a dune buggy. Gould draws major moral conclusions from the role of chance. Evolution fills in gaps, where there happens to be an opening. If intelligence had happened to evolve in small social dinosaurs, we would not be around to know it. Our niche might be filled with the shining cities of velociraptors, commuting in scaly lurex business suits, eating the work force in-stead of downsizing it. Dinosaurs happened to die, perhaps because they happened to be hit by a very large meteorite. Mammals were freed to grow big brains instead. Evolution might easily have missed making human beings, if at any stage another twig on the bush of life crowded out our own.[55]

Chance is a powerful force if you look at details, like what be-comes of you or me. And it is a powerful force in small populations, such as the population that founds a species—our own, for example.

We went through a genetic bottleneck when perhaps a thousand of us started off on a course of our own. We could have missed out completely through one tribal war. For individuals and individual events, chance matters. Some call it history.

However, for large populations, statistical norms and tendencies matter. Life emerged from trillions and trillions of molecules; true cells from trillions and trillions of bacteria; at least thirteen separate lineages became multicellular. That is not one chance but the sum of chances that become probability. Natural selection acts on individuals, but evolution is change in populations. A slight tendency to be better adapted, on the whole, is likely to succeed, on the whole, over time.[56]

The implication of all five arguments—lag in tracking the environment, genetic constraints, spandrels, stasis, and the role of chance—is that perfect adaptation is not the single, blow-by-blow explanation of all we see. Note the qualitative ring of this conclusion. You can perhaps pin down what it means quantitatively for a few assemblages of fossils—but it leaves open the question how strongly adaptation affects human nature. As with the fossils, the answers must come from looking at the species that interests you. How much does its behavior—our behavior—fit with Darwinian theory?

The adaptationist paradigm and its extrapolation, sociobiology, illuminate many aspects of human nature. It does not explain Aunty May's own choice, but it does clarify the common dilemma facing all of a species' Aunty Mays.

Sociobiology and the Status Quo

Let us turn briefly to three fundamental questions that recur in this book, and for that matter, in life. Is instinct fate? Is it right? And does biology justify the status quo?

Is instinct fate? I shall come back to instinct as fate at greater length in Chapter 12. For now, let me give a quick answer. No, it isn't. Throughout Wilson's book, and indeed throughout the study of animal behavior, the theme of behavioral plasticity recurs. It is clear that every genetic tendency depends on an environment to develop. It may be a strong tendency, like growing arms—but not if your environment had German measles, or fed you thalidomide at the wrong moment. It may be a strong but more complex and modifiable tendency, like showing certain kinds of aggression if your environment has been relatively rich in testosterone. It may be a very iffy tendency, like being a great violinist, but only if your parents bought you a violin, not a sitar. All this is banal common sense.

The philosopher Philip Kitcher constructed four fallacious steps of reasoning that can lead to the idea that biology is fate. He calls it Wilson's Ladder; I prefer to call it Kitcher's Ladder. The first step is to argue that a certain trait would be adaptive to a given population. Step two is to see that the trait is prevalent and then assume it was achieved through natural selection. Step three is to say that since selection can only act on genes, there must in fact be a genetic basis for the trait. Step four is to say that genetically based traits are hard to change, and so we are stuck with them—past evolution is current fate. Not one of the links between steps is logically sound.[57]

But try arguing it backward, and try putting the steps as questions, not flat statements. Usually the thing we know most easily about a human trait is whether it is modifiable. How easy is it to change a behavior? Most human traits are easily modified by changing a child's upbringing or by changing current payoffs. Take a trait that is not really surprising, the statement that men tend to be more physically violent than women. Does this mean that all men are always violent? That boys cannot be raised more or less violently? That men do not figure the consequences of their actions? That a man cannot go through life, after schoolboy tussles, without ever actually hitting another person? Clearly in this case, the biological tendency is not fated to be expressed. But if one had a controlled experimental arena, or a large set of statistics about which sex goes to war or gets arrested, on average men are more violent than women: the trait is present, though variable.

The next step is to see if there is a genetic basis. Here, turn to all the animal data about the sequence from Y chromosome to hormones and from hormones to behavior. Human data are more haphazard, but there are now plenty of controlled studies of testosterone secretion and behavior. (There is also widespread folk knowledge about the 'roid rages that accompany artificial testosterone taken by muscle-builders.) The genetic and physiological bases are overwhelmingly probable, though their detailed proof depends on work in other species.

The next step is to determine the prevalence of the trait, from cross-cultural data on war and on local violence. Both are male-biased, in all human populations. It could be massively cultural if one considered only data from this step, but you have already found a likely genetic basis in the step above. One can then conclude that the male bias is not a low-level oddity, maintained by the occasional mutation, but something strongly favored by natural selection, at least in the evolutionary past.

The final step is to ask why it would be beneficial to men to be aggressive, and indeed risk violence. For that, go back to basic sexual selection theory. Lo and behold, men are like other male mammals.

I have taken a trait that few would argue over—at least not now, now that we have emerged from the wish of the seventies that little boys and girls might be androgynous. But I have also chosen a trait that emphatically is not fate. Change the circumstances, and violence need not happen, not even violence by men.

Is instinct right? George Williams remarked that "natural selection maximizes short-sighted selfishness. T. H. Huxley's term 'moral indifference' might aptly characterize the physical universe. For the biological world, a stronger term is needed."[58]

If a trait is natural, is it justified? Are men to be excused or even praised for violence? Are infanticidal stepfathers excused? Is the manipulation of a child for the parents' benefit a fine thing, even if it leaves a bitter and emotionally crippled adult? Should well-fed women have twelve or more children at natural, two-year intervals? What about abortion?

The philosophical questions are bound up with whether we admire the "natural," currently a term of almost pure praise. In many past centuries, however, "natural" meant raw, naive, brutal, uneducated, even mentally deficient. Being natural was no excuse. It just implied lack of civilization.

Our feelings about the "natural" are closely bound up with our feelings about instinct as fate. If "he couldn't help it" is a justification, then behavior that is seen as instinctively based can be blameless. We are still left with all the usual judgments about how modifiable the behavior is. No one can be blamed for breathing, even if they are using up air, say, in a trapped submarine. However, if a murderer "cannot help it" and does not control his actions because he cannot understand their consequences, we judge him legally insane.

In short, what is natural is not necessarily right, in general or in particular. Sometimes it is; sometimes it isn't. We must argue it through case by case in the situation at hand.

Does biology justify the status quo? It is tempting to say no and stop there. If instinct is clearly not fate, but often highly modifiable, and if it does not necessarily provide moral justification for action, then isn't biology off the hook?

Well, a social status quo is a summary of actions of a large number of people who will, on average, be acting averagely. That is just the sort of situation where biology is relevant. If people act in a given sphere in a highly polymorphic fashion, you can say they have differing genetic tendencies, or a common tendency to be highly susceptible to environmental factors. Whether people live in straw huts or skyscrapers seems to be quite variable. But whether people live in some kind of shelter seems not to be. Even nomads carry tents. We have remarkably consistent behavior to keep out cold, wind, and lions.

We all talk. We all nurse children, or substitute bottles. All societies have more male than female violence. We all have status differences and status symbols. We all wear ornaments, we all fix our hair, and virtually all of us wear clothes covering the genitalia. We all have warfare and loving partnerships and we all have gods. In any other species, we would assume that such regularities in behavior result from strong innate tendencies.

So yes, biology relates to the status quo. It can help explain human universals and why they evolved. (Not tell us what they are—that comes from observation of the species in question.) It may not justify them, but it does help explain.

However, if circumstances change—if the patriotic warfare that did well for chimpanzees and protohumans becomes maladaptive in the nuclear age, biology says the wise thing to do, to safeguard future generations, is to change the status quo. If the tendency in question is relatively strongly innate, that may be difficult, but by countering it with another, such as the capacity to consider the future, we can indeed change course.

A banal conclusion—as it should be. Evolutionary interpretations neither change our behavior nor explain it away. The fascination of sociobiology is not the number of repugnant actions that it can explain as rewards for short-sighted selfishness. The fascination is in understanding how loving families and supportive communities could grow from such unpromising ground. Sociobiology reduced to the mere interest of the individual is a trivial pursuit. Its loftier goal is to show how a *society* could emerge, one in which altruism toward a stranger may be the highest gesture of all.

II
Wild Societies

Figure 7.   Haru, lead female of a bonobo group in Wamba, Congo, beckons to gain group members' attention. Haru sets the timing and direction of group movements. In the background, an adolescent male whose mother was killed by hunters sits alone, young juveniles groom an adult male, and two females hoka-hoka. (Body form after de Waal and Lanting, 1997; gesture after Hashimoto and Jolly, observation)

6
Women in the Wild

Having outlined the fundamentals of natural selection and sexual selection and why biologists presume that these principles apply to people, I will now turn to a more historical approach. It's a good big chunk of history, going back to a common grandmother, beaming from under her brow-ridges at some grandchildren who became chimpanzees, while their sisters (or brothers—be fair!) gave rise to human beings. It is difficult to deal scientifically with our own cousins. Primatology is a relatively recent field of study, taking off mainly since World War II. It is one where women have played a leading role.

People often ask me why so many women study primates—the order of mammals that includes prosimians (lemurs, lorises, tarsiers), monkeys, apes, and humans. People outside primatology regularly confuse apes with monkeys and indeed lemurs with lemmings, but they believe that women watch whatever it is. The question rapidly splits into four different ones. First, are there indeed a surprising number of women out in the field as notebook-toting members of monkey troops? Second, do women have some special insight and affinity for nonverbal creatures?

The third question is much more threatening. Do we as observers see only what we are prepared to see because of our own gender and historical context? That question has been sharply put for female primatologists because Donna Haraway's 1989 book, *Primate Visions,* has become a landmark in postmodern critiques of science. Haraway shoved women monkey-watchers right into the firing line of the Arts and Sciences' territorial squabble at the copper beech tree.

The fourth question is related to the third. Have our views of sex roles in other primates changed during the decades of the rise of feminism, so that the content of accepted theory reflects historical trends? Does a silverback gorilla get to treat his mates as he pleases, or is he also subject to the dictates of Derrida?[1]

There are actually about the same number of men as women in the professional primatological societies. Perhaps the misperception that women dominate just means that we have finally achieved equality! Even equality means that women are more in evidence than in many other fields of science. The public thinks that Jane Goodall is the chief primatologist of National Geographic, and Dian Fossey is immortalized by Sigourney Weaver in *Gorillas in the Mist,* so women must run the field. Why so many of us?[2]

Do women dote on other primates as substitute babies? Hardly—it is difficult to babify an actual adult gorilla or baboon. Even apparent fubsys like ringtailed lemurs slash one another with razor-sharp canines and arrange mortality for the infants of their competitors. Primates are hardly conducive to mawkishness.

But maybe women are just "naturally" more sensitive to other animals? I think there is something in this, though I would give much credit to the way we have been socialized. The patience to observe nature without needing to mess it about with experiments was long considered a feminine virtue rather than a masculine one—acceptance rather than control. Way back in the domain of the innate, it is possible that women evolved more patience and interest in the averbal, emotional minds of young children. This is subtly different from the need to maternally bond with children. It is more like a capacity to understand, learn from, and teach them—certainly not confined to women. It would be quite plausible, biologically, to start with a sex difference in this capacity and then heavily channel the talent by later social expectations. In other primates, such an innate bias is clear. Young females are more likely than young males to play with babies, except that where adult males of a species also hold infants, so do their growing sons.[3]

An opposite, and equally important, rationale is that far from being gentle and patient, the kind of woman who is likely to break for the wild wants to do science without being messed about herself by the controlling atmosphere of a hierarchical lab group. In the forest she can be as competent as she likes with no one to deny her. The lone scientist who eventually directs a camp full of students, trackers, support staff is more a feminist ideal than a feminine one. Of course, none of these possible motivations excludes the others.[4]

The best description I have read of our actual feelings is Sy Montgomery's introduction to her triple biography of Jane Goodall, Dian Fossey, and Biruté Galdikas. In her own case, it was not even primates but Australian emus—man-sized racers, with the alien minds of birds. Montgomery testifies to the love of wild animals that grows out of respect and awe at their difference from ourselves. It is a love that does not expect anything in return but their tolerance of our presence and our curiosity. Far from treating them either as babies (which they are not) or as sentimental projections of an ideal of nature (field biologists see nature raw) or as grist for one's own career, wild animals are wondrous to us precisely because they are wild. For those who are endlessly fascinated by them, our love is for the Other. This is not a trait of women or men; it is a trait of naturalists.[5]

The Haraway Wars

"Simian Orientalism!" cries Donna Haraway. She opens her book *Primate Visions* with a chapter on monkeys and apes as the ultimate

Others, the screen on which we project all the prejudices of our own lives and societies. She stresses that the screen is not blank. There is someone out there, being seen, but also someone who sees. The watchers' eyes and minds do not transmit undistorted rays of pure truth. We see through our binoculars, darkly. Even the awe and delight in nature are historically rooted in childhood nature walks and science news and experiences of *being* "the Other." Love of nature is perhaps older than humanity, but love of the untamable wild is not a given. Petrarch was apparently the first European who admitted climbing a mountain just to look at the view (and write poetry), though Po Chü-i of ninth-century China was way ahead of him.[6]

The term "orientalism" is Edward Said's. Said fumed about the West's "orientalizing" the Middle East. Nineteenth-century Europeans and Americans painted setpieces of virile sheiks brandishing curved swords and silken, inviting houris—a vision of romance, and above all sex, having much more to do with western wish-fulfillment than eastern reality. Said and others have since risen up to take control of their own story. Monkeys are ever so much safer to orientalize. Monkeys don't talk back.[7]

Haraway's *Primate Visions* raised almost universal acclaim in the humanities and rejection from scientists.[8] Matt Cartmill staked out the scientists' position: "This is a book which clatters around in a dark closet of irrelevancies for 450 pages before it bumps accidentally into its own index and stops; but that is not a criticism . . . because its author finds it gratifying and refreshing to bang unrelated facts together as a rebuke to stuffy minds . . . A deconstruction of a text or concept is a reading that calls into question its underlying assumptions, its supposed objectivity, and its authority. Deconstruction is not a friendly act, and Haraway's approach to science in general and to primatology in particular is an unfriendly one, which makes no effort to understand or to sympathize with the intentions of scientists . . . She has no interest in the objective 'facts' about primates themselves, for the simple reason there are none. Facts, reality, and nature are, in her eyes, constructs cooked up by Western scientific elites to justify and enhance their power over the oppressed—chiefly women, colonized third-world peoples, and the working class."[9]

Haraway felt that this grossly misrepresented her. Cartmill equates her with the strongest deconstructionists, the few academics who spin careers out of a stance of solipsism. There are indeed critics who argue that the external world is an agreed-upon myth. Haraway claims instead that she treats science *as if* it were narrative fiction.[10] She has never denied that science is about the search for truth; she agrees that you find out more about a monkey's life if you actually go out and look at a monkey than if you sit at home making it up. However, she wanted to write the history of primatology as it is performed by real flesh-and-blood people—not as a chronicle of bloodless intellects and not as the hero-stories scientists sometimes write to represent themselves.

Scientists are human. We live in social climes. We need money. We have gender, and (shhh!) we even have sex. Haraway wished to tell the whole, thick mixture of feelings that go into the making and funding and fights of being a scientist—not ice-clear published conclusions but primatology as a kind of passion-fruit mousse. "How are love, power and science intertwined in the construction of nature in the late twentieth century? What may count as nature for late industrial people? . . . In what specific places, out of which social and intellectual histories, and with what tools is nature constructed as an object of erotic and intellectual desire? How do the terrible marks of gender and race enable and constrain love and knowledge in particular cultural traditions, including the modern natural sciences?"[11]

She begins her account of primatology with Carl Akeley's expeditions to collect specimens for the American Museum of Natural History in New York. It was Akeley who first told us of the majesty of the mountain gorilla, Akeley who asked to be buried on the shoulder of the Virunga volcanoes, Akeley who persuaded the Belgian government to make the Virungas a national park for the gorillas' sake, Akeley who looked in the face of a silverback male and minted the term "a gentle giant." After, of course, he had shot him. Today, that silverback is in New York's American Museum of Natural History, with females and babies, beckoning you to dream yourself into the mists of the volcanoes beyond.

The macho-male safaris of Akeley and Teddy Roosevelt (and Akeley's not-so-convinced wives) gave birth to the magnificent dioramas of the American Museum. In each, Haraway sees that one animal looks straight out at you. Its eyes invite you into a whole imaginary world, with birds on the bushes and even, if you crane to see around the corner, the thrill of a bushfire distant on the savanna, visible only to those who almost fall into make-believe Africa. There are no people. Not in Akeley's African Hall, not until you walk on down the corridors to Ian Tattersall's Hall of Human Evolution. Africa is where our ancestors always belonged, but Akeley's Africa is Eden without Eve or Adam. His magician's hand is on the nape of our necks, making us see visions of truth in what is only bones and fur hung up in a painted cave.

Haraway launches into the career of Jane Goodall. She emphasizes one view of that career: the public image, told through the pages of *National Geographic* magazine and the plummy voiceover of Orson Welles. That image, to Haraway, is the lone white woman, a gentlewoman, who reaches with healing hand into Africa's darkness. Africans become alter egos for the bristling chimpanzees. The iconic photo of Jane's hand intertwined with David Greybeard's became the trademark for Mobil Oil's support of science, and Jane became, according to many polls, the best-known scientist in the Anglo-Saxon world after Albert Einstein.[12]

A different biography would focus on Jane Goodall as one of the most determined and forceful people of this century in taking control of her own fate. Louis Leakey helped her to begin her work, but Goodall herself accomplished those first three years. She watched chimps at the Gombe Stream at hundreds of yards distance, shivered with malaria, lived at first with her equally indomitable mother, then with African aides whose help she always acknowledged. As Haraway emphasizes, the practice, though not the image, changed continually. Gombe in the late 1960s became a "collective, international research site—perhaps more so than any other primate research site established by Western observers . . . with a dense web of social and intellectual ties among former Gombe workers. The entire structure is systematically invisible in the *National Geographic* version of the secret life of chimpanzees."[13]

Goodall for years ran her research entirely through Tanzanians— local men, from local villages. They follow chimps twelve hours a day to take notes and make videos of behavior, and they patiently train visiting students in what to look out for. Haraway's book tells that story, too, better than any other published source except the newsletters of the Jane Goodall Institute. Goodall's current campaigns for humane treatment of captive chimps and her educational Roots and Shoots clubs for young people in Tanzania and elsewhere are all the kind of deep involvement with the consequences of human behavior missing from that early tale of Eve alone in Eden.[14]

Many primatologists reacted to Haraway's book with baffled fury. Like Cartmill, we felt attacked, diminished. Amid 450 pages of research and scrutiny of our lives, where was the shining edifice of scientific understanding that we try to build? Where was even an acknowledgment of that vision, in the welter of intersections of class, history, and gender?[15] Of course, practicing scientists are shaped by all our experiences, including gender—that weak statement is acceptable to anyone. It is uncontentious pabulum. The strongest view, that there is no shining edifice, only a sand castle crumbling into the waves of new historical epochs, Haraway herself denies. But she hybridizes the two, the science and the making of science, as in her metaphors of chimerical cyborgs, simians, women, and OncoMice.[16]

It is all profoundly unsettling. Two separate members of my department (of the male majority) have stormed out after hearing her speak, raging, "That was the single worst lecture I ever sat through in my entire academic life!" Two others (women, as it happens) remarked that they quickly stopped trying to follow analytically. They just sat forward and enjoyed the mind-leaps and poetry, as T. H. White described the boy Arthur learning from Merlin: "left to leap along in the wake, jumping at meanings, guessing, clutching at known words, and chuckling at complicated jokes as they suddenly dawned . . . with the glee of the porpoise, pouring and leaping through strange seas."[17]

At a recent conference uniting primatologists and what one of us dubbed primatologist-ologists, including Haraway, one conclusion leaped out at us. *Each* of the field scientists claimed that our best-known discoveries were forced on us by the monkeys, apes, and prosimians themselves. In fact, many of us at first fought what we were seeing, because it contradicted our previous mindset. Of course some mix of gender and funding, family and national background, got us into the field and prepared our minds for change, but we still *saw* something new. And of course, when we did, that was what became famous, for it surprised others, too.[18]

Matt Cartmill says that Haraway "has chosen not to acknowledge a truth that Orwell, like Marx, always insisted on: that there is a world antecedent to human ambition and desire, and that the powerful and arrogant are occasionally constrained to acknowledge that reality by

having their noses rubbed in it."[19] Reality exists. Only, being primatologists, we are less likely to have our noses rubbed in it than to have
it plop down on us from the trees above.

So I opt for the weakest admission of historical influence. (Of
course, I am a primatologist, not a postmodernist.) This said, three of
the biggest changes in our understanding of primates clearly reflect
the influence of particular people's backgrounds, of rising feminism,
and of society's changing pressures. The first change is the recognition of animal minds and personalities. The second is the understanding of the importance of females and female bonding in group
structure. The third is usually called conservation biology, but it
could also be called the study of future evolution.

Appreciating Individuals

Jane Goodall was one of the modern pioneers in championing the
importance of animal personality. When she published her first scientific monograph, a spin-off of her doctoral thesis, she was discouraged from using the names she had given the animals. (There is some
discrepancy in the stories about this—it is not clear who did how
much discouraging.) The establishment certainly thought it more
scientific, at the time, to generalize about "the adult male," not David
Greybeard and Goliath and Faben and Figan. "The chimpanzee
mother" sounded more objective than Melissa and Flo. Goodall stuck
to her guns, saying a large part of the scientific interest of chimps was
their extreme individuality.[20]

This seems, to some, "feminine"—women wallowing in personal
incident (soap opera again) while real men categorize. There's a problem with this story, however. Ever since World War II an independent
tradition of primatology has been flourishing in Japan. The Kyoto
School, now in its fourth generation of researchers, was launched by
Professor Kinji Imanishi, whose central concern was to understand

how individuals create species-typical societies and how societies in turn reflect the nature of the species. This was explicitly not an evolutionary approach, neither Darwinian nor sociobiological. It was and still is about how levels in the hierarchy of cooperation interact from individual upward to social group, not individuals downwardly subdivided into genes.[21]

To do that, the Japanese have to know and follow individuals, even in provisioned Japanese monkey troops of more than 500 monkeys tumbling together out of the forest. Japanese scientists stem from Zen Buddhist and Shinto traditions where humans and nature are a continuum; they have no Cartesian taboo that holds the human mind apart. Junichiro Itani wrote in an early journal article: "In January 1961 Jupiter, the No. 1 leader of A-Troop, died, and Titan took over. Two years have passed since. We were able to observe some changes since Jupiter's days. First, the individuals kept closer together. Secondly, males made fewer attacks on females, especially during the breeding season. Though it seemed to be only temporary, yet it was quite a remarkable change. An increase in the population may be one reason. Another important reason may lie in the personalities of the leaders. Jupiter was violent to the point of cruelty, while Titan was quite tolerant. Personalities cannot be ignored."[22]

Shock! Horror! Anthropomorphism!

In more recent decades, Japanese and westerners translate better and better to one another. But one thing is clear: the Japanese approach is not "feminine" in their own terms. It is still extremely difficult for women in Japan to become scientists. The handful of present-day women primatologists look eagerly to one another for support but have to be personally aggressive, in contrast to the men's network, which is explicitly modeled on the household.[23] It is equally clear that a massive shift in western science now permits us to speak of individuals.

Female-Bonded Groups

A second sea-change in the "pure science" of primatology paralleled rising western feminism. The article that focused it was written by a man; it was Richard Wrangham's 1980 "Female-Bonded Groups." Wrangham was one of the Gombe network. He studied male chimpanzees' ranging in relation to their food supply. He could not help noticing that female chimps, unlike almost any monkey, spent most of their days alone. He began to wonder why a female would ever travel with others. Being in a group just means that she has to share her food supply. Why don't all the other primates live as sensibly separate as chimpanzees?[24]

Chimps are big. In the tiny reserve of Gombe, they have few predators. That lifts one constraint. Female chimps do not need company for defense. A second reason to travel alone is their finicky diet. Chimps (like other apes and humans) prefer ripe fruit, not green fruit. One chimp can quickly strip a small fruiting tree. Monkeys, on the other hand, eat more leaves and green fruit, and so there is enough food for a crowd. Wrangham argued that monkey females band together in troops because a big troop can chase little troops away from the food supply. Within the limits of tree size, bigger troops are better.

If you are committed to sharing with a troop, it's best not to share with strangers. More of your genes will wind up in the next generation if you share with kin. As monkey bands evolved, the females stuck more and more with kin—daughters, mothers, sisters, aunts. The food-defense band became a *female-bonded* society. You can tell a female-bonded troop in the first day or hour of watching: females sit and groom one another. One picks through the fur while the other lolls under her friend's massage. They carry and cuddle one another's babies—at least their own nearest relatives' kids.

The basic premise is that society is shaped by environment: food, predators, shelter. This was not a new idea. Marx also made it his basis for human social life. But two new insights mold Wrangham's argument. First, *females' ecology* underlies group structure. Second, *kinship across generations* is as important as the cross section of the group at any given moment.[25]

Females depend on food to nourish their young. Each extra bite at the margin makes more babies, or healthier ones. Females spend their energy as adults gaining resources first, and more or better males second. Males, by contrast, use any extra ounces of strength to mate with more females, or to invest time and care to ensure paternity with their "official" mates. It is females who pattern their lives according to the distribution of food. Males just go where the females are.[26]

Focusing on females, like any idea, had a history. I think particularly of Sarah Hrdy's emphasis on the different agendas of male and female langurs, which could even lead to infanticide by males. She extended the gender agenda to all other primates, including ourselves. Another major source was Jeanne Altmann's article showing how most of our early sampling techniques privileged active, showy males and undercounted the behavior of females. She followed up with her landmark book, *Baboon Mothers and Infants,* quantifying the physical limits to motherhood for baboons on the harsh, drying savanna below Mount Kilimanjaro. Barbara Smuts (whom Haraway credits with writing the first sociobiological study of primates) had long discussions with Wrangham—his first footnote thanks her. Still, it was Wrangham's article that crystallized the central social role of female ecology.[27]

The second major insight is that kinship structure across generations is as important as mating structure. Of course any anthropologist should expect that. Anthropologists go on and on about human kinship. Early students of monkeys and apes, though, could not ask kinship questions of animals who didn't talk. We perforce looked at cross sections of the troops. How many males can we count, how many females? This approach fit with the 1950s and 60s assumption that males called the tune. The question was how many females could a male mate with, and did he have to share them with other males?

A kinder assumption mixed with that one. If males and females fully cooperated in group life and parenting, you did not have to worry about the females' separate interests. Our stance was male-biased; it was biased toward a cooperative view of the sexes, and it was also biased totally toward mating relations.

But links across generations? I remember being both thrilled and astonished to hear that Jane Goodall's adult chimpanzees recognized and related to their old mothers and their adult siblings. Soon it became obvious that even lemurs have family relations throughout life. In all female-bonded groups, females stay with their parents, while young males leave; this avoids inbreeding. Most primates are like that. The few who are not include our own ape lineage and the chimpanzees that first set Wrangham wondering; both have a very special ecology which allows or forces the females to separate.

This shift of insight turned primatology inside out. Now, suddenly, there was a whole new way to see things—female-based, not male-based, and with bonds across generations, not just between mates. It was no coincidence that this followed the feminist uprising of the 1970s and that many contributors to the change were outspoken male and female feminists.

Perhaps it needs saying that we needn't credit the changing views just to vociferous studies of females by females. Many people who contributed to the sea-change and who write about primate females are undeniably male. Also many females superbly describe males, like

Goodall's series of alpha males in the Gombe, including Humphrey the bully, Figan the wily, and clever Mike, who terrified the others by banging and clattering Goodall's empty kerosene tins. It seems to me that identifying oneself with a group movement (such as feminism) is important if it matters to the scientist involved. But to say that only females are likely to appreciate females—or males, males—does not tally with the experience of primatology.[28]

Embracing Future Evolution

In recent decades primatology has become increasingly multifaceted and multicultural. The growing number of primatologists from the tropics has not yet overcome the dominance of Europe, America, and Japan, but they are on their way. Scientists from rich countries are now deeply involved with conservation, animal well-being, and politics in their own and other countries, especially collaborating with third world friends and colleagues. The intellectual change is inextricably mixed with politics, just as the focus on females mixed with 70s feminism.[29]

It is clearer and clearer that saving biodiversity is not a game for lemur lovers in their own neck of the woods. Biodiversity involves rights to land, to valuable timber, to watersheds, and to the distribution of income between rich loggers, poor peasants, and hotel owners with overseas connections. Also, of course, governments, which are not exactly servants of the people, just other bunches of people. "Pure science" is not only a search for the ecological present and evolutionary past. We now exercise our intellects to embrace the continuing evolution of ourselves and other primates.

This too involves women as well as men. Primatology's most famous martyr is the tormented, tragic heroine Dian Fossey. There will be others, as the love of nature and the hunger to understand nature confront the future.

She sat among dark lives on mountains,
among those who do not change.
To choose another species over one's own

isn't natural, or admirable exactly—
but who doesn't have some flaw?
Hers was the size of the gap between human and
    gorilla.

When she was on the TV Show
With the famous, well-dressed comedian,
she was relaxed only when she spoke gorilla.

"Naoom, m-nwowm, manauum-naomm, naoumm."
The gaze of the mountain gorilla simply rips your
    soul open.
She's buried among gorillas.

It was a rage she was in.
It was a love she was in.
To give one's life to sit down among animals is a strange,
    great thing.[30]

7
Lemurs, Monkeys, Apes

In Paris 40 million years ago, we clambered about in the trees. It was not yet a city, only the ground that would make the future basin of the Seine, but there was a fine warm rainforest for our home. There were several sorts of us. Some came out by day, some by night. Some were paunchy and clinging and ate leaves. Others leaped elegantly between trunks with spines roughly vertical like the sifaka of Madagascar today. Some were tiny, weighing no more than 100 grams, and crunched up insects for supper.[1]

We were not confined to the future City of Light. Distant species inhabited Wyoming and Texas. It looks as though the first primates of modern aspect evolved in Eurasia and colonized the whole northern hemisphere. They already had the essentials. Their eyes faced forward, not sideways. This was useful for stereo vision to leap between branches, and also particularly useful for hunters: squirrels can jump in trees with their eyes at the side, but cats and chameleons and primates look forward at prey. Furthermore, having your eyes very close together is especially good for focusing on prey about where your hand can catch it, and not so much for judging parallax of distant branches. Their hands had divergent digits. This does not mean a true opposable thumb, not one that could swing around under separate control, but a thumb that stuck off to the side and often held on round the opposite side of the branch. It could well be that they had already embarked on a reproductive trajectory of few infants, long birth spacing, prolonged parental care. Two of the larger species had big males and smaller females. That means that perhaps they already lived in social groups, with more females than males.[2]

Those ancestors looked remarkably like their present-day descendants, the group called prosimians or "pre-monkeys." T. H. Huxley—Darwin's contemporary and defender—noted that the prosimian brain "is far more widely separated from that of any Simian, than the latter is from the Human brain."[3] However, what struck Huxley much more was the continuity within the primates: "Perhaps no order of mammals presents us with so extraordinary a series of gradations as this—leading us insensibly from the crown and summit of the animal creation down to creatures from which there is but a step, as it seems, to the lowest, smallest, and least intelligent of the placental mammals."[4]

Who Are Primates?

Primate social life evolved in each of four great modern lineages, now totaling around 270 species which live right around the tropics, north to Japan, and south to Argentina and the Cape of Good Hope. This helps immensely for understanding the background and constraints of social behavior. You can treat the four great lineages like an experiment with controls. If some species in each one have settled on the same family structure, such as monogamy or harems, you can ask what environment and lifestyle led to convergence on the same solution. That means we gain a fairly firm basis for speculating about social behavior. It also helps to tease out the peculiarities of our own group, the apes. The four lineages are apes, Old World monkeys, New World monkeys, and the prosimians, those distant cousins who look most like the great-grandparents from Paris.

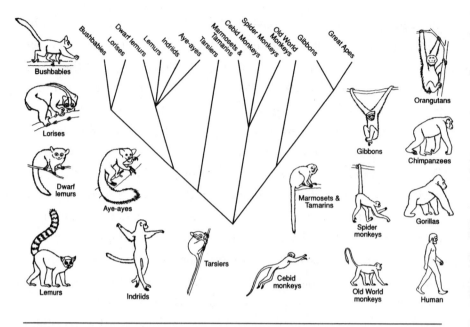

Figure 8.   The families of living primates. Prosimians have radiated into six different body types. Tiny tarsiers are intermediate between prosimians and monkeys. There are three families of monkeys in the Americas, only one in Africa and Asia. Gibbons are "lesser apes." The great ape family has only five living species: orangutans, gorillas, common and bonobo chimpanzees, and humans. (After Fleagle, 1988)

Prosimians.   Modern prosimians are the leaping bushbabies and slow-poke lorises of Africa and Asia and the lemurs of Madagascar. Prosimians range from the pygmy mouse lemur, a 31-gram midget discovered only in 1994, up to gorilla-sized sloth-lemurs who disappeared in historic times, less than a thousand years ago, after people settled Madagascar.[5]

Lemurs apparently reached Madagascar on floating logs long after that island broke loose from Africa. Monkeys and apes evolved too late; they never made it. That left the lemur Robinson Crusoes free to evolve diurnal habits and social life. They gave birth to bizarre species like the dancing sifaka, the wailing indri, and the aye-aye with its bat ears, beaver teeth, and skeleton finger that winkles insects out of tree holes. Some thirty species and sixty subspecies of lemurs are alive today. Perhaps fourteen larger species became extinct in the past thousand years. The giant lemurs died just a little too soon to become the darlings of conservation.

In Africa and Asia, prosimians are all small, nocturnal (since monkeys took over the daytime), and therefore difficult to study, but that doesn't stop them from being interesting. The slow loris of southeast Asia is one of the very few poisonous mammals. Its bite kills mice, and secretions on its fur nauseate predators, which is why it can afford to be both slow and smelly.[6]

New World monkeys. New World monkeys probably evolved from an Old World stock that somehow managed to cross the ocean. They are more like Old World monkeys than like prosimians, but people argue whether their ancestors island-hopped from Africa or took the long way through North America. The smallest are the marmosets and tamarins, tiny creatures who scamper through the woods like a cross between a bird and a squirrel. They have clawed hands like squirrels. But like birds, they sport flashes of red and gold, head-plumes and Kaiser Wilhelm mustaches, and trilling voices.

The larger, more familiar New World monkeys are the only group of primates that hang by prehensile tails. They include the cebus (the intelligent monkey sometimes trained to help people with disabilities), gangly spider monkeys, ear-splitting howlers, and much rarer creatures of the deep forest: sakis, uakaris, woollys, and big blond muriqui.

Old World monkeys. Old World leaf-eating monkeys are superbly graceful leapers, who range throughout tropical Africa and Asia. The infanticidal hanuman langurs are among these species, exceptional in living outside the forest in desert and bazaar. Many more langurs live in Asian jungles. One group is lumped as the "odd-nosed" monkeys: the Pinocchio-snouted proboscis monkey of Bornean river edges and the snub-nosed golden monkey of China, which migrates up and down snowy mountains in troops several hundred strong. Their African cousins are colobus, some black and white, some fantastically red. Black-and-white colobus are all too well known as tourist rugs, made of the pelts and white capes of eight or a dozen unfortunate victims.

Macaques and baboons are also Old World monkeys. These brownish, stocky omnivores are as at home on the ground as in the trees. Macaques used to range right across the northern hemisphere of Eurasia. At one end of the range live the so-called Barbary apes of Gibraltar and Morocco. At the other are the snow monkeys of Japan. Holy (but holy terror) Tibetan macaques steal food from pilgrims at Mount Emei in China. They have frightened a few people into plunging from the cliff paths to their deaths. Rhesus macaques range from the Himalayas to the bazaars of middle India; crab-eating and pig-tailed macaques inhabit southeast Asia; toque macaques, with their centrifugal hair-do, pose on the temples of Sri Lanka. Macaques have not been nearly so hard hit by humanity as true forest monkeys, because they are weed species adapted to disturbed habitats.

The baboons, their near relatives, range from Saharan oases to the Cape of Good Hope. Thoth, monkey-god of Egypt, magnificent in gray mantle and by tradition prodigiously oversexed, is the hamadryas baboon of Ethiopia and the upper Nile.[7] The macaques and baboons were familiar to Old World human cultures throughout history. It was probably a tailless Barbary ape that led the satirist Ennius to sneer, in the second century BC:

> A nasty beast
> Is the simian.
> How dreadfully is he
> Like to man.[8]

Apes, including people.   Apes appeared about as early as Old World monkeys, at least 35 million years ago. There were many more kinds of apes in the Miocene, when the world was hotter and wetter, than there are today, probably much more confined to high forest while monkeys explored the forest edge. Today, just six genera remain. Lesser apes are gibbon species and the larger siamang. They swing hand over hand through the jungles of southeast Asia, warbling and bellowing in ear-splitting duets. The orangutan diverged next. The three African apes—gorillas, chimpanzees (with two species), and prehumans—were the last to separate, between 5 and 10 million years ago. It is now clear that the common chimpanzees and the bonobo (pygmy) chimpanzees are more closely related to ourselves than they are to gorillas. We are their nearest relations. At the molecular level, we are essentially chimpanzees, and not all that far from gorillas.[9]

Sitting in a restaurant eating a quiet meal, I'm often
   struck
by how much human beings resemble gorillas.
The way that woman over there
just now lifted her small daughter's hair
to examine some scratch or bite on the back of her
   neck,
just above the shoulder,
and then let the tender dark hair fall back down,
and patted her on the head.
Or the way my son,
young adult male of this species,
picks up french fries one by one,
making a casual but satisfactory choice each time,
dips each one into a pool of ketchup
and raises it to his mouth,
resembling the way the chimpanzee
raises the twig crawling with termites to his.

All this quiet lifting of food to mouth
around the dusky room—
the steadiness and murmur—
there's something beautiful about it—
something ape-like and sane.[10]

Primate social groups. The social group is made up of those who live together, who move around and feed in the same space. This may or may not be the same as those who mate. Cuckoldry happens to "monogamous" primates as it does to monogamous people. DNA analysis now shows that it happens much more often than either the primate-watchers, or probably the socially ostensible spouses, expected.

Richard Wrangham's insight that females are the fundamental sex for understanding the ecology of primate society gives us a way to start classifying group structure. Do females tolerate the company of other breeding females? Do they live with an assortment of female friends? Or do they live only with other females who are their own kin?

This classification amuses me, because when I first began studying primates, we used the reciprocal classification: do *males* live alone or with other males? If you cross-cut the tolerance of males and females for living with kin and nonkin, you come up with the classical descriptions of society, such as monogamy. It does, however, sound a bit different when you start from the females' point of view.[11]

Sex and the Single Female

I am calling a "single female" one who does not tolerate other breeding females. She may or may not associate with males. Some live quite alone.

Living alone. Solitary primates include lots of little prosimians, which you might expect, and one of our own ape relatives, the orangutan. One key seems to be a food supply that comes in very small packets. There is not enough of most insects to share, even with your best friend, and prosimians like bushbabies tend to be insect eaters. Pottos even specialize in noxious insects that other animals will not eat. They ooze along with a peculiar, slow gait that does not tremble the branch, and clamp one forceps-like hand on the creature's back. I have seen captive pottos catch cockroaches this way without alerting the insect's supersensitive vibrissae. Anyone who has battled cockroaches in a New York kitchen must admire the potto's stealth—but it is hardly a social skill.[12]

Oddly, orangutans seem to have the same constraint. There is so little fruit on most rainforest trees that a big male orang strips them in a sitting and then must trek across the landscape to find more. Rainforests are huge and varied, but often grow on poor soils and are not rich in fruit. So, for the orang of either sex, the best feeding strategy is to gobble alone. When the forest is exceptionally rich, orangutans come together and share the same fig tree.[13]

The other constraint is predation. Solitary animals can hide from predators better than a group can. Again, solitude is what you might expect in nocturnal prosimians. But why orangs? Perhaps orangutans have become more cryptic because of human hunting. Organgutans are native to southeast Asia, not Africa. When *Homo erectus* walked out of Africa and settled in southeast Asia, he possibly turned to hunting orangs. Then *Homo sapiens* arrived, also out of Africa. The orangutans did not have five million years to co-adapt to the invaders, as chimpanzees and gorillas did in Africa. It may even be that the double onslaught of humanoid hunters drove orangutans to retreat from a more social life they had enjoyed previously.[14] Orangutans reared by humans or kept in zoos with others of their species have a full set of social skills and seek out others' company. In the wild, young females rendezvous with friends and with their mothers. Individuals know one another, even if they do not meet for years on end.[15]

In both orangutans and bushbabies, when single females come into estrus, they seek out dominant males for the period of mating. Among aye-ayes, many males converge on the female, fighting with one another to possess her—a pattern more like diurnal lemurs. But in all species, females actively associate with preferred mates.[16] Both orangutans and bushbabies also have small, subordinate males, who may be tolerated by the dominant males but rarely get to mate. As we have seen, small orangs sometimes grab nonestrous females in what seems like rape. So far as is known, small male bushbabies do not even get that far.

Monogamy: one of each.   Monogamy among primates means a single adult male living with a single adult female, though not necessarily in absolute sexual fidelity. DNA analysis now shows that gibbon mates sometimes cheat on each other, and probably the many species of monogamous lemurs and New World monkeys are equally untrustworthy.[17]

So why do these species go through the motions of monogamy? At least four reasons, which can all operate at once. First, the basic distribution of food may be so spotty that a female is not willing to share it with another breeding female, let alone the other females' brats.[18] In marmosets and tamarins, females hormonally suppress their daughters' maturation. In captivity a mother may even kill a daughter who succeeds in reaching estrus but is left in the same cage. In the wild, of course, the daughter would leave, perhaps chased away by her mother.

But then why tolerate a male? Perhaps for his help in watching for predators. Marmoset and tamarin males assume more than their share of vigilance for the family, running the risk of exposure on higher branches where they can watch for hawks. This matters to such tiny animals—they, like songbirds, are favorite prey.

Having a male around may be even more important to guard against other males, that is, against infanticide. Suppose originally females were solitary, likely to meet a male alone for mating. Any other male she encountered would know the infant could not be his. If he killed it, she might soon come into estrus, and if he could keep track of her, he might be the next father. Under these circumstances a female would have strong incentive to stay with one mate, to avoid victimization by the next.[19]

Turning this round to the male's point of view, he would also have strong incentive to stick around after mating to guard his own off-spring from other males, even though this curtailed his chance to roam across the forest looking for more females. One should admit that infanticide has hardly, if ever, been seen in monogamous species, even though the idea is now fashionable. One should beware of the circular argument that it has not been seen because monogamy works so well that the infants are indeed protected. However, there is a logic here: for the system ever to originate, it should offer some advantage to both males and females. Probably the males stayed originally as paladins, and then gradually made themselves useful as nannies.[20]

Monogamous males often also help with child care. They are fairly sure of paternity, so they are good fathers. (There is a skeptical view, here, that they are not so much helping their last children as currying favor for mating and fathering the next child.) Gibbons and titis and indri defend feeding territory by song and actual battles against their neighbors. Admittedly, it is not clear whether they are defending resources for their children, for their mate, or just to keep out other males, but the effect is the same: their family has more to eat. Indri males actually defer to their females, who eat higher up in leafier branches—dietary chivalry.[21]

Marmosets and tamarins are even more chivalrous. The male carries the twin young from shortly after birth until the infants' combined weight equals the father's. This frees the female to conceive again. For much of her adult life she is simultaneously lactating for twins and gestating the next set. She could not possibly do this without male help, any more than a robin can rear its young without both parents' full-time efforts.

Polyandry: the help of several males. A few single females live and mate with more than one male. The only nonhuman primates practicing polyandry are tamarins. Sometimes the males are brothers, or father and son; sometimes they are unrelated. The breeding female mates with both. (Mortality and migration are high enough so the son is usually not her son but a stepson.) Both help carry the young. One male may stay under cover with the infants in the bushes while the other feeds in higher exposed branches and simultaneously watches for predators. Then they shift the infants over and change roles. In the Manu Park of the Amazon, red-bellied tamarin groups never actually raise their young unless more than one male shares the burden. This system is an extension of monogamy evolved to depend on an enormous input of male support.[22]

Human polyandry is best known among Sherpas and other hillpeople of Nepal. In humans it seems always to be fraternal polyandry: several brothers marry one woman. It relates to the extreme lack of land on the high Himalayas. It takes the work of several men to raise one family, just as it takes several tamarins. Usually, only men with as close a tie as brothers are willing to share one wife. Even if a child is not genetically their own, at least it is a niece or nephew.[23]

Female Friends

In a true harem, one male guards a group of unrelated females, who center their attentions on his important self. When early primatologists counted a bunch of females with just one male, they used to automatically call it a harem. It turns out that just two out of all known nonhuman primate species live in true harems.

The hamadryas baboons of Ethiopia have been the life work of the Swiss primatologist Hans Kummer.[24] Hamadryas are a feminist's nightmare. Males attract females, often as young juveniles, who cuddle against the male's silver cape and transfer their need for protection directly from mother to future mate. In adult life when a female is frightened, she runs directly to her overlord even if he is the one who threatened her. Males threaten and punish a female who strays near another male, biting her on the nape of the neck while she loudly screams.

Kummer's experiments show that males have a code among themselves which prevents most fights. They respect one another's rights to females. If a strange male and female are introduced and given 30 minutes to set up a relationship, another male placed in the enclosure will sit far away, looking in any direction but the bonded couple's. This is so even if the second male is dominant in a one-on-one situation with the first. It is only when a male becomes exceedingly old, visibly losing force, that others use their daggerlike canines in earnest to take over his harem. Otherwise many harems coexist in a single band. They all sleep on adjacent narrow ledges on Ethiopian cliff faces.

Recently, Barbara Smuts has argued that human females' subordination to males evolved as a result of this same kind of agreement among males to respect one another's rights to females. That is, males became dominant over females because of their reluctance to challenge one another's rights to particular females. Kummer refers to hamadryas females as "wives"—and for hamadryas, there is good sense in the term.[25]

The other harem-living primates are our near relatives the gorillas. They do not congregate like the hamadryas, but males in different groups still know one another. Silverbacks, the dominants, challenge other leaders or intruders from the bachelor groups, with chest-thumping and bluff charges but only rarely real fights.

Females make their own choice of mates. An emigrating female can pick up and run to a new group across hundreds of meters of hillside. There is little the original leader can do about it, nor should he, if the female is his young daughter. Once they reach adulthood, most females stay for years with their silverback, grooming him assiduously. Gorillas share our own body language, our own timing of glances in the eye and even more polite refraining from a direct stare. (Life within the group is so placid most of the time that I have wondered whether gorillas attract such high-strung primate-watchers because the apes serve as therapists for the humans.)

The gorillas of the dense tropical rainforest are much harder to study than the mountain gorillas who live in the giant celery and nettle pastures of the Virunga volcanoes. It seems that they eat more ripe fruit and have more frequent contact between groups, and even a tendency for subgroups to form—something like the fission–fusion patterns we see among chimpanzees, rather than the cohesive group-ings on the mountains.[26]

Why do mountain gorilla females bother to join a group instead of wandering alone, like orangs and chimps? Food is abundant, at least in the Virungas, and so there is no need for females to limit group size in order to reduce female–female competition. What females gain by sticking close to a male is protection. Gorilla males are highly infanti-cidal. There is all too much grisly evidence to doubt the fact. If a female with an infant strays far away from its father, or if her group is taken over by a new male, she is likely to see her infant killed.[27] Males also protect the group from predators. Leopards could eat gorilla juveniles, but humans may always have been the gorilla's greatest threat, long before the modern carnage of Rwanda. Our lineage evolved alongside theirs. Today, it is the males who charge or (fortunately) often bluff-charge observers. It is males like Dian Fossey's famous Digit, or Carl Akeley's "gentle giant," who die defending their groups.[28]

Group marriage. Sometimes several unrelated females live with several unrelated males, especially among the leaf-eaters. It is not a common system. Why not stay with kin instead? The standard answer is that leaves are a much more evenly distributed resource than the omnivorous diet of most monkeys, so there is less competition within the troop. However, monkeys make strong choices even when eating mature leaves. Some trees provide more protein, some are more highly defended by plant poisons. The question is open why these groups do not recognize the advantages of kin bonding. Really open—we do not actually understand why.

It is not so hard to see why they are in groups, rather than solitary, though. They are preyed on by leopards and ocelots and eagles. Red colobus are ferociously hunted by chimpanzees. They are also often subject to infanticide by incoming males. There is every argument for females to live in a group where males will take on their defense against predators and their own kind. It just isn't clear why they do not stay with their relatives. One possibility is that females need to be able to migrate away from groups that are at risk of takeover. They usually join groups that have several males, or up-and-coming males. If infanticide is a major danger, this could outweigh the support of female kin.[29]

This group structure is very rare in humans. A few people have tried it—hippie communes of the 60s and earlier religious idealists like the Oneida Community. One of the longer-lasting experiments were the Shakers, who chose group companionship rather than sex. Experience seems to show that group marriage does not succeed well in our species. Paradoxically, in the last century many philosophers have taken it as axiomatic that humans began with group marriage and only moved to one-male harems or monogamy with the advent of civilization. Friedrich Engels wrote of the progression from group marriage toward "monogamy supplemented by adultery and prostitution . . . women, but not men, are increasingly deprived of the sexual freedom of group marriage. In fact for men group marriage actually still exists even to this day. What for the woman is a crime entailing grave legal and social consequences is considered honorable in a man, or at the worst, a slight moral blemish which he cheerfully bears."[30]

Darwin did not concur: "It seems probable that the habit of marriage, in any strict sense of the word, has been gradually developed; and that almost promiscuous or very loose intercourse was once very common throughout the world. Nevertheless, from the strength of the feeling of jealousy all through the animal kingdom, as well as from the analogy of lower animals . . . I cannot believe that absolutely promiscuous intercourse prevailed in the past."[31] Darwin is right again. It would be fascinating to trace the concept of group marriage, and why Rousseau thought it the lost Eden of innocence, or indeed why it became a 60s utopian ideal when it is one of the few primate systems that our species seems unable to adopt for long—and neither does any other ape.

Female Families

The large majority of primate groups are families of females with one or more males. Rarely, a group of related females lives without males. Bands of rhesus macaques, Japanese macaques, and ringtailed lemurs in the course of fission can persist for weeks or months, moving coherently through their range, led by adult females alone. This seems to be only a temporary state, however. If the group is viable, one or more males move in.[32]

Gigolo groups. A female family with one male looks like a harem but isn't. Gelada and hamadryas baboons underline the contrast. On the grassy clifftops of Ethiopia's Simen plateau live geladas. The male is a huge heap of windblown brown fur, quite as spectacular as the silver hamadryas with his harem. The gelada male, though, is only a visitor for a few years within a matriline, a group of related females who groom and interact with one another. If you want a pejorative term to match "harem," call it a "gigolo group."[33]

This kind of society is common in primates, among langurs and guenons of the Old World, cebus in the New. Female kin bonding is, in fact, the norm among mammals, but it is rare or unknown among humans. A man in a polygamous culture may well marry two sisters, as when Jacob (in the Bible) married Leah and Rachel. He does not, I think, marry sisters and their mother and their aunts and their nieces and go to live in the household thus formed. We do not have gigolo groups.

Coping with several males.   Many female-bonded groups attract several adult males. They are among the best studied of all primates. One reason is that they frequently range in the open and on the ground, where you can see them. A better, but related, reason is that we humans range on the ground, so perhaps they—especially the baboons of the African savannas—hold clues to our own adventure in terrestrial life.

Baboon and macaque groups consist of several matrilines of amazingly nepotistic organization. The day a baby female baboon is born, you can predict her eventual slot in the hierarchy. She will rank under her mother and over her older sisters. If she later has a younger sister, eventually the mother will help that baby in turn into top slot among the daughters.[34]

There is a little play in the system, however. Some females just seem to be ambitious. By aggression and alliances they climb above the rank of their birth. This is especially so if accidents of mortality thin out the matrilines above them, so there is less enforcement of rank. But even more surprising is what happens to a young female whose mother dies after she reaches two years old—that is, old enough to survive on her own, and old enough to have learned where she "ought" to belong in the troop. The orphaned juvenile baboon is young and subordinate for three more years, until she too is an adult. Then she systematically threatens the females of matrilines that ranked below her dead mothers', until she reaches her birth rank.

There are privileges of rank: feeding privileges, drinking first, and above all raising more infants. The discrepancy in reproductive success is clear, particularly when food supplies are uneven. The most extreme example was artificial. Japanese scientists who had been feeding sweet potatoes to the macaque monkey troop on Koshima Island had to cease provisioning. For several years the subordinate matrilines raised no surviving infants, and newborn females took seven years, rather than the previous four, to reach sexual maturity. Only the two dominant matrilines successfully reared young.[35]

The effect is hardly mysterious. It depends on the distribution of food. A troop in Ryozen on the mainland had more natural food available in the forest than there was on Koshima Island. When provisioning stopped, their reproduction actually became more equal among the matrilines. Given piles of sweet potatoes, the dominant females hogged them all; in the forest everyone foraged apart, with more equal shares.[36]

Male–female relations can be extraordinarily complex and subtle. Males have a variety of dominance styles: the aggressive bully, the cagey forger of male alliances, the suave type who forms "friendships" with particular females. Wild male baboons with type-A personalities have higher stress hormones, same as humans. The lowest basal stress hormones are found in dominant baboons during stable periods of the social structure, and especially if the male is relatively socially suave, responding with threat or aggression only when truly threatened himself. Unstable dominance relations, or a personal style of flying off the handle, plays hob with baboon stress hormones.[37]

Baboon males learn social skills as they grow older and have longer tenure in the troop. Often a young immigrant offers to beat up all comers. Older, subtler males join a new troop by making friends with females—and gain more matings as a result. A large part of making friends with a female involves grooming and carrying her infant. Burly males with gleaming canines walk with an infant riding jockey-style, or groom and lip-smack to tiny wrinkle-nosed young.

Males in these groups are larger than females. In altercations, they can beat up and bite females who try to evade them. However, females have a surprising degree of choice in their mates. They can rely on the help of long-term male friends, or incite one male to challenge another, or simply escape behind bushes. DNA analysis of baboon and macaque reproductive success usually shows a correlation with the male hierarchy, but sometimes that is not clear at all. Instead, females often choose immigrant males who are, apparently, exciting because they are new.[38]

Who is *not* chosen matters, too. Females grow bored with males who have been around too long—in macaques, typically this takes about four years, just the length of time for those males' oldest daughters to mature. Spurred by this female indifference, the males move on, even though they may be dominant and in their prime, to some other troop where incest is not an option.

New World female-bonded monkeys seem to operate rather like baboons and macaques, though with less extreme differences between the sexes. Ringtailed lemurs, though, offer another variant. Females completely dominate males, whether for food or space on a branch or choice of mates. There have been various attempts to explain it, the most convincing being that females sexually select for deferent males. The bottom line, though, is that we do not really understand what is going on. The multimale, multifemale ringtail groups have exactly the mating structure that we imagine has fostered dominance among male baboons—but the ringtails didn't read the textbooks.[39]

And what about humans? We recognize many of our own personal relations among the baboons—friendship between males and females outside of estrus, male baby care, male jostling for different styles of dominance, life-long bonds between female kin, and females playing off one male against another. However, we have probably converged on the same social subtlety from the opposite direction, from a group of bonded males.

Male Clubs

Our closest relatives, the common chimpanzees and their rangier cousins the bonobos, live in groups of bonded male kin. So do the spider monkeys and muriqui, largest of the New World monkeys. Females are likely to forage alone. Sarah Hrdy outraged some feminists by pointing out that humans have inherited female competition as much as female bonding. From what we know of our ape relatives, it is males, not females who stick together.[40]

All of us apes share a taste for ripe fruit. This makes apes dependent, often, on very small food resources. One or a few animals eat the ripe fruit from a tree, leave the green fruit, and move on. One way to solve the dilemma of being a large animal with a small food patch is to live alone like orangutans. Chimps found a different solution: they live in fission–fusion communities.

Chimpanzees. The whole community or group numbers 20–60 animals. When food is scarce, chimpanzees split to forage separately. In the Gombe Stream Reserve, in fact, female chimps spend most of their time alone, with just their dependent young. Among chimps, a female has little to gain by banding together with her kin. She is big enough to defy most predators, and she does not appreciate the competition.[41]

Males are much more social. They groom one another, hug one another, and hold one another's scrotum in reassurance. They have elaborate systems of shifting alliances, which involve a real politics of power. They have elaborate means of reconciliation after disputes so they can continue to live together. They even have patronage: a male who has captured a prey animal doles it out to those whose favor he needs. Ntologi, lead male for fifteen years in the Mahale Mountains, systematically gave tidbits to third- and fourth-ranked males who would back him, not to a second-ranker who might challenge him. Males also hand meat to estrous females in exchange for sexual access. The males, in short, are both rivalrous and clubby. They can even join as a war platoon, with some of the comradeship that implies.[42]

Chimpanzee males defend a joint territory against other groups— and along with the territory they defend the females who have chosen to live there. Chimp males go "on patrol" together to their borders (it is only half a joke to call it a platoon). They travel in tense silence, quite unlike their normal rowdy progress. If a juvenile with them makes a noise, older animals have been seen to put a hand on the juvenile's mouth. Neighboring communities may live in peace for ten or twenty years (as humans in Europe did from 1920 to 1935) and then erupt in a major confrontation. The Gombe chimpanzee "war," during which the main community exterminated the males of the southern group, was the most thoroughly observed, thanks to a corps of Tanzanian watchers, but it is clear that similar group annihilation has happened in the Mahale Mountain Reserve and probably in the Tai Forest of the Ivory Coast.[43]

Richard Wrangham argues that the key to this demonic behavior is the fission–fusion nature of chimpanzee society. If foraging bands vary widely in size at any given moment, then it is possible for a big gang to meet, corner, and safely attack a smaller group or a lone wanderer. The rule for chimps is to attack only if you vastly outnumber your victims.

Unfortunately, chimp gang warfare is all too much like a great deal of human tribal war—classically among the Yanomamo of the Amazon or the precolonial Tandroy of the region where I work in Madagascar. If numbers are equal, don't fight, just shout insults. If the enemy is out-numbered and off guard, kill the men and steal the women.[44] The evolutionary basis for such behavior began from proto-humans foraging on the savanna in flexible groups, some digging for roots, some hunting, some gathering nuts. Our feeding ecology promoted the dangerous life of fission–fusion communities.

What about relations between the sexes in male-bonded, fission–fusion groups? New World muriqui are amazing—they have virtually no aggression between adults and no possessiveness about sexual partners. Males essentially queue up to mate with an estrous female; they are all related, so each gains some fitness no matter who fathers the infant. There is competition, of course—at the level of the sperm. Katie Milton wrote that when she first saw her study group of muriqui, she thought they had some fearful elephantiasis; their huge scrota just could not be normal. Well, they are normal—for males who are willing to stand in line to flood the female with semen, while laying a reassuring blond arm across a comrade's furry shoulders.[45]

Common chimps are tougher. Their males dominate females and emphatically contest dominance with one another. They display by barreling through the woods flailing branches and flinging rocks. If a female annoys them, they flail her too, or even toss her out of a tree. The alpha male does a high proportion of mating, in full view and unchallenged, but all males get a chance and, like the muriqui, are well stocked for sperm competition. A female often mates with six or eight males in quick succession, but most chimpanzee infants at

Gombe are conceived "on safari"—perhaps a prototype for humans' desire for sexual privacy.[46]

Even in chimpanzees it seems that females have more choice than observers (and presumably their own males) suspect. DNA tests of chimps born in Christophe Boesch's troop in the Tai Forest of the Ivory Coast showed that 7 of 13 troop offspring were sired by males outside the community. This means that the females were successfully sneaking away unseen to find partners elsewhere. This is, of course, one troop, and any one troop has special circumstances. Two of the alpha males, with a total of 11.5 years tenure, sired no surviving infants, so it is possible that those females simply did not like who was most available at home. However, it suggests that all the herding, corralling, fighting, and seducing by chimp community males still may not succeed in limiting females' choices.[47]

Bonobos.   The bonobo, or pygmy, chimpanzee was not recognized as a separate species until the 1920s, and even now very few scientists have been intrepid, or lucky, enough to see them in the wild. They live only in the rainforests of the Congo, south of the great arc of the Congo (or Zaire) River. They are not actually smaller in height and limb length than common chimps but are rangy, gracile, and often bipedal—in fact, unnervingly close to the body shape of Lucy the australopithecine, our probable ancestor of 3.5 million years ago. They are neither nearer nor farther kin to us than common chimps— the chimp lineage diverged from ours 5 to 8 million years ago. The two chimps themselves split up perhaps 4 million years ago, when Lucy's lineage was already heading for more open country.[48]

Bonobos are undeniably the sexiest of primates. We used to think that humans were, but bonobos have taken on sexual contact as a way of life, especially when there is any tension in the group that needs defusing. Or indeed any play going on. Or when waking up, or meeting again, or consoling an infant, or some other excuse. Males mount and intromit females. Two females rapidly rub their clitorises against each other, called in English G-G or genital-genital rubbing, though I am all for adopting the local Mongandu term of hoka-hoka. Adult males mount one another or rub backsides. Infants get into the act, both in play and as partners for adults. In this society it would be child abuse to deprive infants and juveniles of sex.[49]

Not surprisingly, males can tell when females are fully swollen near estrus and are more likely to mate with them then, though females keep a partial estrus swelling for most of the month and are willing to play-mate throughout. Males compete for access to females in maximal swelling. Here appears a whole new role for a female's mother-in-law. If an older female manages to hoka-hoka a desirable female partner, she may then simply hand her over to her own son.[50] Professors Kano and Furuichi, who have studied bonobos longest, were at first perturbed that of all the relationships in the group, a few closely bonded males and females never, ever mated—unlike absolutely everybody else. In their second and third decades of research the scientists figured out that these were dominant mothers and adult sons.[51]

Although bonobo groups, like common chimps, are fission–fusion, male-kin societies, the strongest personal bonds may be between mothers and sons and, secondarily, between female friends. When an adolescent female is timidly trying to join a new group, she tries to find an older female best friend and mentor who will help her among the others, and sex is part of this mentoring arrangement. Female coalitions can also displace individual males. On the whole bonobos are much more cohesive than common chimps. Bonobos feed on herbaceous undergrowth in the forest, not just ripe fruit, so they do not have to split up so often to feed. It is not known whether there is ever the kind of intergroup antagonism found in common chimps and in ourselves. In the only group meetings seen, males and females wound up in a frenzy of cross-group mounting—making love, not war—although the encounters clearly include tension and threats.

When I look at a male bonobo, however, I see immensely powerful arms, with veins like weight-lifters' snaking under the skin—a very different build from the females. That, and their powerful canines, suggest they are equipped to fight something, whether leopards or one another. I fear that, in Sarah Hrdy's phrase, "the other shoe hasn't dropped yet about bonobos." Meanwhile, they offer a model of an exuberantly affiliative society.[52]

Do they resemble us? Well, anyone you know?

Honor Thy Grandparents

We are the only primate species that regularly keeps kinship bonds throughout adult life with both sides of the family. We do not do so equally. In the vast majority of cultures, if a newly married couple settles in or near an extended family, it is the husband's. This has been elevated to a principle by the anthropological guru, Claude Lévi-Strauss. He wrote that men in Amazonian Indian tribes, and elsewhere on the planet, make peace between groups by exchanging sisters and daughters.

A feminist view of this barter emerged in Gail Rubin's article, "Traffic in Women: What Right Have Men to Trade Us Off?" Primatologists have yet a different view. Female gorillas, chimps, and bonobos, all our nearest relations, pick themselves up at adolescence, and sometimes again later in life, and trek off to join a new and interesting bunch of males. True, bonobo females have to make friends with their future mothers-in-law. True, a chimp female may often range largely alone, and so does not always associate with any male. But they do choose their groups.

As for gorillas, one might as well say they fall in love, with enormous personal preference for one mate over another. You can see it on the mountains. Zoos are also full of stories of females placed with a male the zookeepers preferred but who managed somehow to mate with their own choices through the bars. My favorite tale is the gorilla in the Los Angeles Zoo who managed to scale a twenty-foot wall, get down past the moat on the other side, and join the male of her desire. The puzzled curators tried to decide if this presented a public hazard. They finally concluded no gorilla had ever done it before, and she was where she wanted to be, so if they just left everyone alone they would stay put. That was thirty-odd years ago; she did stay put.[53]

But chimp and gorilla behavior also has some flexibility. In an isolated chimpanzee group in Guinea, West Africa, males as well as females have left their group in search of new partners. In the Mahale Mountains, a five-year-old male transferred with his mother to the victorious group in a chimp "war." He played with and mounted young females and followed the victorious males—like other young male chimps, he hung out whenever allowed with the male club. He did this in spite of the fact that his "heroes" had apparently killed the males of his own group and he had seen them kill and eat his own infant brother.

Conversely, not all chimp females leave their natal groups. Three of the best known ape families—the Flo family and Melissa family of Gombe chimps and Beetsme's group of Virunga mountain gorillas—have adult daughters who remained home with their mothers.[54] When old Melissa built her night-nest some 10 meters from her daughter's, her grandson Getty "climbs away [from his mothers nest] through the branches, a tiny figure outlined against the orange-red of the evening sky. When he reaches a small branch above Melissa's nest, he suddenly drops down, plop, onto her belly. With a soft laugh his grandmother holds him close and play-nibbles his neck and his face until it is his turn to chuckle. He escapes, climbs up, and plops down again for more. Then again, and again." Though ape females usually do not live among female kin, female emotional bonds remain, between mothers, daughters, and even grandchildren.[55]

This is what humans do. We have a strong tendency for male bonding and female migration, but most families, and most cultures, keep track of both sons and daughters, mothers and fathers through the generations. It takes negotiation—his family for Thanksgiving and hers for Christmas. Links break down, like any other human relation. But overall, we are the ape that honors and loves grandparents and grandchildren on both sides of the family.

8
Human Apes

Lemur, monkey, and ape societies reflect many bits and pieces of our own behavior; humans can live in almost any grouping. Conjugal families are always framed by a larger group, with strong elements of male bonding. This gives us the social subtlety to play off allies and rivals, the way chimpanzees do. The really rare mating structures are a strict solitary life, polyandry, and group marriage—all usually thought rather odd by surrounding society.

Women in any given human population are only 85–90 percent as large as the men. A baboon male is twice the female's size, and so are gorillas and orangutans. The baboon lives in a multisexed social group; the gorilla has a true harem; the orang is alone except when his bellowing attracts amorous mates. Body size ratio clearly does not reflect any particular group structure, only that males are proportionately larger in species where a few honchos impregnate many females, leaving other males with none. The female's size relates to efficient foraging and reproduction, in other words, to her ecology. She has not grown smaller; it is rather that competing males have evolved to be big and dangerous.

Conversely, in monogamous animals, males and females are essentially the same size, with the same size antlers and tusks. They may be different colors or have small fluffy headdresses, but they are physi-

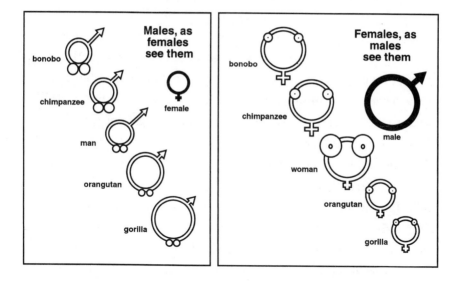

Figure 9. The bigger a male body is compared to females of his species, the more mates he is evolved to have. The bigger the sexual attribute of either sex, the more mates a female is evolved to have, although humans also choose monogamy and other lifestyles. (After Short, 1994; bonobos added)

cally a match for each other. The variance in reproduction between sexes is relatively slight. Adultery happens, but most reproductive effort is bound up with the partner's success. Gibbons are the closest of our cousins who are monogamous. Gibbon partners have equally large canines, as each fends off rivals of her or his own sex. In mammalian societies generally, male-to-female size ratio correlates closely to male-to-female ratio of mates: the bigger the size difference, the more mates the males "expect" to have.

If we turn to men and women and apply this formula, we conclude that human males "expect" perhaps two or three mates. In other words, the difference in size between men and women probably evolved through mild polygamy, on average, in spite of our behavioral scope for men to take either one wife or several hundred.

What do we conclude about the number of mates if we look at another revealing trait—the ratio of testis weight to body weight? Again, humans are intermediate, at least when compared with our primate cousins. Human testes weigh about the same as those of a 450-pound gorilla, which suggests that men compete with other men at the sperm level more than gorillas compete with other gorillas. Because gorilla groups are widely separated, a gorilla female is not likely to go off among the giant nettles to mate with a neighboring male unless she opts to change groups completely. It would be taking a huge physical risk to sneak off for an adulterous liaison and sneak back, hoping to escape detection. Occasionally there may be two adult males in a single harem, and both may even be allowed to mate; but then the younger one is generally the favored son of the harem leader. Gorillas compete by muscle power to establish a harem; they do not need to compete by quantities of sperm.[1]

Common chimpanzees and bonobos, by contrast, who live in multimale groups, are well equipped for sperm competition. Humans are partially so, both in testis size and even by quantity of ejaculate, depending on the length of time the pair have been apart. This implies that men hedge their bets against likely infidelity. They are evolved to defend at the sperm level against their mates having another lover in the same fertile period—not many others, but one or two.[2]

In other words, body size suggests, on average, that early human males might have had two or three mates at a given time, while testis size suggests that women also found more than one simultaneous mate over the eons that our bodies were evolving.

Extended Sex and Pair Bonding

In prehuman societies, during most of their reproductive life females were not cycling, being almost permanently pregnant or lactating. When our foremothers evolved, we spent only two or three years of the life span having menstrual cycles. Modern women, in contrast, may spend as many as 35 "reproductive" years neither gestating nor lactating. Birth control technology is the major reason for the change. In industrialized countries today, family size is commonly two children, or in Spain and Italy one or none. In Africa, completed family size was six children 25 years ago, whereas it has now dropped to three. A second factor is age of menarche: in the West it has dropped from about eighteen to about twelve during the last century, due to improved health and diet. Finally, when modern women do become pregnant and give birth, they tend to breastfeed for just a few months or none at all, as compared with several years that our foremothers spent nursing each child. The physiologist Roger Short has pointed out that the upshot—year after year of monthly bleeding—is a highly unnatural life course for modern women.[3] In earlier times, when human females were evolving, menstrual cycles were a rarity.

Among gibbons, the absence of cycling leads to a nearly sexless monogamy that lasts for much of the three or four years between infants. Among baboons, male "friends" shadow and guard females, and often carry their infants, again over years before the female returns to estrus. Pregnant or lactating protowomen, too, might well have figured out a way to attract the help and interest of males without having sex with them, but we found another route. Our solution was to extend sex to become a social, not just a reproductive, function.

Sex for passion, for intimacy, for relief of tension, even for material reward, for all the reasons people now make love—sex in early humans became a social lubricant. It seems to serve this function predominantly between sexes. Although people have much capacity for homosexuality and bisexuality, we are a long way from the equal-time arrangement of bonobos. Given their reproductive interests, it seems reasonable that protowomen's extended sexiness evolved in such a way as to attract primarily—but not only—males.

Sexual eagerness evolved in women along with their other social needs. This is extended, not constant, interest. As Frank Beach, who has devoted his professional life to studying sex in animals, remarked:

"No human female is 'constantly sexually receptive' (Any male who entertains this illusion must be a very old man with a short memory or a very young man due for bitter disappointment)."[4]

Sex outside estrus is rare among mammals. It does happen. Even in chimpanzees, half of the mating goes on when the female is already pregnant. The chimp's labia continue monthly swelling while the fetus is already growing. Hanuman langur females give sexual invitations to a new and potentially infanticidal male whatever their reproductive state. They judder their heads from side to side and present their hindquarters, which is their species' sexual invitation. Some are definitely pregnant already, but perhaps the new male, having mated, might accept the infant and let it live. It is only humans and bonobos, however, who have managed to prolong female sexual appetite over months on end, not just a few days of real or apparent estrus.[5]

Concealed ovulation.   In multimale primate species, sexual swellings signaling estrus have evolved at least seven separate times. The bright red advertising balloons flaunted by chimps and bonobos appear in species that attract many males at a distance. Human chimp-watchers can spot the pink flash of a swollen bottom on an opposite hillside—much more so an eager chimp male. Male chimpanzees and baboons are exquisitely sensitive to the flamboyant sexual swellings of females. Dominants allow subordinates or even subadults to mate in the days before and after the peak of a female's ovulation but claim their rights at the peak itself.

Orangutans and gorillas have almost no visual sign of ovulation, but pheromones and behavior cue in the males to the relevant days. Visually concealed estrus is widespread in multimale species, and it is common in single-male and monogamous species as well. It does not tell you much about the mating system—only that the female lives in a group where scent and behavioral cues serve to notify all the males she needs.

Human females, by contrast, conceal the moment of ovulation. Throughout human evolution, the moment when conception was likely was unknown even to women, let alone to their mates. In recent times, some women have learned to take their temperature daily and monitor their cervical mucus for clues to ovulation. However, we are apparently alone among mammals in needing technology to ascertain our own state of fertility.[6]

Instead of hormonally varying swellings, human females have evolved breasts and buttocks—permanent advertisements for receptivity, even if cycling is absent. Along with "faking" fertility, women can also fake desire and orgasm. Men can't. A female chimp can't either. She is not above trying it on a bit: estrous chimps are more likely to mate with males who give them meat, and much of the male's hunting seems to be centered around having tidbits to give to estrous females. All the same, a chimp prostitute could earn a living only a few days in the month.[7]

There are two main theories why concealed ovulation and extended sex evolved in humans, and at first they seem diametrically opposed. Jared Diamond has called them the "daddy-at-home" theory versus the "many-fathers" theory.[8]

"Daddy-at-home" has a long paternity, including the gleefully mischievous ethologist Desmond Morris, who in 1967 wrote a paean of tribute to explicit but monogamous sex in *The Naked Ape*.[9] A woman who wants to keep a husband interested in her and their children (and away from other females) can jump him any time she likes. All the better if neither she nor he knows just when they may actually produce a child. This can exploit a man's justified paranoia about fatherhood. If nobody knows just when the right time is, he had best keep at it, not leave his wife to find other men in the intervals.

This is certainly not the function of bonobo free-for-all sex. Sarah Hrdy opts instead for the "many-fathers" hypothesis. Perhaps extended sex is a way for a female to interest many men, not just to nail down one. If females live among males so jealous they actually kill rivals' children, then there are two ways out. The gorilla route is to get a powerful male committed to defending you and his own offspring. The bonobo route is to mate so widely that rival males must all treat your children as possibly their own. What better way than being ready to mate whenever a good opportunity comes? Bonobos clearly mate to defuse tension, especially over feeding. Males often swap food for sex, though both parties act as if they appreciate both the food and the sex.[10]

Setting up "many-fathers" and "daddy-at-home" as a contradiction ducks the fact that both apply. David Pilbeam of Harvard has been pointing out to students for years that evolution does not have to have an either-or explanation for every trait. Two good reasons are better than one. If human women can have sex just about whenever we want, and can fake an interest even when we do not want, we can have our cake and eat it too. Men are quite right to feel threatened on occasion and deeply and warmly and wholeheartedly loved on other occasions. We play it both ways.

Concealed sex. Seeking privacy to mate is just as extraordinary a development as ovulating in private. There is almost no primate parallel. The nearest one is the "safari" of chimpanzees, when a male chimp inveigles a female to follow him in silence for days away from the group until her estrus has finished. Of course, many primates, like other mammals and furtively mating birds, are onto the principle of hiding momentarily from interference. Dominants and legitimate nesting mates, however, do not need to seek privacy.[11]

The trait seems to have developed along with the evolution of publicly recognized mating partners. As our ancestors' groups coalesced into recognized liaisons, each couple tried to keep away from interference by the others. This implies as its obverse that each couple also needed the others. After all, if they could afford to break up into separate territories like birds or gibbons, they would not just seek privacy for sex but for a lifetime.[12]

Our peculiar system comes from strong personal bonding within a larger group. It is the safari of the chimpanzees extended to a two-week honeymoon and then into spending nights in the couple's own nest or hut, yet in the daytime foraging and hunting in cooperation with the rest of the group. This strategy strengthens paternal certainty and support while allowing participation in a larger, more integrated social group, including friends who help with child care.

Is concealed estrus critical to making the system work? Privacy might be harder to maintain if females advertised with overwhelming smell, or brilliant bottoms, to all the men of the tribe just when the moment comes. However, you could argue that it is just as hard to restrain indiscriminate coupling with permanently sexy females like ourselves and the bonobos. Bonobos don't have restraints; we do. Sex in private is human.

Male control of females.   How did humans turn toward fidelity? It is often the hope of both parties, but sometimes it is forced on the female by male coercion. Why the asymmetry, where women take far more pains to conceal their secondary lovers, if any, than men do? Recall the crack of billionaire James Goldsmith: "A man who marries his mistress creates a vacancy."

It isn't enough to go back to the male–female difference in overall mating strategy, even as ratcheted up by mammalian pregnancy and lactation. We take on *obligations* to our principal mates. The verbal contract is one that only *Homo sapiens sapiens* could make, but the limiting of obligations to so few partners, and the asymmetry between partners in their degree of commitment, is perhaps much older.

One primate illuminates our own situation: the hamadryas baboon, the silver-maned Ethiopian species with its true harems. One afternoon I watched a film of an experiment conducted by Hans Kummer, in which a brown savanna female was introduced to a male hamadryas. I happened onto a particularly antagonistic version of *The Taming of the Shrew* that same evening, which played up Kate the shrew's feminist independence rather than her sexual chemistry. The brown savanna baboon kept wandering off to feed; in her own species' troop she would go where she pleased. A hamadryas female, in contrast, would have known her place, following her harem over-lord but also expecting his support and protection. The silver, scarlet-bottomed hamadryas male seemed furious at the brown female's be-havior; he herded and bit her ever more viciously, while she seemed to become more and more disoriented by his behavior. At last, on a narrow cliff path, she stopped fleeing and began to follow him. She learned her new role through straight aversive conditioning. It spoiled any possible pleasure I might have taken in *The Taming of the Shrew.*

I am pleased to report that after two or three days the brown female evaded her hamadryas lord and ran away. I should also add that Kummer, one of the most sensitive as well as clearest-minded of primatologists, used as few animals in these experiments as possible to understand the situation and now says he might not even have embarked on such experiments if he had foreseen how devastating it is to wild primates to be taken from their social context.[13]

The hamadryas provides all-too-familiar a model of those human couples where males enforce the pair bond and females submit. The crucial factor is not just the difference in muscle power between the sexes but the pact of mutual respect between males. This makes it difficult for hamadryas or, as Barbara Smuts points out, human fe-males to escape to another protector or to live in personal inde-pendence.[14]

The converse of the male pact is male jealousy. Some have sug-gested that jealousy in a man, especially "irrational" jealousy, is a mark of low self-esteem. It may, of course, reflect an accurate assess-ment by the man of his own undesirable status. And if his wife shares this view, he may soon have grounds for jealousy—which at that point may not be "irrational" at all. Throughout the world, male jealousy is invoked as a legal defense for so-called crimes of passion, even murdering a wife. The phrase "rule of thumb" derives from the eighteenth-century English judgment that beating one's wife with a stick is a crime only if the stick is thicker than the husband's thumb.[15]

The evolution of two-parent families and marriage. If the down side of pair bonding is coercion, the up side is that, in humans, men want to care for the kids. All single parents know how badly they need help with finances, housing, personal safety, just somebody to take a shift when the baby is howling with earache or has vomited over the last pair of sheets. Sometimes being on your own is just too much to cope with. Of course we want long-term partners. Both sexes do. Our love for each other has risen out of an evolutionary need for co-parenting. This does not mean that the pair must be biological parents of the children: present and future bonding matters more than the past—and an extended family of grandparents, aunts, and uncles also helps mightily. Male–female asymmetry remains, but a woman who is the only wife, or the official wife, or the favored wife, can enjoy her husband's support. Marriage—the formal recognition of partners with rites and rights—is a human universal, and so is love.

> Raising up
> from my weeding
> of ripening cane
>
> my eyes
> make four
> with this man
>
> there ain't
> no reason
> to laugh

but
I laughing
in confusion

his hands
soft his words
quick his lips
curling as in
prayer

I nod

I like this man

Tonight
I go to meet him
like a flame[16]

Birth Spacing and Child Care

Human infants and children are more dependent for longer than any other primate. At birth, they don't even have prehensile toes to support themselves on our fur, if we had enough fur for them to hold. They may be thirty and just emerging from graduate or professional school before they are financially self-supporting.

However, an ape-child's physical demands on its own mother are even greater. Orang birth spacing in Tanjung Puting in Borneo (before the fires of 1997–98) has been about eight years. The mother carries her baby for much of the first two years. She suckles it for five or six years and only resumes cycling after seven. She is also the sole tutor for the child in the complexities of rainforest food supply and travel routes and its sole defense against clouded leopards.[17]

Chimpanzee females, who often range alone, have a similar challenge. The birth interval is slightly shorter, five or six years. They meet more often, and siblings stay till about age eight, so the children do play with aunts and uncles and older sibs. Six-year-old males can tag along after adult male "heroes." But by far the major energy demand for care, carrying, and lactation falls on the mother. Among the great apes, only gorilla females, in their tight harem groups, have constant companions to watch for predators, and playmates always at hand. Gorilla birth spacing in the giant nettle meadows of the Virungas also adds up to five or six years.[18]

One measure of the mother's role in infant survival is what happens to orphans. Among chimpanzees, four-year-old orphans die; five-year-olds go into depression, fail to develop and learn, and also die except in the rare cases when a competent older sibling or adult essentially adopts them. Among bonobos, the limitation may not be so strict. After Shijimi of Wamba lost his mother to hunters as a tiny three-year-old, he was raised by the group at large, especially grooming and cuddling with the three big group males. However, even Shijimi has suffered. An extraordinary BBC film followed his growing up and recorded a jealous adult female biting off one of his fingers. His own mother might have defended him from such an attack, as well as providing milk and comfort.[19]

Birth spacing is a sum of the physical and energetic needs of the mother and child. In zoos with plenty of food and no work, great ape mothers give birth every two to three years. Human Hutterites and other well-fed women who do not use any birth control except breast-feeding have the same interval. Our physiological potential for reproduction is thus much like that of other apes, but no known population of humans takes as long as wild apes to conceive again.[20]

Consider the archetypal society of the !Kung San, studied by Marjorie Shostak. It is archetypal not because it is the only way hunter-gatherers have lived but because the San until recently kept their traditional lifestyle, roaming the Kalahari Desert. Among these groups, the birth interval amounted to four years, during which a mother nursed her toddler and carried it on the long daily trek for food. But even among the !Kung San almost any adult would look after an infant for a while. Older children—the four-year-olds who, like Nisa, were displaced by new baby siblings—ran round the camp watching adult behavior and began their own apprenticeship in adult skills. Meanwhile a four- or five-year-old chimpanzee or orangutan would be still almost wholly fixated on its mother. The contrast becomes even sharper if you look at the Hadza, hunter-gatherers of Tanzania. During the 1980s when they were studied, they lived in richer gathering grounds, with fewer predators than the San. Even babies and toddlers could be put in the care of older children, so mothers nursed less often, carried their offspring less far, and conceived after a mere 2.5 years.[21]

It is perfectly true that human young need care from birth to puberty and beyond. However, there is no human society where all the care comes from the mother. That is what I mean by the relative independence of human children. Mothers could not bring them up without help. Even if our kids left home at eight, they would need help from someone.

Throughout the species, help is given. In modern human groups, much of the care comes from other women and older children, particularly girl children. Some comes from husbands; almost no physical or material help is offered by men who are not the mother's father, brother, or mate. It seems that human childhood, and all the dependence it entails, evolved within the mother's support group. The support group may have been kin, especially female kin, and friends, especially female friends but also male "friends" in the sense of potential or long-term mates.

One primate social system that we do not adopt, except for madmen and anchorites, is solitary life. This is odd since our female chimp relatives may range alone for weeks, and orangs for many months. Our need for support is such that "the heart of a person is only a jelly-fish in the human ocean, and when the ocean is removed from the jelly-fish, the latter dries up."[22]

A Scenario

Here is one possible scenario. The ancestors of all the African apes lived as a group of related males, like chimpanzees, bonobos, and the minority of gorilla groups where young males stay at home to inherit the territory and females. Young females came to join the males, crossing the forest on their own to find attractive mates. They also made friends with older females already in the group, though possibly with some continued tension toward these mothers-in-law.

First the gorillas split off. Their groups grew smaller, more focused on the dominant male, until most were true harems. The chimp–human line instead foraged on resources that led groups sometimes to fission but often to feed in large assemblages. Possibly females even then were beginning to extend the period of sexual interest, mating with many of the males.

Again two lineages split. One, the chimp–bonobos, grew large sexual swellings that advertised receptivity over a limited time to any male in eyeshot. The protohuman line continued extending the timing of sexual receptivity. They grew enlarged breasts and buttocks, permanent advertisements to any male around. Females of this second lot were not only attracting several males for their own choice but, with their sexually selected displays, were themselves subjects of choice.

The chimp line split a final time. Chimp sex swellings were originally temporary, but bonobos converted theirs into the equivalent of buttocks. By keeping them for a large part of each cycle, they showed their own prolonged sexual receptivity, although they were still aware of changes in the intensity of the signal. Meanwhile, the protohuman line went still further in blurring their signal. More and more, the

193

time of ovulation was concealed until nobody—not even the woman herself—could be sure of it. And any occurrence of ovulation was rare, perhaps only once or twice over the course of two or three years.

The protohuman females, in their multimale, multifemale groups, mated with many males and befriended many females, including, at times, female kin who remained home. The males protected them all. They lived in the kind of forest-edge or savanna habitat that encouraged a fission–fusion society and mutual attacks between groups. Within the group, however, personal links coalesced. Males made compacts with one another, unconsciously at first and later consciously, to respect mating rights. Females also encouraged particular males to think themselves the father of particular infants, and they grew jealous of their own mates' attention to other females. They enticed the males to bond, heading off harm to the infants.

On the whole, peace within the group would help them raise the needy infants. This was, of course, a conditional strategy: you could change partners, or just plain stir up trouble, then as now. Sex in private was doubly helpful. Females hid like any other primate if mating with a non-legit male. But hiding with the main or recognized mate also helped deflect jealousy. For the taboo against public sex to have become so strong and universal, it must have been a very important factor in keeping the peace, for all parties.

One crucial part of the story is almost pure speculation. This is what happened between neighboring troops. Among most primates there is an uneasy truce, punctuated by standoffs between females over food and territory. Males do fight, sometimes lethally, to take over females, as witness the "demonic males" among Wrangham's chimps. It matters if they have kin in other groups; immigrant vervet monkey males fight less hard against their own kin.[23]

Chimplike warfare intensified among the evolving human apes. Still, the females had kin in other groups—the mothers and brothers they, as adventurous adolescents, had left behind. In spite of the tribalism of modern chimps and humans, for most of the time human neighbors live in peace. When, and how, did humans start keeping in touch with both mother's and father's kin? My scenario would be that it was long, long before conscious "traffic in women" by males swapping sisters between groups. Perhaps, just perhaps, it was the mediation of the sisters, who maintained links between their husbands and brothers. With, of course, the agreement of the men, who found ways to extend the male pact of mutual respect to their in-laws.

Menopause and the Value of Age

One final group to consider: the old. Life does not stop with fertility. It is extraordinary that among women, fertility stops long before life or even sex does. This is not true in many primates. I remember meeting the oldest female of the Arashiyama Japanese macaque troop. She was 33—silver-haired, rheumy-eyed, arthritic. It hurt just to watch her hips move. Her infant clung to her back, bright-eyed, wrinkle-snouted, with the newborn's part in its topknot. It was just the age of her great-grandchild born the same spring.

In fact, most female mammals decline in fertility as they age. Women, pilot whales, and some elephants and chimpanzees stop cycling completely, as do individuals of many other species. Why do we give up reproduction so long before our time? Or, conversely, why do we live so long afterward?

It could just be a by-product. If genetically influenced breakdown of reproduction—and other bodily functions—are delayed long enough, animals will die anyhow of disease and starvation. In other words, natural selection acts to favor living vigorously and reproducing robustly up until the point when the animal is likely to be knocked off anyway. There is little selective pressure on genes to keep you healthy after that point, say until age 200, if the lion or the TB germ would have got to you first. Craig Packer and his colleagues examined the extensive records for wild lions and baboons and concluded that they live just long enough past their last child's birth to "fledge" that child. Both lions and baboons do offer child care for their grandchildren. However, having a grandmother alive, or a grandmother alive but not herself reproducing, does not seem to influence the grandchildren's survival.[24]

Lion mothers need only live on for a year, baboons only two, to launch their own offspring with as good a chance of survival as they are likely to get. Human mothers need more. Further, human mothers overlap their broods: we do not wait until one is "fledged" before starting on the next. This means that a risky final childbirth, in an older mother, may also put her existing children at risk. The "by-product" of putting off the effects of genes that lower reproduction until late in life may turn into a "spandrel" for us, which lets women stop having children in order to take better care of the ones we have. (Men seem to be in no particular danger from late reproduction, except the real possibility of a heart attack during intercourse—apparently worth it.)[25]

A second tier of benefits accrues to grandchildren, though the statistics do not show that it aids survival. Lions and baboons and even little vervet monkeys lick, cuddle, and protect their grandchildren. Human grandmothers do all that and share food as well—surely a part of the aid that allows our longer childhood and relatively shorter birth spacing than other apes, and a contribution to survival and success of the family.

The other aspect of postmenopausal life is not the stopping of fertility but life span itself. We are all too well aware in our aging society that old women are likely to live longer than old men. John M. Allman and his students have compiled life tables for a wide variety of primate species, in the cossetted conditions of zoos: not survival at the margins but, for a primate, life in the middle class. Strikingly, in species where the females do the bulk of child care, females live to be older. In the firmly monogamous species—gibbons and siamang, little New World owl monkeys and titis—males also live to be old. In other words, males who carry, play with, feed and defend their babies are chosen by natural selection for a ripe old age. There is a politically correct moral here, though unfortunately helpful men can only expect to reap the benefits over evolutionary time. There is a politically uncomfortable conclusion, too: that females in our prehuman past did most of the work to care for their young.[26]

For humans, of course, the benefits given by aged parents and grandparents are not mere food and care. Sometimes they share wisdom. When Jared Diamond was inquiring about edible wild plants on the island of Rennell in the Solomon Archipelago of the Pacific, the tribesmen led him to their ultimate authority. "There, in the back of the hut, once my eyes had become accustomed to the dim light, was the inevitable, frail, very old woman, unable to walk without support. She was the last living person with direct experience of the plants found safe and nutritious to eat after the hungi kengi [the catastrophic cyclone of 1910] until people's gardens began producing again . . . Her survival after the 1910 cyclone had depended on information remembered by aged survivors of the last big cyclone before the hungi kengi. Now, the ability of her people to survive another cyclone would depend on her own memories."[27]

Again, this is foreshadowed faintly among other mammals. There was an aged mangabey female who led the group to a distant swamp at the height of a drought when all other water sources were dry, as, on a grander scale, do aged elephant matriarchs. Having at least a few elders to be the troop's memory could mean survival for children and grandchildren, and the tribe of all one's kin.[28] It is better for everyone that women should live to be old.

The two of us sit in the doorway
chatting about our children and grandchildren.
We sink happily
into our oldwomanhood

Like two spoons
sinking
into a bowl of hot porridge.[29]

9

Apish Intelligence

We humans tend to assume that intelligence is always an advantage. This is not always true for human individuals and hardly ever true for other species. Let us follow on from the notion sketched out in Chapter 2, that intelligence is like sex. It is a dangerous gamble to let other people's ideas mess up your mind. But there are potential benefits. When things go right, intelligence gives both more stability against damage and more flexibility to deal with change. In this chapter we ask what are the specific benefits for a wild ape? What were the benefits for prehuman apes? Why did they take those first steps that lead us now to suppose even politicians should use some intelligence?

Gender bias lurks in every discussion of ape and human intelligence. Nineteenth-century theorists found evolutionary reasons for men to become human, while the other sex rode along on their coattails.[1] Even Darwin concluded that selection for intelligence worked only on males: "The chief distinction in the intellectual powers of the two sexes is shewn by man's attaining to a higher eminence, in whatever he takes up, than can woman—whether requiring deep thought, reason, or imaginations, or merely the use of the senses and hands . . . To avoid enemies or to attack them with success, to capture wild animals, and to fashion weapons, requires the aid of the higher mental faculties, namely, observation, reason, invention, or imagination. These various faculties will thus have been continually put to the test and selected during manhood . . . partly through sexual selection— that is, through the contest of rival males, and partly through natural selection,—that is, from success in the general struggle for life."

"Thus," Darwin concludes, "man has ultimately become superior to woman. It is, indeed, fortunate that the law of the equal transmission of characters to both sexes prevails with mammals; otherwise it is probable that man would have become as superior in mental endowment to woman, as the peacock is in ornamental plumage to the peahen."[2]

This view is totally unacceptable today, of course. We know that women support themselves in dealing with the environment, except in the minority of human cultures where the woman's environment is largely her husband. We know that women do the crucial teaching of small children. We know that sexual selection cuts both ways in humans: both sexes choose and are choosy.

Darwin's argument was, in fact, quite logical. If only one sex did deal with life and the other were totally dependent, all the second sex would need is the limited repertoire of a good parasite, like a *Boniella* male settling down for life on the female's tongue. What Darwin got wrong was the anthropology. Most of our ancestors were not Victorians.

Outsmarting the Environment

Manipulation and tools.   When I was in graduate school in the 1960s we learned that the leap to human status was making tools: we were "Man the Toolmaker." Then Jane Goodall watched chimpanzees pluck straws and twigs, bite the end neatly, strip off leaves and branches, carry the twigs as much as half a mile away, and settle down with these shaped objects to fish for fat, prereproductive termites. Louis Leakey exclaimed, "We will have to change the definition of tools, or the definition of man—or else agree that chimpanzees are human!"[3]

Of course we changed the definition of "man." Toolmaking fell from favor, to be replaced by language and culture. Even more so after Bill McGrew pointed out that the really skillful termite-fishers among chimpanzees are the females, who spend long hours at a task that gives high reward in protein, while the infants play alongside. Fishing female chimps get more termites for time spent than the males do.[4]

It may be that "technical intelligence" played a crucial role not so much in the difference between ape and human as between monkeys and apes.[5] The great apes least interested in tools are gorillas. However, almost everything gorillas eat is defended by hooks, spines, or stings. Researchers venture out in rubber rainsuits, gloves, and Wellington boots and still nurse welts raised by the Virungas' giant nettles. Meanwhile, gorillas chew up the nettles with placid impunity. They actually perform an elaborate sequence of maneuvers. They slip leaves off the stem with a one-handed swipe, fold in the stinging leaf edges into a neat packet with the other hand, and eventually pop the leaf wedge into the mouth—the whole process with minimal stinging to hands, lips, and tongue. Giant bedstraw, with its barbed hooks, gets a different handling sequence. Gorillas differ individually in details—and family members share these details. It looks as though young animals copy the moves of their parents. Their food handling qualifies as a protocultural tradition.[6]

The search for "embedded food"—food that is encased in something else, such as an insect in a curled leaf, termites in a termite hill, nettles with stings—means that something else besides the food itself must be tackled first. The animal must be able to deal with a goal two steps removed, not just one step. Animals that are clever about objects are those that search for hidden food, such as raccoons and aye-ayes, cebus monkeys, chimpanzees, and of course human beings.[7]

That does not take us very far, though—not even into the difference between aye-ayes and chimpanzees, let alone humans. Most plants have not invented complex means of defense. Plants up the ante by making the nutshell harder to crack, or the nettle sting excruciating, or the poison more toxic. They have not usually invented a sequence of challenges—a nutshell that opens with a lever inside one that opens with a drill bit inside one that bashes with a stone.

Apes, on the other hand, have thought of sequential solutions. The nettle-stripping and folding sequence is actually a string of about six different moves performed in order. Apes have a capacity for solving problems in a multistep series that shows up clearly in the manipulative sphere—which probably also becomes a stepping stone to thinking in ordered sequences.[8]

Sequential thinking matters in many other domains besides the domain of technical intelligence. The ability to see more than one step ahead and to remember or even plan a sequence of acts is fundamental to many human skills, including erecting the sequences of language. Meanwhile, chimpanzees in the wild have been known to use whole toolkits of different sticks and stones to reach an elusive goal.[9]

Mental maps.   Mapping is another sphere of environmental intelligence. Even howler monkeys map sequences of moves to minimize their travel distance between food trees. (Howlers are New World monkeys, lethargic leaf-eaters, who save energy by resting among the branches for hours each day while their monkey-watcher fidgets underneath.) The nervous, wider-ranging spider and cebus monkeys map even longer sequences in their visits to patches of fruit.[10]

Orangutans are probably, again, the primate champions. An adult orangutan eats around 400 different food items in the rainforest of

Tanjung Puting in Borneo, from at least 229 different plant species, plus fungus, insects, and wild honey. Most fruits mature at different rates, on trees where each member of a species may fruit asynchronously, on schedules several years long, and spread out in space over five square kilometers of tropical rainforest for females, much more for the large and itinerant males. Our own ancestors, ranging in savanna, presumably covered even more square miles than orangutans do and paid an even higher price in energy and dehydration if they started off in the wrong direction.[11]

Is map memory a narrowly specialized knack or something that could have been a wider stimulus to human brain power? Spatial memory can sometimes be a specialized skill. The true champs are not orangutans, or even ourselves, but Clark's nutcrackers. These California birds, with a teaspoonful of brain apiece, cache up to 30,000 piñon or whitebark pine seeds in the fall, remember them using visual landmarks, and find at least the 3,000-odd caches they need over the next six months.[12]

Primates, instead, apparently work from much more general principles of categorization and causation. Food-finding helped hone our ability to juggle categories, causes, and simultaneous spatial organization. Primates evolved more general abilities than the nutcrackers, though not so exquisitely adapted for surviving all winter on piñon seeds.[13]

Clever huntsmen. A third key to intelligence, in addition to tool use and mental mapping, was hunting. Anthropologists have nearly come to blows round their intellectual camp fires debating how much prehuman men hunted, how much they scavenged the kills of the big cats, how much they merely played about with a rare if classy supplement to the basic vegetarian diet.[14]

Nowadays, we recognize that men were hunters, like modern chimpanzees. Further, it was and is *men* who hunt, not women. Even enterprising Inuit women mostly trapped small game while men were out on the ice bagging seals and walrus. Men hunt, and men share the resulting meat. It turns out that among chimpanzees the meat frequently goes to estrous females, and among humans to women other than the man's official wife or wives. Some things never seem to change, at least not in the last five million years or so.[15]

Hunting spurs on intelligence by rewarding quick reactions, concentration, foresight, plotting the prey's routes of escape, and avoiding stronger animals that fight back in defense (adult male red colobus monkeys rush to head chimpanzees away from their juveniles). Chimpanzees even have a recognizable division of labor in the hunt. In the Tai Forest, some male chimps act as beaters while others wait along the prey's flight route as killers; there is true collaboration. Prey meat is divided among all the hunters, not just dominants, which means that it profits the males in Tai who cooperate in the chase. (This is not so true of chimps in the Gombe, where the forest canopy is so much lower that single males often succeed in catching prey, and the division of meat is not so egalitarian.)[16]

In primatologists' excitement over documenting the heroic pursuit of big game, they have paid less attention to snaring and outsmarting prey. Craig Stanford tells the only tale I know of a primate's trap. Stanford saw a party of male chimps in the Gombe Stream drive a group of red colobus monkeys into a tall, isolated palm tree. The colobus males threatened downwards, until most of the chimps gave up and left. Only one, wily Frodo, hid just beneath the crown of spreading palm fronds, out of sight of the monkeys above. Gently, Frodo pulled a frond downwards, making a bridge between the palm and the next tree. The colobus apparently did not see the hairy hand below. They ran toward safety. After seven or eight had crossed, a mother with clinging baby tried the passage, only to have the waiting chimp lunge upwards at her—but short. Clever or not, Frodo missed. The mother and child got away.[17]

Lions and wolves cooperate in driving and ambushing prey, much as chimps do. House cats crouch patiently by mouseholes. Spiders build traps as elaborate as any of ours. The really big difference is in the sharing of meat after the kill. Chimpanzees play politics artfully with their prize, encouraging allies as well as seducing females. Hunting may well employ speed and complexity of decision, and trapping demands patience and foresight, but like tools and maps, primate hunting seems to encourage thought because primates do it thoughtfully.[18]

The complexity of the environment. Before moving on from these environmental uses of intelligence, we should beware of our own bias. People in developed countries believe the physical environment is mostly under control. We see it tamed. Humans closer to nature, on the other hand, endure fires, floods, droughts, plagues of locusts, diseases that kill children, predators in ambush—and also enjoy hard-earned bonanzas. As our early abilities to categorize stimuli and to think in terms of cause and effect evolved, being able to predict even a little what the environment would serve up next was a great advantage. It may well have rewarded far more sophisticated reading of the signs around than city-based people can imagine.[19]

I once listened to an ex-colonial official, a ruddy-faced Briton, describe being taken out hunting by !Kung San in Botswana. He fired at a distant antelope but only wounded it. The trackers followed its trail. At intervals they announced that it had changed direction, that it was a young male, that it was hurt in the left hind leg but not badly, so they would have a long pursuit. The official kept saying, "I never felt such a bloody fool in my life! I would look and I could see *absolutely nothing*. Whatever signs they were following, I never saw *anything*."

Not only do people in developed countries underestimate the complexity of the environment, but traditional psychologists in laboratories simply hate it. (Rat psychologists almost by definition are control freaks.) When we turn in later chapters to talking of social intelligence, we will find that much of the data we have consists of anecdotes about the most surprising acts the watcher ever saw. Even quantitative social data come from wild settings, where animals show off what they do naturally, and do best. In contrast, laboratory tests of spatial memory and planning are often based on a priori criteria of what we think important, not on the animals' aptitudes. Such experiments can substantially underestimate capacity in the wild.[20]

But even in the wild we tend to boil down capacity to saying, for instance, that ringtailed lemurs eat in perhaps ten major sites in a week, within an area of only 5–20 hectares, where they go round and round a repetitive daily course. We rarely tell the environmental anecdote. One day in 1992 Lisa Gould's ringtailed lemur troop picked up, left home, crossed a dry riverbed, and scampered in a straight line through scrub and farmers' fields for two kilometers until they reached a single mango tree in fruit. In the rainy season with so many other trees in fruit and blossom, did they smell mangoes that far away? Did they know the tree was there all along, with a tradition passed from one female to another or brought from afar by an immigrant male? How much else do they know that we miss, through not picking the one right day to see it or not jogging as fast over two kilometers of scrub as a lemur troop heading for a gourmet meal?

Gould thinks this one is socially preserved knowledge, repeated from year to year. "Two years later, in 1994, I was looking for them in their home range all afternoon without success, and just at dusk they came charging back into the reserve, after being out for at least half a day, with orange mango faces and very sticky hands."[21]

Machiavellian Intelligence

It took a surprisingly long time to realize that social relations are a major key to the evolution of general intelligence. Society is the only evolutionary pressure that provides accelerated feedback *within* a species. As rivals, and friends, grew smarter, there has been ever more selective pressure to be smart enough to keep up.

In 1982 Frans de Waal published *Chimpanzee Politics,* a rip-roaring tale of coalitions, deception, coups, and counter-coups among Yeroen, Nikkie, and Luit, three male chimpanzees in the Arnhem Zoo. The males were both playing to, and supported by, a chorus of females, led by determined old Big Mama. De Waal made it clear that the maneuvering males acted with considerable subtlety, as though they had read Machiavelli. Machiavelli, he said, "presented rivalries and conflicts as constructive and not negative elements. Machiavelli was the first man to refuse to repudiate or cover up power motives. This violation of the existing collective lie was not kindly received. It was regarded as an insult to humanity."[22] De Waal also pointed out that such maneuvering need not be consciously directed toward its goal. Human parents rage over who picks up the adolescents' dirty socks, without quite realizing that the socks betoken a coming shift of power in the family.

In 1987 Dick Byrne and Andy Whiten again invoked the Florentine realist, in *Machiavellian Intelligence.* The authors rightly claimed they had hold of an idea whose time had come, and this would be the last book that could summarize the field. Since then, we have seen an exploding literature, much of it fortunately condensed in the sequel, *Mach II.*[23]

Why did it take so long to shift the main (or fashionable) emphasis from dealing with environmental variables—tools, maps, embedded food, hunted prey—to dealing with social life? Even people who are not especially feminist must see that this has meant shifting away from focusing on the challenges of an external, or male, world, toward a sphere that has long been thought female. This is an odd admission, when the most complex case of primate social maneuvering we know is male chimpanzee politics, with all its parallels to male human ploys. But somehow the resistance to social life as a major force in intellectual evolution involved a feeling it was all too soft, too vague, too empathetic, too intuitive, too accessible to be real science. Not just female, but a lot of adjectives feminists recognize.

It is worth telling what sparked Byrne and Whiten. Some credit should go to a juvenile baboon called Paul, on a grassy slope in South Africa. He watched an adult female dig up a root from the hard ground—a root that his juvenile strength would never let him unearth. Then Paul screamed. He screamed just as though the female had attacked him. His mother, who was out of sight, came belting over the hill and threatened the other female, who fled, dropping her prize. Paul's mother went back to her own concerns. Paul ate the root. It looked to Whiten and Byrne very, very much as though Paul "lied" on purpose.[24]

The problem was that this was only an anecdote—inadmissible and unbelievable in the scientific canon. Paul did it again when his mother was out of sight, again to a female subordinate to his mother. That made two anecdotes.

Instead of giving up, Byrne and Whiten wrote to every other primatologist they could find, asking for all the stories we never dared tell about cases where primates had apparently deceived one another. They assembled 225 such accounts. They concluded that there is no evidence of deliberate deception in lemurs and other prosimians, and only a few cases among New World primates, but overwhelming evidence for Old World monkeys and the great apes.

There are different levels of sophistication in deceit. Simple concealment, or hiding, is widespread—especially for females mating with a subordinate male out of sight behind a bush. Looking away, so as apparently not to register the hostility of a dominant or the importunity of one's offspring, is also common. Delaying one's own response takes more self-control—to pass by a tempting food with one covert glance and return later. Dandy, a young male chimp in Arnhem Zoo, did this so successfully he even fooled his human watchers.

Without breaking stride, he passed a place where de Waal had buried a grapefruit, so the observers thought they had buried it too deep to detect and their experiment had failed. Hours later, when the adult chimps finally went to sleep, Dandy got up and went straight back to the site, dug up the fruit, and ate it in peace.

Jane Goodall told similar stories from the wild. Figan, as an adolescent, spotted an overlooked banana in a tree above the head of a much larger male. Figan went round behind the research house, as though aware that he could not help giving away the game by his eager glances, and only came back when the male had left. Then he went straight to the banana.[25] Young Figan would also stride off into the forest (probably toward a distant food source), followed by a trail of gullible adults, then circle back himself to feed on overlooked bananas. Once when he tried this, Goliath turned up in the meantime and was gobbling the fruit when Figan returned. Figan lay down and had a tantrum.

In the laboratory, David Premack tried to teach chimpanzees to react differently to a "good" trainer, who gave them food they pointed out, and a "bad" trainer who ate the food himself if he found it. Young chimps learned to turn their backs on the bad trainer, to hide their tell-tale gaze, but only an extremely test-savvy adult named Sarah actually pointed the bad trainer to the *wrong* place.

Perhaps the clearest observation of all again came from the "chimpanzee politics" at Arnhem. Luit, the dominant, was losing power to the younger Nikkie. Sitting with his back to Nikkie, a fear grin spread over Luit's face. He took his fingers and pressed his lips together— three times, until he finally succeeded in wiping the grin off his face. Only then did he turn around to confront his challenger.[26]

Using a third party as a tool is fancier still. This was what Paul apparently did, if he really did con his mother into attacking the female with the food. Hamadryas baboon females in the Zurich Zoo developed "protected threat" in which they made faces at another female while standing directly in front of their harem overlord. If the victim had the temerity to threaten back, she would apparently challenge the overlord—not a good idea for a hamadryas. This particular behavior was never seen in the wild; it seems to be a product of the social hothouse of the zoo.[27]

The accumulation of these and many other accounts added up to a new proposition: outsmarting others is adaptive. Machiavellian intelligence had arrived.

Machiavellian cooperation. Whiten and Byrne never meant their term to cover only the nasty elements of social intelligence. Deceit was a good place to start in trying to understand communication of any sort, because communication can be best understood when you look at what blocks it, at why and how information is withheld. Smoothly integrated communication can happen by just about any means, from pheromone to telepathy; so deceit just gives you a better handle.[28]

Machiavelli himself did not think deceit is all there is to success. "For a prince . . . it is not necessary to have all the [virtuous] qualities, but it is very necessary to appear to have them . . . [It] is useful, for example, to appear merciful, trustworthy, humane, blameless, religious—and to be so—yet to be in such measure prepared in mind that if you need to be not so, you can and do change to the contrary."[29] We are right back to the calculated relationships of sociobiology. Yes, love your family. Yes, cooperate with your friends, to the extent that you and your genes benefit in the end. That can also be said more encouragingly: to profit yourself, think first of others. Betray only when there is no better alternative.

Primates do love, support, cooperate, reconcile after fights. Read the paeans of praise for prehuman morality in Frans de Waal's books *Peacemaking among the Primates* and *Good Natured*. He chronicles the subtle and continuing search for harmony within primate troops. In spite of all the rivalries, these animals must live together, and have

Figure 10. Cooperative intelligence. Nikkie, a chimpanzee male in the Arnhem Zoo, Holland, has climbed a dead tree and broken off a large branch. He now holds the branch for Luit to climb past electric fencing into the branches of a live tree. Luit, in turn, will break and throw down leafy branches for the whole group to eat, especially Nikkie. The males sometimes used such cooperation to defuse tension over status rivalry. (After de Waal, 1982)

evolved the wish to do so. Reconciliation between former enemies has not been seen much in lemurs, but seems to exist to some extent in all monkeys and apes.[30]

This is hard-headed stuff. Rhesus macaques hardly ever initiate reconciliation with a weaker opponent. In their hierarchical world, it is the weak who curry favor with the strong. But stumptailed macaques, close relatives of the rhesus, live in much more egalitarian groups. Like bonobos, stumptails invite heterosexual and homosexual mounting. Strong stumptails invite mounting by the weak, as a way to lubricate their social lives in the troop. Chimpanzees, both common and bonobo, seem the most elaborately—I nearly said deviously—cooperative of all.

Perhaps the acme of social manipulation yet seen in primates was peacemaking by Big Mama, the old and powerful female in Arnhem Zoo. When the males were tense after serious fights, Big Mama sometimes sat grooming one of the rivals. She would shift her bottom along slowly, the male following to stay within reach of her clever hands. At length she could also reach the other male, who might be staring at the sky, or anywhere but at his opponent. Big Mama would groom him too—and of course the males groomed her back. At length, Big Mama would slip out from between the two of them, leaving them willy-nilly grooming each other. Mission accomplished.[31]

Goodall and Nishida say that they have never seen this kind of mediated reconciliation in the wild, but reconciliation of sorts is a daily occurrence. Reconciliation would not be necessary if there were no conflict, but it is as much a part of life as conflict—in other primates and in ourselves. Reconciliation and the formation of social alliances could equally well be called "social intelligence" to avoid the cynical connotations of *The Prince*. But Machiavelli summed it up: appear to be virtuous—*and be so*—unless it is necessary to be otherwise.[32]

Social Intelligence Evolving

Back in 1966, after my own first field study, I noted that social behavior is widespread in mammals, including lemurs. I concluded that social life probably preceded and shaped the development of monkey-level intelligence. We had no way then to rank social complexity. All I could say was that ringtailed lemurs lived in groups with several males and several females, like the troops of most monkeys. Lemurs apparently faced the same challenges in dealing with dominance hierarchies and kinship (though we were not so aware of kinship then). I also said that I had not seen Kummer's "protected threat," so perhaps the details of social interaction were simpler in ringtails than in his hamadryas baboons.[33]

Thank goodness for the weasel-words, that is, for due scientific caution. Thirty years later it is clear that lemurs are indeed less socially sophisticated than most monkeys. Ringtailed lemur relationships are almost wholly black or white: they consistently like or dislike each other animal in the troop. The result is that they have scant use for reconciliation with enemies and little ambivalence unless a relationship is changing.[34]

Their networks of grooming and affiliation are close between mother and daughter and somewhat so between sisters. Lemur friendships do not ramify over eight or ten relatives as among macaques (though even for macaques the level of support may fall off fairly rapidly for more distant kin). Juvenile rhesus macaques, when they are threatened, give different calls for threat from relatives and nonrelatives. Mothers respond much more strongly to being told a nonrelative is threatening their child, and rightly so, for nonrelatives are much more likely to escalate to violence. Lemurs, in contrast, need no such subtlety. In fact, mother lemurs rarely support their children—at least not enough to guarantee them a nepotistic slot in the dominance hierarchy, although often a daughter supports her old mother.[35]

Lemurs have "targeted aggression." When resources are scarce, dominants harass subordinates to the point of expelling their unfortunate cousins from the troop. Monkey troops can split, too, but monkeys do not go in for this extreme personal bullying of their cousins. My explanation is that in a monkey troop, many animals cooperate to defend the troop against others. For monkeys, the cousins are a positive benefit. In ringtailed lemurs, most of the intertroop vigilance and warfare falls on one or two dominant females. Subordinates who do not defend the troop are optional extras for those dominants—useful for passing on family genes and for distracting predators but expendable if resources are tight.[36]

There do not seem to be tripartite relations—no complicated three-body problems to juggle. It is true that lemurs cuddle into lemur-lumps, purring as they try to reach and lick new infants with six or eight black-and-white ringed tails dangling from the furry conglomerate. It is also true that they often attack others in concert—but in that case it seems to be to the advantage of each individual. Michael Pereira has borrowed Mark Twain's description of when an auctioneer lies: "I never saw a lemur attack another unless it was strictly convenient."[37]

Are there also different levels of social sophistication at the next great taxonomic division, between monkeys and apes? One must specify which apes: little monogamous gibbons and solitary orangs probably should not count. But yes, deliberate deception is much commoner in great apes. Sizes of coalitions, particularly among chimp males, are larger (hardly ever more than two or three in monkeys). Use of a third party as a "social tool" is more frequent and more complex.

I still stand by the earlier argument that social behavior is more widespread and challenging than manipulation of objects. Living in a troop of twelve or twenty others means dealing with twelve or twenty separate relationships on a daily basis, even if you have fairly simple rules of thumb to handle each one. It still seems likely to me that troop life preceded the development of more complex relations, and that these in turn shaped the evolution of intelligence.

What has changed is that now there is something to evolve. Social intelligence itself is not a given but has mutated, as minds became larger and more complex. Simply counting the number of animals in a troop does not tell you all the permutations of their relations: twelve lemurs are indeed a much less challenging combination than twelve chimpanzees.

Brain size and society. Increased number of animals in a troop does correlate with larger brain size, and especially with larger size of the neocortex. The capacity to keep track of social relations may even set an upper limit on troop size. In a few weird habitats where the food supply allows, some primates live in huge conglomerates but still interact meaningfully with only a small set of friends. This is what we do in Manhattan; it is apparently what Japanese macaques do in the thousand-strong provisioned troops at Takasakiyama. Perhaps it is what mandrills do in the horde of over 600 counted in Gabon rainforest, including 25–30 fully adult males, with their red-white-and-blue muzzles and matching rear ends.[38]

Establishing a correlation between brain size and troop size is highly complex. First, you usually try to discount the effect of big bodies, since larger animals need larger brains simply to run the machinery. Even this is open to dispute: perhaps the interesting measure is absolute brain size, whatever it is running. There is also some doubt which bits of the body should "count" as machinery. Leaf-eating monkeys apparently have smaller brain-for-body weight than fruit eaters. But does that mean that it takes relatively more brain to find patchily distributed fruit, or does it just mean that eating leaves takes relatively more body to digest such coarse food? (Leaf-eaters have big paunchy guts!)[39]

Then you need to check that the two species you are comparing respond separately to selective pressures; closely related species may be just variations on an inherited theme. Then you try to imagine confounding variables: does a large brain simply reflect greater visual acuity?

After controlling for all conceivable objections (that is, the ones so far conceived), there is still a solid result. The size of the neocortex compared to the rest of the brain is larger for animals living in large groups, among both primates and carnivores, and possibly even bats. No environmental variables correlate at all—not diet or day range or even diurnality, once other factors are removed. Vision, however, is highly important in neocortex size and probably also relates to social life.[40]

Robin Dunbar suggests that troop life selected not for sequential processing but instead for parallel processing to deal with the simultaneous social challenges of large primate troop. Perhaps parallel processing evolved originally to deal with visual data and then immensely enlarged under the stimulus of social life. This would mean that the realm of social intelligence set the stage for our own massively parallel intellect.[41]

Social rules of thumb.   How clever does social intelligence have to be? To calculate from first principles in most situations, we would have to have minds of almost unlimited computational capacity. Instead, we often use rules of thumb. Simple rules may not always achieve the "optimal" solution, but often the "optimal" solution is only fractionally better than one that is good enough, and in fact, worse, once you subtract the time and energy needed to discover the optimum. Herbert Simon, starting from computers rather than people, called good-enough rules "satisfycing."[42]

Ringtailed lemur females do a good deal of disposing of their rivals' infants. At first sight it seems as though they sometimes deliberately prevent a subordinate mother of their own troop from retrieving a fallen infant. They drive away mothers of other troops, too. If the infant falls during group combat, the winners may carry the baby away into their own home range, where it is eventually abandoned to die. If you look at the details of behavior, it seems to be put together from three very simple rules of thumb. First, go to any crying baby. Second, pick it up if it is your own or your close friend's but sometimes let it crawl on you of its own accord. Third, treat the mother as you always treat her, that is, harass if she is a troop subordinate or attack if she comes from another troop. The third rule is, in fact, a nonrule: just behave as usual. The sum of the three yields many variants of separating other mothers from their infants, but there is no need to imagine elaborate goal-directed plotting—just an evolutionary advantage.[43]

Going back to the hoary question of human mate selection, the rule of "pick the first candidate above some threshold" may often work a lot better than "check all conceivable partners for the best." By the time you decide on who is the finest in all the land, and call back, he/she may be busy elsewhere. This may be one reason why neoclassical economics, with its mathematical calculation of marginal cost and benefit and its search for absolute optima, seems so counter-intuitive, not to say wrong. People just do not work that way, at least not when dealing with other people. We do not simply factor in time lost in the search (which economics can do, too). We always factor in rivals and alliances.

Too Smart

We all know people who have been too clever for their own good. The eleventh century poet Su Tung-p'o sardonically remarked on the birth of his son:

> Families, when a child is born,
> Want it to be intelligent.
> I, through intelligence,
> Having wrecked my whole life,
> Only hope the baby will prove
> Ignorant and stupid.
> Then he will crown a tranquil life
> By becoming a Cabinet Minister.[44]

However, it seems even more striking that intelligence is useful, and that people seem even to be overendowed with the trait. A few of us compose symphonies or create quantum physics. Most of us seem willing to step into a device made of two tons of metal and aim it at 70 miles an hour down the New Jersey Turnpike, while twenty-ton heaps of metal hurtle by even faster—yet the majority of drivers come home safe at the end of the day. How do such "extra" capacities arise which could not possibly have evolved in the same context? There are four answers, all of which contribute to solving the puzzle. First, we systematically underestimate the intellectual demands of others' lifestyles, even apes'. Few human city-slickers would survive a week in a chimpanzee habitat, even if we cheated and took a prefabricated hand-ax to start off with.

The second point is much more fundamental. None of us descended from ancestors who merely coped with average conditions. At some point, all lineages that survived did so by surviving the extraordinary: the predator, the falling tree, the drought, the child stuck up in a tree whose mother solved the problem of getting it down. The extraordinary is not just what is bad; it is also the bonanza of what is good—the fruit tree whose higher than normal branches could be reached only when one chimp thought of holding a stick as a ladder for his friend.[45]

The extra capacity appears in all domains. We shall see in Chapter 19 that humans have increased our body mass by over 50 percent and our longevity by 100 percent in the past three centuries. This is much too fast to be a genetic change. It is expressing a genetic potential already there, as world nutrition levels have improved. There have been a few times in the past when such body stature was widely expressed; we are just today returning to the height and size of the Cro-Magnons. For stature, stamina, and intelligence, what is selected is not the mere survivor but the superfit.[46]

The third point, which recurs over and over in this book, is the importance of differences. If sex itself evolved to randomize genes among one's offspring, a distribution of traits and skills would result. Perhaps there is a random component in the architecture of the developing brain which would on occasion give rise to new and innovative capacities.[47] In humans, a random genius who turned out to be Newton, Mozart, Einstein (or the inventor of the hand-ax, the tamer of fire) would be worth a good deal of the risk of leeway in construction of the brain—at least to his kin, even if he died childless.

The fourth and last point is that skills which evolved in one context may be used in another. Spandrels for intelligence, in short. If we evolved the capacity to deal in sequences and categories in order to find fruit in the forest, and fast reflexes to deal with the branch that broke under us or the lion that surged from the long grass, we can use those same abilities to deal with the behemoths of the Jersey Turnpike and to map the route through its spaghetti junctions. Even more: if there is a bit of brain not too involved with the mechanics of life, it can be enlarged and co-opted for general intelligence. Which brings us to the question: is there general intelligence?

Demons and Domains

A library's worth of psychological literature has argued whether intelligence is an accumulation of specific skills within specific domains or rather some general multipurpose ability that crosses all fields. A "vertical domain" is a context for intelligence, such as tool use or route mapping or Machiavellian social juggling. A horizontal domain is a skill useful in any field, like memory or perception. General intelligence (sometimes called the g-factor, to make it sound like mathematics) would cross both vertical and horizontal domains.[48]

As animals, we evolved with specific needs in specific contexts. Like any other animals, our brains adapted to meet those needs. Eventually we evolved some spillover, such that we could use mental skills over a wider field, even though they were originally honed within single domains. Even now, though, as Francis Crick remarked, "The brain does not look even a little bit like a general-purpose computer."[49]

The complexity of thought in the mind, as in the computer, is achieved by routines that mobilize subroutines. In computerese, a little operating routine may be called a "demon." (This comes from one attempt to make artificial intelligence by having many small demons vie for success in a computer world called Pandaemonium.) A demon can be any program that slots in whenever its cue appears, such as "Capitalize 'I' when it stands alone." Each demon has its own protocol which embraces still smaller demons, down to the smallest imps that turn individual neurons on or off. Without nested hierarchies of demons, it would be impossible for the brain to operate or to have evolved. This was Herbert Simon's basic principle in the parable of Tempus and Hora: every complex organic structure is a hierarchy.[50]

The brain is not a single, simple hierarchy. Its parallel processing ensures that active sets of demons compete and cooperate inside the brain, rather than answering in lockstep to a single satanic overlord. Can one still picture the various uses of intelligence as something like executive demons, so that near the top, demons become domains? Perhaps. We retain a clear legacy of bias toward dealing more effectively with relations in social than physical spheres.

In other primates there is such a bias. Lemurs rarely play with objects; infants, juveniles, and adults play with one another. They do not solve spatial monkey-puzzles, unless a pull on one bit of the apparatus will do the trick, like hauling in fruit on the end of a branch. However, they are exquisitely sensitive to one another's spatial positions, with the potential for threat or support. Vervet monkeys do not seem to associate python tracks with the possibility that they may stumble over a python, or a fresh leopard kill with the thought that there might still be a leopard about—both, one might suppose, rather useful inferences to make. However, they do associate a juvenile's distress call with the particular mother that ought to be answering that call. Children (and monkeys) show concepts of ranking and transitivity in the social sphere much earlier than they can rank inanimate objects. The examples can be multiplied almost indefinitely.[51]

Sometimes a psychological test becomes a kind of type specimen, the example to standardize what people mean by a concept. In this case the test is the Wason Selection Task.[52] Here are four cards, laid out in front of you. Their backs say:

$$\boxed{4}\ \boxed{5}\ \boxed{A}\ \boxed{B}$$

The psychologist tells you each number is matched to a letter on the other side. There is also a rule to test out: if 4 is on one side of the card, there should be a B on the other. What two cards do you turn over to find out if the rule has been broken?

Now try it this way. You are an employee. Your firm has a rule that if you work on a weekend day, you should get a day off during the week. The cards read, "Worked weekend," "Didn't work weekend," "No day off," "Day off." What cards do you turn over to find out if the rule is observed?

Now try it this way. You are the firm's employer. Same cards as above, same rule as for employees. What cards do you turn over to check the rule?

The problems are logically identical. The right answer is the same in each case: the first and third card. However, in one run of this test, only 10 percent of participants got the letter-number version right, 75 percent of "employees" got the social case right, but a mere 2 percent of "employers." Leda Cosmides argues this is all about a special domain in our minds to check for *cheating*.[53] The employee wants to know, "Is the boss chiseling me?" "Worked weekend" and

221

"No day off" tell him that. The employer asks, "Are the staff scrounging free time?" She turns over "Didn't work weekend" and "Day off." Few people bother with logic, which matters much less.

We have the exact bias that Trivers predicted in his thoughts on reciprocal altruism. Human alliances are very important but vulnerable, so they must be repeatedly tested. It seems that cheat-testing is a universal human mental bias. There are versions of the Wason test about a chief giving out cassava in return for tribal face-tattooing, where Amazonian Indians opt to spot the cheaters rather than follow abstract logic, just as we do.[54]

One of the most convincing arguments for social intelligence was made by Nicholas Humphrey in 1976.[55] We anthropomorphize whenever possible. In fact, we even vivomorphize. We have built-in programs to decide what kind of movements mean that a thing is alive. Tiny babies, like adults, conclude that dots of light moving in synchrony are a single "organism." Toddlers, like adults, know when two such dots are helping or hurting each other, chasing, hunting, supporting. If in doubt, it's alive. And if it is alive, we suspect it has intentions—the stuck automobile, the purring pussycat, the automatic bank teller, the thunderstorm. In the beginning of this book, I talked about our conviction that nature itself has purposes: just the idea you would expect from an intelligent social primate.

Over millennia since the founding of astronomy and astrology in China and Babylon, we have looked for purpose even in the stars. We look out at them; we see them looking back at us. They watch us with primate eyes.

III
Developing a Mind

Figure 11.   Athlete Maria Patino had the courage to say that one gene is never the whole story. (After photos courtesy of Maria Patino)

10
Organic Wholes

In the first two parts of this book, we progressed from evolutionary principles through bacterial sex and marched right on to the intelligence of apes. To continue further, we need a new approach: the development of people as individuals. Each of us starts from a handful of inherited genes, a string of DNA cushioned in surrounding cytoplasm. How does this tiny blob, with its compressed string of coded information, turn into a large and visibly organized blob such as Arnold Schwartzenegger? The overtones of thinking about development are very different from considering natural selection. Our bodies are forms most beautiful and most wonderful, in Darwin's phrase. Their complexity and their efficiency are a delight.

In Chapters 2 and 3, we listed five characteristics of living organisms: self-maintenance, reproduction, a boundary, cooperation between the parts, differentiation of parts from one another. Looking at development rather than evolution gives a different take on all of these. What you see in growing bodies is an extraordinary degree of cooperation between the parts, even as they differentiate from each other.

Genes within a mammal like us can reproduce only if their organism survives to reproduce. They cannot take off on their own, like bacterial genes hitchhiking on viruses to lodge in wholly different bacteria. They cannot reproduce until their organism does. Our own genes' interests are wholly bound up with the success of their community. Truly selfish genes do arise, in the sense that they reproduce themselves at cost to the other genes in the genome. As I said in Chapter 5, talking of multilevel selection, the surprise is how few of these anarchically reproducing genes there are. Many more genes are simply deleterious, accidentally damaged chemicals that are bad for their organism's lifestyle but at no special benefit to themselves.

When either a deleterious or selfish mutation arises, it may be selected out by the failure of its organism's lineage. If it persists, though, other genes will be strongly selected to compensate. The other genes silence the defect or counter its action. There are many other genes in the genome. Their collective presence is likely to produce a counter-mutation quite quickly. All their Darwinian self-interest lies in cooperation, not competition, to create the organism that carries them on. This mutually supportive effect has been called "the Parliament of the genes."[1]

The dour evolutionist points out the price. Billions of sperm and a half-million eggs in your own two parents never survived, and probably many fertilized eggs were lost before they grew into an embryo. Uncountable myriads of would-be ancestors left no progeny even in your own lineage—all the cousins who died without issue since the beginning of life. The embryologist looks at the results: the cooperative genetic architects of the triumphant being that is you.

The organism's boundary also looks different from this standpoint. A living entity does not exist without its environment. The DNA's string of information works in concert with its surrounding cytoplasm. The embryo's placenta needs its mother's womb to grow. The child needs teachers and feedback from its own experiences. In talking of natural selection, we emphasize boundaries: this organism lives, those die. In talking of development we emphasize how permeable the boundaries are. Where do you draw the line between you and the air? As it goes up your nose, or into your lungs? Or when its molecules cross the barrier to your blood? Where is the line between your ideas and mine as you read? The living, growing system is not walled off from the world; it is a node of interaction with the world.

Reductionism and Emergence

Populations of discrete entities that we define as interchangeable are subject to statistical laws. The study of natural selection as changes in gene populations is mathematically tractable.

Studying development is harder work. Individuals that differ from one another, and each make their own unique contribution to building an organic whole, make different demands on understanding, even if those individuals are just different genes building a body. When you throw in the environment, that adds a potentially infinite sphere of interacting influences. This is still science, but it attracts a different kind of scientist.

Unfortunately, it also attracts a lot of wishful thinking. People repeatedly turn to individual development, with its sunny view of cooperation in nature, as a kind of relief from Darwinism. Cautious versions of this point of view are illuminating. It is true that too many biologists ignore the cooperation among organisms' internal parts and the external environment. However, stronger versions imply that holistic explanations somehow replace or contradict explanations by natural selection.

This is a classic battleground. The most recent case, for me, was hearing a lecture by Brian Goodwin on emergence of order. Bath water and other fluids form a pretty whirling vortex if drawn through a narrow plughole, a shape that looks almost organic. If you heat oil evenly from the bottom, it breaks into convection currents spaced at regular geometrical intervals, almost like separate cells. If you rhythmically drop magnetic particles that repel one another onto a central point, while also drawing them symmetrically away from the point, you can mimic the mathematical spiral pattern of the seeds in a sunflower head, an embodiment of the Greek proportion of the Golden Mean.[2]

This is wonderful stuff. These are all examples of fairly complicated organization derived from simple laws of physics. They are patterns in the world that emerge without natural selection. Not very complex organization, not like a real biological cell or sunflower head, but still they are physical patterns that could offer raw material for evolution.

Then Goodwin somehow skipped from order in physics to rejecting Darwinian competition along with free-market capitalism. The emergence of naturally ordered structures without natural selection became for him an alternative morality—structure achieved without rivalry or cruelty. This led on to hope for human society based on cooperation. Admittedly this was a lecture, so he had no time for the middle steps of his argument. In a way, though, it made the goal all the clearer: to remove or weaken Darwinian competition as a driving force in creating order. (I glowered in the background, while his audience of computer programmers cheered.)

Many other modern schools pursue related agendas. The complexity theorists of the Santa Fe Institute, like Goodwin, see physical structures simply falling into ordered patterns through mathematical necessity, in examples ranging from cells in a developing embryo to the distribution of galaxies to cycles in the stock market.[3]

A Viennese school of genetics, disciples of the Austrian Rupert Reidl, also champion biological order and emergence. They do not deny natural selection; they just think that there is far too little emphasis on gene context. Each gene's actions ramify through the whole body and are influenced in turn by the genome as a whole.[4]

Feminist biologists, especially Ruth Hubbard and Evelyn Fox Keller, also stress the importance of context for gene action. They see the egg's cytoplasm too often coded as being female, passive, mere machine tools and food supply. In fact it interacts at every stage with the nuclear genetic program. The nucleus itself is derived from both parents, but a subtext codes it as the organizing male, because its genes can be treated mathematically as a population of digital elements.[5]

Somewhere in the emotional background to all of this is a reaction against Darwinism. It is rarely so explicit as in Goodwin's exposition, but it is there—that structure is good, or even female, while particulate analysis is the evil war of all against all, presumably male.

Reductionist explanations in science involve analyzing larger, more complex entities into simpler parts and principles. Eventually we imagine all science as a web of understanding, with biological organisms explained in terms of chemical molecules, and molecules in terms of physicists' atoms. At each stage reductionist explanation deals with lower levels as populations of identical elements. If you actually look at the elements, they may in fact have some individuality, but that does not matter to the mathematical treatment. For instance, population ecology may lump together all frogs in a pond, human demography all humans in a country. We can then project population trends, but not by consulting the individual propensities of each person or tadpole.

However, there is an equal need to understand higher-order complexity. A group of equivalent elements are just a population, but if they once start differentiating from one another, cooperating, developing boundaries, we are into the realm of ordered wholes, which have different properties from their smaller parts. We are back to Koestler's holons, the assemblies that are both individuals and communities of individuals.[6]

Holons have emergent properties. Emergent properties belong to no one part, not even a simple sum of a population of parts, but to an interlinked architecture of parts. You do not have to seek even as far as a living cell to find the importance of architecture, as Goodwin's lecture emphasized. Organic chemistry is all about the different arrangements of remarkably few elements, mostly hydrogen, oxygen, and carbon. The whole power of organic chemistry is that complex shapes build up from simple elements. Even water makes life possible by the odd angles hydrogen and oxygen set up when they bond, not simply because two of one join one of the other. Water's emergent property of "aquacity" could not be easily predicted from the properties of its constituent elements. Every level of structure, whether chemical, cell, or organism, has emergent properties that could hardly be guessed from the level below.

Reveling in Complexity

Both art and science in the twentieth century over and over have reduced wholes into separate, hard-edged, digitized parts. William Everdell, in *The First Moderns* (1997), traces the trend in many intellectual spheres. He begins with the mathematics of real numbers, progresses through the components of the atom, and arrives at the colored pixels of a Seurat painting and the juxtaposed viewpoints of early Picasso—and of course the independently segregating genes of Mendel. Digitizing is not just a technique of science and information technology but a widespread trait of modern thought. In furniture, in apartment blocks, in assembly lines we tend to reduce larger wholes to modules.[7]

I have personally never been able to see a logical conflict between reductionist and emergent descriptions of nature. To me, the Janus face of the holon can see itself *both* as individual and as a member of a population. We can rejoice in complexity without denying that in living creatures complexity has been reached through cumulative natural selection on populations of genes.

Still, arguments flare up over and over again, not only because of intellectual difference, or even differences over ethical implications, but through fundamental emotional conflict. Basically, some people *like* reductive analysis into component particles. Others revel in complex wholes. This split is not masculine versus feminine, or science versus art. It is inside every field—and at times in every thinking person. It is what philosophers contrast as analytic versus synthetic, what my artist mother called controlled formulation versus baroque exuberance, what Elizabeth Sewell, a poet, called number versus dream. It is even inside mathematics. Husserl, in his *Philosophy of Arithmetic*, contrasted the idea of sets of discrete elements as against the consciousness of continuity. To him, some mathematical constructs like the continuum of all real numbers must be conceived as partless, indivisible wholes.[8] The influence of Husserl's later work spread to Heidegger and Sartre. If you study philosophy in France or Belgium today, you expect to read Husserl as an undergraduate. This stream eventually flowed into Roland Barthes' and Jacques Derrida's postmodern attempts to ground "objective" science in subjective human consciousness.[9]

Meanwhile, a particulate, analytical, Anglo-Saxon school of philosophy hopped (not flowed) from Bertrand Russell to Gödel to Turing and thence into the computer revolution run by people who think in terms of digitized bits. Alfred North Whitehead wrote to his friend and colleague Bertrand Russell: "There are two kinds of people: the muddle-headed and the simple-minded. I, alas, am muddle-headed. But you, Bertie, are simple-minded."[10]

For examples, let me take Sewell's juxtaposition of two French symbolist poets, Stéphane Mallarmé and Arthur Rimbaud, at her poles of number and dream. Mallarmé aspired to the purity of absolute void, *Le Néant*. He expected the first thing to strike his reader would be the blank spaces on the page, onto which he dropped his words like separate pebbles of type. His poem "Un Coup de Dés" is more or less about chance and probability and the statistician's paradox that the number on the dice, once rolled, becomes a certainty. The last line is: "Every thought emits a throw of dice."[11]

For Mallarmé even stars are merely bits, elements of a cosmic calculation:

A CONSTELLATION
cold with forgetfulness and lack of use
not so much
that it does not number out
on some vacant and superior surface
the successive shock
sidereally
of a total sum in formation.[12]

Rimbaud instead wants everything at once. The drunken boat, center of the poem that is his masterpiece, leaves human constraint in the first four lines, as shrieking redskins take its towmen and nail them naked to the color-masts. It careens among rutting Behemoths and rotting Leviathans, drowned bodies sinking backward, singing fish so beautiful it would like to show them to children, and flowers with the eyes of panthers and the skin of men. At last the boat soars up to stars which are not Mallarmé's numbered points but whole tropical archipelagos:

I have seen sidereal archipelagos! And isles
Whose delirious skies are open to the sea-farer!
—Is it in these bottomless nights that you sleep, you
exile yourself,
Millions of golden birds, Oh future Energy?—[13]

The two forms of thought, Janus's two views of any object as both sum of parts and richly constructed whole, struggle within our minds and our emotions. They are both right. A tadpole is one element of a population in the frog pond, and at the same time its own, urgent, wiggling self. This is a good deal easier for most of us to understand than how light can be simultaneously a particle and a wave. But small wonder that for biologists the two emotional stances take shape as warring theories over the desirability of seeing from small to large or large to small, irreconcilable as the Lilliputians' war over which end to open an egg.[14]

11
A Gendered Body

$\mathcal{N}$ettie $\mathcal{S}$tevens was a research assistant at Bryn Mawr when she discovered the Y chromosome, the Mendelian segregator for sex. Her mentor, the great geneticist T. H. Morgan, wrote in 1903 that he thought a fertilized egg might become an adult of either sex, depending on environmental conditions, but he supported Stevens' application for a Carnegie Institute Research Grant.[1]

She studied *Tenebrio,* the little black beetles that infest dry flour, convenient beasts to keep in laboratories. *Tenebrio* has only ten sets of chromosomes. Stevens painstakingly stained cells as the chromosomes paired up before germ cell division. In 1905 she published her finding that males, but not females, showed a tiny chromosome paired with a larger one. She reasoned that this mattered—a genetic determinant of sex. Stevens eventually won a research professorship in her own right, but she died in 1912 of breast cancer and did not live to occupy her chair.[2]

Embryonic Sex and Gender

The Y chromosome—indeed, one small part of it, called the *SRY* gene in humans and *Sry* in mice—is the particle that launches the process

234

that makes mammals into males. A genetically female mouse embryo given an *Sry* gene develops as a completely functional male.[3]

It is an extraordinary gene. It is essentially the only switch gene we know that identifies a large alternate caste of people, without apparent disadvantage in survival or reproduction, who have predictably, multiply different bodies and behaviors from those without it. It also exemplifies what we mean by "a gene for" something. It is a switch gene, but like any other gene it cannot work alone.

We, and all other mammals, have females with two similar large chromosomes, called X and X. In birds, it is the males who have two like chromosomes, called Z and Z. The other sex has one large and one small chromosome in the pair: XY male mammals but ZW female birds. The sex with two like chromosomes is the default setting. An animal missing one chromosome, so that it is genetically just XO or ZO, will develop as the default, respectively a female mammal or male bird. The activity of a single gene on the small, odd chromosome, the Y or W, turns on a cascade of hormones that shapes the alternate body plan. It is a potent gene, working even against much opposition: people or mice who are XXY or even XXXY still become male.

At this point we get another of those funny linguistic extrapolations, like the active macho sperm and the stolid passive egg. For mammals the female body plan is the default setting. Does this make females the Ur-mother, the Earth Mother? Are females somehow the way we all should be, the ultimate form, Eve evolved complete, with not a rib but a gene tweaked aside to produce a second string of Adams? Or to take the other tack, should we see the female as inert and unformed clay without the touch of that Master Potter, the Y chromosome, to shape her into an ideal male?

Fortunately, we have the birds as counter-example. It is instantly obvious that it does not matter which sex is the default, which sex has the switch gene. The bird lineage happened to start off one way, mammals the other. If you look still farther, the genes that determine sex in insects and roundworms are different yet again. Genetic sex determination evolved independently several times. The only thing that matters is that somewhere there is an adequate instruction as to which fork in the road an embryo will take, to produce two complementary sexes in the end.

What does the switch turn on? The analogy of the switch is important. Almost all the actual machinery that makes the switch effective is coded in other genes on other chromosomes. This means that it is present in both male and female embryos. Think of a light switch. The household wires, power cables, generating plants, dams and turbines and oil wells and uranium rods are elsewhere. In evolutionary or historical time they were indeed built and adapted to meet the needs of a set of houses and factories, but right now, for this household, they exist whichever way the switch is turned. The SRY gene does not build a male from scratch any more than flicking on the desk light builds a power plant. Indeed, much of the power plant operates for both sexes.

What does the switch actually do? One section of the protein produced by SRY apparently alters the expression of other genes by bending their DNA into a sharp angle of 70–80 degrees—in other words, changing the architecture of other genes. There are perhaps a half dozen candidate genes targeted by this very early action of SRY.[4]

The cells that will differentiate eventually into eggs or sperm appear very early. They do not start off within the embryo's body but in the embryonic yolk sac—perhaps set aside as early as the 8-cell stage. The future "germ line" is segregated right at the start, before it can be messed up by development. These prospective gametes then migrate into the body and settle in the appropriate place: a "genital ridge" of cells that will become the adult gonads. The fetus grows not one but two sets of ducts leading away from the genital ridge, and the primordia of what will be external genitalia. At this stage, the only apparent

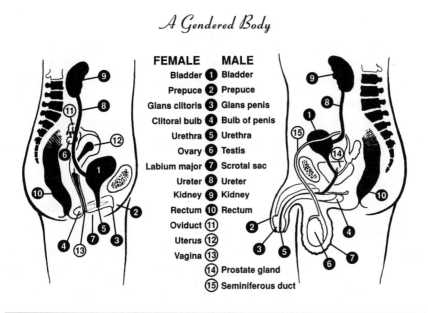

FEMALE   MALE

Bladder ❶	Bladder
Prepuce ❷	Prepuce
Glans clitoris ❸	Glans penis
Clitoral bulb ❹	Bulb of penis
Urethra ❺	Urethra
Ovary ❻	Testis
Labium major ❼	Scrotal sac
Ureter ❽	Ureter
Kidney ❾	Kidney
Rectum ❿	Rectum
Oviduct ⑪	
Uterus ⑫	
Vagina ⑬	
⑭	Prostate gland
⑮	Seminiferous duct

Figure 12. Most male and female organs develop from the same, undifferentiated primordia (nos. 1–10). The embryo also forms two different sets of internal ducts. In most females, one persists to become oviduct, uterus, and vagina; in most males, the other gives rise to sperm ducts and associated glands.

difference is that male embryos are developing slightly more quickly than female ones. Male specifications need to kick in early; otherwise, the undifferentiated structures might start down the default path toward being female.[5]

Testosterone is the earliest sex hormone embryologists can identify. Testosterone reaches the genital ridge, which takes it up. Some of the ridge cells become a specialized lineage—the Sertoli cells. Sertoli cells align themselves as "cords," the future sperm-making tubules. The cords lie in parallel rows, so the tiny organ looks stripy under a microscope. Voilà! The first visible male trait: stripes.[6]

Sertoli cells secrete a second hormone, Mullerian duct inhibiting factor, under the influence of a gene that seems to be a direct target of *SRY.* One set of embryonic ducts disintegrates, the ones that would persist in a female to become fallopian tubes, uterus, vagina. The same hormone also disrupts any of the early germ cells that have begun division. This matters, for if these cells developed into primordial ova, they themselves could trigger hormones that would maintain the gonad as an ovary.[7]

Sertoli cells then promote a second kind of testis cell: Leydig cells. Leydig cells secrete three different hormones. F-alpha-reductase virilizes the external genitalia. The nubbins that would otherwise become labia and clitoris develop instead as scrotum and penis. Estradiol, similar to estrogen, targets the fetal brain. In laboratory rats, mice, and monkeys it primes the brain for later masculine behavior. What evidence there is (mostly analogy from rats, mice, and monkeys) suggests that estradiol similarly primes male brain and behavior in ourselves. The Leydig cells also secrete more testosterone. Testosterone influences the second set of ducts to become the sperm ducts that lead from testicle to penis.[8]

In cold-blooded reptiles, the gonads remain within the body. Hot-blooded mammals need to keep their sperm at cooler temperatures, outside. During fetal development, mammalian testes descend from the body into the scrotum. In some individuals, one or both testes are undescended at birth. Unless they descend by puberty, either naturally or through surgery, they become infertile. Testes take a direction worked out bit by bit in the course of evolution, not the route that foresight might have chosen. This leaves the sperm ducts still routed into the body cavity and out again, looped around the ureters (in Steven Pinker's phrase) "like a hose snagged round a tree."[9]

All this differentiation acts in concert with many other active genes. Estrogen, for instance, is absolutely essential for both sexes. Mice that lack the receptors for estrogen have multiple deficits and cannot develop either sperm or eggs.[10]

The principle is clear. The most dramatic switch we have, the one for maleness and femaleness, is exactly that: a single switch within a hugely complex matrix of development. In fact, most of what we do know is how the switch stops males from being females. The basic plan, the female, is so complicated that scientists have only begun deciphering how she arises.[11] One candidate gene, called *Dax-1*, seems to oppose testis formation, but Amanda Swain, who has analyzed its action, suggests that *Dax-1*, and the *Sry* gene itself, are both superimposed on some far older and more general sex-determining mechanism which is common to all vertebrates.[12]

Strangely, until very recently anatomists were not even very sure what it is that arises. The clitoris is much larger than the textbooks declare. The anterior parts are about 2–4 cm long, with two arms projecting backward for another 5–9 cm, enclosing the urethra up to the front wall of the vagina. The whole structure is composed of richly innervated, erectile tissue. It obviously functions in sexual arousal, both "clitoral" and also "vaginal." It also probably closes off the urethra during sex, much as the urethra is closed off in the aroused penis. The fact that the gross anatomy was not fully described until 1998 might be because most dissections have been done on cadavers long past menopause, in which the structure has diminished. The part by part correspondence of male and female organs now extends to a detailed correspondence of penis and clitoris.[13]

Female X inactivation. Females have an extra problem. Most genes contributed by a single X chromosome are sufficient for overall development of either a male or female body. Females have a double dose. The solution is that one of the Xs goes silent in each cell. The silence is usually random—different Xs prevail in different cells within the female's body. Sometimes swathes of adjacent cells inactivate the same X, either the maternal or paternal one. The classic illustration is a calico tabby cat, blotched white, gray, and orange. Expressing one or the other X gives rise to her orange or gray.

This has direct consequences for girls who carry a gene like hemophilia on one X chromosome. A boy who has the misfortune to inherit a hemophilia gene on his mother's X chromosome is sure to suffer from the condition, because there is no compensating gene on the Y.

However, a girl who has one of each version (allele) of the gene may be completely protected, if by chance the X with the normal allele is active in all her blood-forming tissues. If instead, by chance, the defective X is expressed in some of those tissues, she may have slower blood clotting and heavier menstrual periods than other girls, though it would be a huge statistical anomaly if she were as much affected as her brother.

Inactivation of the X chromosome does not affect all of the genes. Some escape inactivation and are expressed in double dose. The Y chromosome also has a group of genes that have nothing to do with maleness. These "housekeeping genes" work with their counterparts on the male's X chromosome to provide essential developmental instructions. I said a few paragraphs ago that a person with just one X chromosome develops as a female. She will not be, however, an ordinary female. Ninety percent of such people die as embryos. Those who survive tend to have short stature, anomalies such as a webbed neck, and nonfunctioning ovaries. This condition is called Turner syndrome; it is not really an anomaly of sex determination but of genes that need to operate at double dosage in both sexes.

Ambiguous Identities

I spoke of the *SRY* gene as offering a fork in the road. John Money, who did much of the fundamental research on people with ambiguous sex, wrote that development is not one fork but many. The vast majority of people, having taken a turn to the right or left, go on diverging toward clear male or female gender. The paths never diverge completely: most of us share a wonderful assemblage of traits coded "masculine" or "feminine." Some people, though, take a mixture of turns very early in fetal life. In fact, there is such a complex range of transitions between the already wide ranges of "normal" male and female that the radical biologist Anne Fausto-Sterling argues we should simply abandon the notion that everyone's sex is classifiable and welcome the fact of human diversity.[14]

Some definitions: *Intersexes* are people with physical ambiguity of sex assignment. The term is sometimes enlarged to include people with any mismatch between chromosomal and bodily sex. *Hermaphrodites* are intersexes with developed traits of both sexes. Snails are full hermaphrodites; we never are. No humans both beget and bear children. However, it is possible to have both male external genitalia and female breasts, or an ovary on one side and testicular tissue on the other, each with its appropriate internal ducts.

Transexuals are people whose self-identities are the opposite sex from their physical appearance. They may eventually decide for social, hormonal, or surgical sex change. *Transvestites* cross-dress. For them, the excitement is to belong to one sex while flaunting the trappings of the other, which is quite different from wishing to actually change sex.

Homosexuals are sexually attracted to members of their own sex. Homosexuals are more numerous than any of the other groups of people. Members of any of the other groups may or may not have homosexual orientation; homosexuals usually have no identifiable physical intersexuality. None of these groups is clear-cut, of course. For instance, homosexuality and heterosexuality both shade into *bisexuality,* which is being attracted to either sex, depending on the partner involved.

Genetic sex reversal happens sometimes naturally. Some people with a predominantly female reproductive system are XY, while others with a predominantly male reproductive system are XX. The commonest cause seems to be that the crucial SRY gene gets deleted from a Y (in the former case) or displaced onto an X chromosome (in the latter case). There are other genes that can lead to the same effect, including X-linked defects in the receptors for testosterone. XY females do not develop functional ovaries and they have low estrogen. Unlike the XO Turner females, who are short on average, some XY females grow extra tall because estrogen is necessary to close the growth regions of the long bones at adulthood. Castrated boys, like the boy sopranos mutilated to become opera singers in past centuries, grow tall for the same reason—a lack of estrogen.[15]

Hermaphrodites can result from mosaicism. Either the child is formed from cells fertilized by two different sperm, or else one cell lineage loses a sex chromosome very early in differentiation. Oddly, if there is just one ovary, it is usually on the left side.

Hormonal anomalies can lead to intersexual body forms. Congenital adrenal hyperplasia, for instance, is a condition where the adrenal glands overproduce a steroid that mimics the action of testosterone. In the 1930s a group of mothers at risk of miscarriage were injected with progestin, a synthetic progesterone, to help them keep the pregnancy. Progestin, like the adrenal hormones, mimics testosterone in the growing fetus. Boy babies were apparently unaffected, but girls were born with protruding genitalia like underdeveloped boys (their internal organs were unaffected). John Money, in his study of these children, concluded the girls acted like "tomboys." We have no way now to tell if this was because the parents knew of their birth ambiguities, or indeed whether the girls' behavior would have passed unnoticed with the changed social climate between the 30s and the 90s.[16]

From all his studies, Money concluded that surgical "correction" at birth—essentially by removal of the malelike tissue—was the right solution. He believed that children essentially accept their sex of rearing. However, he felt that gender reassignment after the age of two caused profound emotional trauma, so decisions should be made early by well-meaning adults.

One case entered the literature as an icon. Two 8-month-old midwestern boys, twins, were operated on for constricted foreskins. In one, the doctor slipped and removed most of the penis. Money advised the parents to proceed to surgical castration and removal of the rest of the penis, to spare the child the trauma of growing up with a mutilated penis. After the operation, that twin was given dolls and pretty dresses, and she grew up believing herself a girl. She became the oft-quoted case to prove that environment determines gender.[17]

It was only in 1997 that we heard the sequel. Even as a child she spurned the dolls and hated the dresses. She wanted to climb trees with her brother. Little girls at school called her "Gorilla." Between ages 9 and 11 she decided she was somehow a boy. At that point he was told, "All girls think such things when they are growing up." He recalls thinking, "You can't argue with a bunch of doctors in white coats; you're just a little kid and their minds are already made up. They didn't want to listen." At 14 he refused to live as a girl, wore gender-neutral jeans and shirts, rejected estrogen therapy, revolted against the suggestion of vaginal surgery, and fought psychotherapy to reinforce female identity.[18]

"Doctor XXX said, 'Its gonna be tough, you're going to be picked on, you're gonna be very alone, you're not gonna find anybody [unless you have vaginal surgery and live as a female].' And I thought to myself, you know I wasn't very old at the time, but it dawned on me that these people gotta be pretty shallow if that's the only thing they think I've got going for me, that the only reason why people get married and have children and have a productive life is because of what they have between their legs . . . If that's all they think of me, that they justify my worth by what I have between my legs, then I gotta be a complete loser."[19] He thought of suicide. It was only then that his father, in tears, confessed what had been done to him in infancy.

The boy was not devastated but relieved: "All of a sudden everything clicked. For the first time everything made sense, and I understood who and what I was." Supported by local doctors, he received the surgery he wanted to produce erectile tissue, though still lacking much feeling. He remained in the same high school. His schoolmates accepted him as a boy as they never did when he was a girl. He is now a happily married man; his wife had previous children whom he has adopted and raises as their father. He keeps his anonymity in print but strongly supports making his story known to the medical community, to save others from similar trauma.[20]

One case does not prove anything. Or perhaps what it proves is that in the 1960s it was possible to hope for complete environmental determination of gender, while in the 1990s the research community cites the same person as an example of biological influence on gender. One thing this case does do is add ammunition for the small community of politically active intersexuals. They argue that surgery at birth may be far worse than letting a child grow up aware of its own ambiguity, and eventually able make decisions for her/himself.[21]

Although many people experience "gender confusion," a few people decisively feel themselves "born into the wrong body." Transsexuals who attempt full sex change are a small group—people who announce to themselves and the world that they have been somehow trapped in a body that differs from their mind, and who undergo radical surgical and social trauma to achieve their desired identity. The number of people who feel equivocal, who just wonder about themselves, is surely much larger. Even on purely biological grounds, if prenatal hormones influence mindset, some people must receive intermediate doses, sowing confusion even before the ambiguous impact of society.

The known fetal pathways do offer a biological basis for the disjunction of body and mind. If some people receive the brain priming of estradiol, without the f-alpha reductase and testosterone that masculinize the reproductive tract, that would do it. A gene mutation known in a few families in the Dominican Republic led some boys to be born with ambiguous, feminized external genitalia. Though they were raised as girls, when the testosterone surge of puberty arrived they grew unmistakable male genitals and beards, and their voices dropped to tenor or base. They apparently had no difficulty shifting sex roles, though we cannot know how much was due to chemical brain priming, how much due to changing to the more glamorous role within their culture. They might have protested more if they had to change from the freedom of boyhood to the limitations of mere women![22]

I hesitate to impose an outsider's summary onto an experience that I only dimly imagine. Let me quote instead people who explain themselves. Jan Morris's 1974 book *Conundrum* was one of the first to tell the story openly: "I was three, or perhaps four years old when I realized that I had been born into the wrong body, and should really be a girl. I remember the moment well, and it is the earliest memory of my life . . . I was sitting beneath my mother's piano, and her music was falling around me like cataracts, enclosing me as in a cave . . . By every standard of logic, I was patently a boy . . . I had a boy's body. I wore a boy's clothes . . . if I had announced my self-discovery beneath the piano, my family might not have been shocked (Virginia Woolf's androgynous *Orlando* was already in the house) but would certainly have been astonished . . . Not that I dreamed of revealing it. I cherished it as a secret, shared for twenty years with not a single soul . . . A moment of silence followed each day the words of the Grace—"The grace of Our Lord Jesus Christ, and the love of God, and the fellowship of the Holy Ghost, be with us all for evermore." Into that hiatus, while my betters I suppose were asking for forgiveness and enlightenment, I inserted silently every night, year after year throughout my boyhood, an appeal less graceful but no less heartfelt: 'And please, God, let me be a girl. Amen.'"[23]

Biologist C. S. Hall speculates, "I do think the explanation is as simple as prenatal brain priming. Seeing it as 'simple' is very pleasing. For a long time I was caught in a place that allowed no reasonable explanation: heredity couldn't be responsible for the discrepancy between public and private identities, because I appeared to be physically thoroughly female, and environment couldn't be responsible, because my family and others had always treated me as a girl. Awareness of that pair of contradictions probably started when I was quite young, and by the time I knew enough about genetics and development to counter it, I'd been around the argument so many times I couldn't really see it any more. Quite suddenly, and only a couple of months ago, it occurred to me that sexual identity would be a very useful trait to have as a primary characteristic—not necessarily hardcoded, but extremely early and strong—because it is so basic to so many social as well as sexual interactions. And if sexual identity is primarily not derived, no matter how strongly it is normally linked to phenotype one would expect that occasionally the two would be separated. (By simple mutation, or by environmentally induced wrong timing of some prenatal hormonal kicker.)"

"Sidebar here is that since female is the default state, if as I imagine the male-to-female transsexual has all the required hormones, but misses or miss-times the little squirt of stuff required to change the brain part, whereas female-to-male gets the brain squirt but lacks the rest of the hormonal package, there's no reason to believe the two (mtf and ftm) should occur with equal frequency. My intuition would have mtf the likelier condition. In fact, it is so reported; the current literature mostly maintains that the ftm population is underreported, or more underreported, and that the two 'should' be equal, but maybe not . . . I never felt that I had a choice of gender—although some of the time when I was a child I felt that my choices were inhibiting normal development; that is, that if I had been better at sports or cared more about automobiles or not liked word-games and poetry, I'd have developed proper male genitals."[24]

The last word should come from Jan Morris, who spent forty-five years of army life, journalism, and fatherhood before achieving sex change: "To me, gender is not physical at all, but altogether insubstantial. It is a soul, perhaps, it is a talent, it is a taste, it is environment, it is how one feels, it is light and shade, it is inner music, it is a spring in one's step or an exchange of glances, it is more truly life and love than any combination of genitals, ovaries, and hormones. It is the essentialness of oneself, the psyche, the fragment of unity. Male and female are sex, masculine and feminine are gender, and though the concepts obviously overlap, they are far from synonymous . . . Gender is a reality, and a more fundamental reality than sex."[25]

Olympic Sex Testing and Sports Steroids

Women who compete in the Olympics all undergo a test for biological sex. There is in fact only one case of a man known to have posed as a woman in order to compete. Hermann Ratjen bound up his genitals and, as "Dora," entered the high jump in 1936. He was beaten by three women in the finals, and so lost his big chance to bring glory to the Nazi Youth Movement. He went on to set a women's world record, though, in 1938, and kept his secret, living openly as a man until he told the story in 1957.[26]

The masculine appearance of several later athletes raised suspicions, particularly the Soviet sisters Tamara and Irina Press, who competed between 1959 and 1963, winning five gold medals and setting 26 world records between them. In 1966 at the European Track and Field Championships, women competitors were made to undress for a "nude parade" before gynecologists—extended later to internal examination for undescended testes. The Press sisters and various other Eastern Bloc athletes did not show up and soon retired from competition.[27]

In many sports that need muscular strength or short bursts of energy, women have never reached the world records of men. In others—very long runs of several days rather than hours, and very long swims—women's records have sometimes exceeded men's. The whole point of having separate meets and records is an attempt to be fair to each sex. But to do so, someone has to decide who is a woman. The International Olympic Committee does not think this determination can be left to the athletes themselves.

In recent times, the nude parade and gynecological grope have been abandoned for the more "scientific" method of a buccal smear for chromosomes, updated in the 1990s to DNA testing for the presence of the *SRY* gene itself. Nowadays, drug testing requires that athletes urinate into a cup before witnesses, so any genital abnormalities are obvious in any case. But what about more subtle hormonal intersexuality? Does any single test, even for the gene itself, say what is really happening in an athlete's body?

The first person with the courage to challenge the verdict publicly was Spanish hurdler Maria Patino. In 1985, at the World University Games in Tokyo, her buccal smear came up masculine. The 24-year-old had never before questioned her own femininity. She was told on the way to the race that she must fake a foot injury. She would not be allowed to compete. "I could barely comprehend what was happening. I was scared and ashamed, but at the same time angry, because I couldn't see how my body was different from other girls," she says. "I was crying, but not for my foot. I had to sit in the stands, watching the other girls run my race. And I still had another week to spend in Japan with this horrible secret. Everyone from my dorm was sightseeing and having fun, but I stayed alone in my room. I had no one to talk to." Patino returned to Spain, began the long and expensive round of endocrinologists, and went back into training. She learned that she had indeed XY chromosomes, but also insensitivity to testosterone. Her body never responded to the initial messages of testosterone, so she developed as a woman.[28]

Four months later, as the next track season was starting, the president of the Spanish Sports Federation told Patino the verdict. She must fake another injury that would seem permanent. If she ever chose to run, she faced public exposure. She ran. "I was petrified they would stop the race—I kept looking behind me. When I crossed the finish line, I was sick with worry." It took two and a half more years of public battle before she also won her case for reinstatement in sports, with the help of Alison Carlson, a coach and journalist with a biological background, who campaigns for the understanding that biological sex is far more complicated than any single gene. Carlson declares that cases like Patino's force us to reexamine what we mean by sex and by gender. Maria Patino's conclusion is more direct: "I knew I was a woman—in the eyes of medicine, God, and most of all, in my own eyes. If I hadn't been an athlete, my femininity would never have been questioned."[29]

One in 504 female Olympic athletes fails the sex test—six did so in the 1984 Olympics alone. These are people who have considered themselves women throughout their lives. The discovery about their chromosomes can be psychologically shattering, causing them to retire from sport for life. Doubting their own identity, probably few have the courage to pursue the question, to find out what exactly the tests can mean. If there is a mere false positive, as in the case of swimmer Kirsten Wengler, or androgen insensitivity, as in the case of Patino, these women would have no actual advantage over other women. The International Amateur Athletic Federation voted in 1992 to scrap sex testing as useless, as well as degrading, but the Olympic Committee continues.[30]

Meanwhile, the use of steroids in sport spirals upward. Even in the 1960s javelin thrower Olga Connolly stated that sex testing was a farce; the real problem was drugs. In an extraordinary trial in Germany in 1998, Birgit Meineke Heukrodt, who was the best female free-style swimmer in the world between 1980 and 1984, led the suit of former athletes against the state apparatus for grievous bodily harm. East German swimming team coaches passed out little pink and blue pills, called "vitamins," to underage swimmers for the glory of their nation. Dr. Heukrodt, now a successful surgeon, wife, and mother, oscillates between professional fear of the tumor on her liver and the recognition that everyone went along with the system, including herself and her parents, who asked no questions. She testifies in her baritone voice against the coaches and doctors that she once considered father figures for systematically doping her with androgens all through her girlhood.[31]

There is a neat semantic division between so-called anabolic steroids, which primarily build muscle, and androgenic steroids, including testosterone itself, which primarily build maleness. In fact, there is no chemical division. Anabolic steroids build muscles as in males. No mystery: all athletes know what steroids do, whether or not they succumb to the temptation to take them. For men and women alike, anabolic steroids increase muscular prowess. If ever there was proof of the biological continuum between men and women, steroids are it. The more athletes are caught with the drugs, the clearer it becomes that the drugs do confer success.

Steroids have spawned their own pathology—the cult of body building. Body building for beauty is the mirror image of anorexia nervosa. Anorexics starve themselves, convinced that they are fat even as they grow more skeletal. The complicity of fashion magazines and models rub in the slender image that for a few people grows into clinical anorexia. Body builders, in contrast, are usually considered merely odd, not sick. However, at an extreme they look into a mirror and see a body that is too thin, just as the starving anorexic sees one too fat. In fact, body building could be called megarexia—the insatiable quest for more protein, more workouts, and almost inevitably more steroids to build those muscles. Anorexia is primarily a female malady, megarexia primarily male; but neither is confined to just one sex.

One of the most chilling of the student seminars I have heard was given by an undergraduate athlete. He detailed the effects of anabolic steroids: muscle power and stamina, but also acne (the same as adolescent acne from an extra hormone surge), liver and kidney and heart failure, "'roid rages" of uncontrollable temper and violence, and then shrinking of the testes themselves, with total infertility, when the pituitary region of the brain concludes there is enough circulating testosterone and shuts down its natural source.

His punchline was: "Of course, it's worth it." As the class quailed, he explained, "Look. Suppose it means getting a college scholarship or not? Suppose it means getting into the Olympics, or not? A Gold, not a Silver or Bronze? Even just for the money, never mind knowing you won? Of course it's worth it."

Menstruation and Menopause

Beyond our chromosomes, genitalia, and gender identities, there are other differences between living in the body of a man and living in the body of a woman. Two of the most apparent hormonally based effects are menstruation and menopause, and their associated emotional states.

Barbara Ward, a devout Catholic, remarked that her first act on dying would be to ask God why he made women's plumbing.[32] The hormonal bases of the menstrual cycle are well enough understood. But why bleed? Why jettison blood and tissue instead of resorbing it to start over? We do not actually know. The process is prehuman. Old World monkeys and our kin the apes menstruate, so menstruation has been around for a long time.

One suggestion is that the uterus is actually cleaning out bacteria. Marjorie Profet began a provocative paper: "Sperm are vectors of disease." Her argument has been much challenged. After all, the nose, which breathes in a great many germs, manages with mere mucus, and the digestive system has a whole variety of barriers. Why should the uterus have to shed its lining to stay disease-free?[33]

Possibly menstruation protects against cancer. Uterine tissue is extremely active, dividing frequently. When women take estrogen supplements after menopause, doctors often recommend progesterone as well, which by provoking mild menstruation seems to clean out possible precancerous cells. Medical opinion is divided, however. Some doctors (and many women) see no reason to go on menstruating when it is no longer strictly necessary.

A third possibility is simpler. Bleeding might actually save energy, rather than costing energy. When the uterine lining is deep, spongy, and active, it is metabolically expensive to maintain. When shed, far less so. Remember again that women in industrial countries spend years cycling, while women in other cultures are often in lactational amenorrhea. Our hunter-gathering foremothers, like women in most of the world today, spent years breastfeeding, often while themselves underfed and hence less likely to ovulate. In most of adult life they did not cycle. The question then reverses from negative to positive: When is it worth the energy to build up the uterine tissue for a fertile cycle, rather than, Why lose the tissue when that cycle is over?[34]

A final possibility is that menstruation may be just a jerry-built evolutionary solution, not some optimal design. As our lineage developed ever more intimate contact between mother and embryo, with longer, more complex gestation, we needed to prepare the uterus more elaborately. Evolutionary pressure to conserve uterine tissue from month to month may conflict with a stronger pressure to make it ever more fragile, spongy, and receptive for the life it is ready to hold.

Most women are aware of a biological rhythm to our emotions that tracks our menstrual cycle. Most women, though not all. In my classes, when someone gives a seminar on menstruation, I often ask everyone but the person giving the talk to cover their eyes, and women who are personally aware of such emotional changes to raise their hand. It comes out to roughly 90 percent who are. More surprising is the evenness of distribution. In a class of 20, there seem always to be one or two women who notice no mental variation over the month.

Premenstrual tension has been much championed by Dr. Katherine Dalton as a real disorder, not psychosomatic self-pity. She even successfully defended a case of murder on those grounds. The killer was judged temporarily insane, as she became every month. A few such women suffer almost unbearably during the hormonal drop before menstruation.[35] Conversely, some women become irrationally happy—to the point of hypomania—with the estrogen effect of the early part of the cycle and appreciate the emotional lift of estrogen pills and patches after menopause. For most, it is milder, a pleasure to follow the rhythm of one's own tide.

Menopause itself is defined as the cessation of menses. Its physical changes derive from lowered estrogen levels. The loss of estrogen triggers all the rest. Some women scarcely notice the change and wonder what all the fuss is about. Others find it excruciating—unexpected hemorrhages, mood swings, and hot flashes as often as several times an hour. For those who think a little blushing should not hurt anyone, I report that I once did see a man have all the symptoms of a bad hot flash, so much so that he lay down panting, sweating, and scarlet-faced. I privately thought that women cope better, and much more often. A few hours later, his "flash" revealed itself as the first spike of a high malarial fever.

What I have never seen stressed in an anthropology text (and only once in a doctor's office) is the extreme variability of women's hormonal levels and our emotional reactions to them. Anyone who is doing urine analysis of estrous cycles, for instance, to help gorillas and chimpanzees to breed in captivity knows that each animal has her own timing of each hormone. The general principles are the same, but individual chimps differ in level, in timing, and in response to interventions.[36]

Women swap the stories among their friends: a particular birth control pill or dosage suits one but not another. One radical doctor said, about estrogen, "Try it out! This isn't diabetes, where misjudging insulin could send you into coma. Every woman's optimal dose could differ. Many want none at all, some want only a fraction of 'standard dose.' You will only know by experimenting yourself to find the best level."

Quite probably many other bodily functions and emotions are similarly variable. The tidal ebb and flow of women's emotions have allowed or forced women to be in touch with their bodies, in ways that men sometimes believe they can ignore in themselves and distrust or denigrate in women. Admit it or not, men and women both feel what hormones and neurotransmitters do, and recognize the symptoms:

> Let me only glance where you are, the voice dies,
>   I can say nothing,
>
> but my lips are stricken to silence, under-
> neath my skin the tenuous flame suffuses;
> nothing shows in front of my eyes, my ears are
>   muted in thunder.
>
> And the sweat breaks running upon me, fever
> shakes my body, paler I turn than dry grass is;
> I can feel that I have been changed, I feel that
>   death has come near me.[37]

Is All Sex Determined by Genes?

Sex determination by genes is hardly necessary. Many organisms do develop, as Thomas Morgan first thought, according to their environment. Among Mississippi alligators, if the clutch of eggs develops at high temperature, there will be more males, at low temperature more females. An alligator who nests on a levee has more sons than one who nests in a cool swamp. However, a Greek tortoise who lays her eggs in the hot Greek sun produces more females, but in the cool shade of a laurel grove she has more male offspring. Some reptiles have a two-ended curve, with more females hatched at intermediate temperature, more males at the two extremes.[38]

Sometimes the important environmental factor is social. The creature with the greatest known sexual dimorphism is *Boniella,* a saclike worm of the deep sea floor. When a drifting *Boniella* larva lands on the sea bed, it turns into a female. She makes a burrow for her body but grows an exceedingly mobile tongue, a meter long, that lies out on the sea bed to catch the downward snowfall of detritus which is her food. Sooner or later one such mote on her tongue may be another *Boniella* larva. He seizes his chance to develop as a tiny, parasitic male. The luck to actually find a female makes it worth his forsaking all others to stick with (indeed, into) her.[39]

Sex determination need not last for life. Blue wrasse, the cleaner fish who groom the teeth of moray eels, live in territorial flocks on coral reefs. The male is large and brilliant, his females smaller and duller. If a male dies, the most dominant female begins to act masculine within a day. She herds her colleagues and defends territory. Within a week s/he turns eye-splitting neon blue and has fertile sperm.[40]

Other serial hermaphrodites do not wait for social cues but change in the course of normal life. Garden slugs, if small, are male. When they grow into big, fat slugs, they become female. Simultaneous hermaphrodites are both sexes at once; this is true of many plants and a few animals, like the common earthworm and the slugs' relatives, terrestrial snails.

One organizing principle is that the larger sex profits most by being large. A big male wrasse can herd more females. A big alligator can bellow his love songs in the swamp and fertilize more females' eggs.

255

In contrast, a big female slug or tortoise lays more eggs. She gains more by being big than a male would, since their males reach females mainly by scramble, not physical contest. Another organizing principle is rarity and distribution: a *Boniella* who actually finds a female's tongue in the wide, wide sea should stay with his luck.

Organizing an Organism

The interplay of gene and environment builds an embryo, builds a body, and keeps on shaping our form and our feelings throughout life. Gender development is at once our clearest genetic "switch," the SRY, and a beautiful example of the complexity of bodies and minds that are more or less one sex or the other. A very large number of different genes contribute to the end result. In one sense, all active genes do, because if the embryo did not have the full complement necessary to live, it would not survive to have sex or gender.

The genes' Darwinian self-interest lies in cooperation, not competition. The parliament of the genes, for all its imperfections, creates the organism which carries it on.[41] This is the real reason why people who turn to organismic and developmental explanations find biology so much "nicer" than the people who study evolution at a genetic, evolutionary level.

This is also why I see no problem in reductionist explanation in terms of genetic and environmental influences. The logic of evolution itself is that genes will work together, for the most part, in creating the body's wonderful architecture: the Olympic gymnast twirling over the uneven bars, the diver soaring into space and precision fall, the sprinter chesting the tape.

12

Instinct, Learning, and Fate

In 1992 John Tooby and Leda Cosmides wrote a manifesto for the emerging field of evolutionary psychology. Their style was not exactly conciliatory. They accused their opponents of using "naïve and erroneous concepts drawn from outmoded theories of development . . . a faulty analysis of nature-nurture models . . . [and] an impossible psychology." Their own evolutionary psychology, instead, looked for the universal bases of human nature, the traits selected for survival in the environments where our ancestors evolved.[1]

Tooby and Cosmides contrast their own approach with their opponents' "Standard Social Science Model," or SSSM. According to the SSSM, the human mind is almost infinitely malleable. Culture molds it. An individual's instinctive repertoire "consists of nothing more than the infant comes equipped with, bawling and mewling, in its apparently unimpressive initial performances." Culture cannot be the product of such unimpressive individuals; it must be instead an emergent property of the group. According to the SSSM, human personality is born somewhere in the space *between* minds:[2] "It must be emphasized that this claim is not merely the obvious point that social phenomena (such as tulip bulb mania, the contagious trajectory of deconstructionist fashions, or the principles of supply and demand) cannot be understood simply by pointing inside the head of

257

a single individual. It is, instead, a claim about the generator of the rich organization everywhere apparent in human life . . . For most anthropologists today, even emotions such as 'sexual jealousy' and 'paternal love' are the products of the social order and have to be explained by the conditions [of] the social group, in its totality . . . The [SSSM] conclusion [is] that human nature is an empty vessel, waiting to be filled by social processes."[3]

The SSSM they describe is indeed an impossible psychology, one that never could have evolved. If people did rely on culture to teach everything, a single generation's descent into war and chaos would break off any hope of survival. Who would inform toddlers they must attend to parents' speech in order to learn language? Who would drill children in play to build strong minds and bodies? Who would coax adolescents to start noticing the opposite sex? We are a lot safer relying on instinctive hunger to speak, to play, and even to worship teen idols. That list should dispose of the mewling, bawling baby as the only one with instincts. In each stage of development, our instincts kick in when we need them.

Is the SSSM really preached, or is it instead a Silly Scary Straw Man? Sometimes an attempt to sum up a philosophy in one sentence shows its bedrock, as a caricature reveals more than a portrait. Here is an undergraduate course catalogue leader for Women's Studies at Cornell University—the most urgent idea the social scientists wish to tell prospective students: "Definitions of gender—including those that privilege exclusive heterosexuality—are not natural or universal but are instead social constructions that vary across time and place, serve political ends, and have ideological underpinnings."[4]

I can see their point, but only if gender is defined tautologically as what fits the description, and not as those aspects of gender that go along with biological sex in a statistically large proportion of human beings, in a large proportion of cultures. I hope they just mean that all *interesting* questions about gender refer to its cultural aspects.

In sober debate, the SSSM is of course an intellectual straw man. Even cultural anthropologists often admit that human bodies exist. However, they create a second straw figure: the Bad Biological Bogey Man. In this stereotype, what interests me and other biologists is dismissed as either banal or exceedingly dangerous. The divide between SSSM and BBBM is like a duel of scarecrows armed with flaming broadswords. Each imagines one clear stroke of logic will reduce its opponent to a bonfire and a wisp of ash.

But as Tooby and Cosmides pointed out, biologists have been complicit in the lack of serious communication between the opponents. After the bruising battles over sociobiology, many of us simply retreated to our own side of the copper beech tree. We waggled our fingers toward the Humanities' end of the campus and muttered, "Our theories explain behavior resembling lust and jealousy and tenderness in honeybees and chimpanzees. However, you humanists are the experts on humans. You claim humans are purely cultural. *You* go ahead and explain people, if you think you can."

Instinct and Fate

I think that I could turn and live with animals,
    they are so placid and so self-contain'd.
I stand and look at them long and long.
They do not sweat and whine about their condition.
They do not lie awake in the dark and weep for their sins.
They do not make me sick discussing their duty to God,
Not one is dissatisfied, not one is demented with the
    mania of owning things,
Not one kneels to another, nor to his kind that lived
    thousands of years ago,
Not one is respectable or unhappy.[5]

When we turn and look at animals, we do imagine that they are placid and self-contained, instinctively certain how to conduct their lives. In fact, they can vacillate and anticipate. They can surely be unhappy. But not like us. Not like us.

Does that mean we have no instinctive certainties? Or are we, too, animals? It is one thing to admire the interplay of gene, hormone, and structure in the development of the body. It is quite another to accept their influence on our emotions and behavior. In the last chapter I took the commonsense view that they do influence us, trusting that readers who are women, or athletes, or intersexuals, or are acquainted with some member of these groups, will already know they do. When the thought becomes formalized, though, it is frightening. Is our precious identity a mere product of our genes?

A problem with talking about this is that people cling to the idea that an innate *capacity* to express a certain behavior is the same as being *forced* to behave that way. Undergraduates dutifully produce for examinations a clear statement that genes and environment interact and that the interesting question is which genes are expressed in which environment. (They know that's what professors want to hear.) A few elaborate some version of Steven Pinker's statement that mere interaction is like claiming a computer is so sophisticated that it does not matter what you type in. "Having a lot of built-in machinery should make a system respond *more* intelligently and flexibly to its inputs, not less."[6]

Then a day later, in conversation, students jump right back to instinct as fate. I can only suggest that what people mostly mean by "instinct" is not much like the biologists' version. People usually do mean a kind of fate—a character trait that they personally will never escape in any range of environments they personally are ever likely to meet. Arizona State Senator Ann Day wrote a law to prohibit insurance company discrimination based on genetic testing. She quoted James Watson, co-discoverer of the double helix: "People used to think their destiny was in the stars. Now we know it's in our genes, in our DNA."[7]

An increasing number of psychologists point out that a *tabula rasa*, the mind as a blank slate, would not give us total freedom of thought. It would just be so inefficient that a child could hardly grow up, let alone that its ancestors would have grown and mated and left surviving children in the course of evolution. If we really had to learn everything from scratch, we would not even know how to start learning, how to suck at a nipple or pay special attention to human voices. Some basic programming is essential.

Return to Pinker's metaphor. Suppose someone gave you a superb computer, with gorgeous gigabytes of memory and all you desire of RAM. Do you really want to have to tell it what to do in machine language? (Neither mind nor machine can start with a complete blank slate, though machine language is fairly near a *tabula rasa*.) Wouldn't you prefer word processing, spreadsheet, e-mail, and Solitaire preinstalled? Now you can write stories and make business plans and chat up your friends, not fumble for years to build the basic programs. Instinctive programming allows more creativity, not less. In fact, when innovators in artificial intelligence try to make machines do simple instinctive things we take for granted, such as recognizing a rectangle seen from an angle, it turns out they must program in a very large number of "innate" assumptions about regularities in the real world to get anywhere at all.[8]

Instincts do not merely write programs. They also give us reasons to use the programs. William James, father of psychology, pointed out: "To the broody hen the notion would probably seem monstrous that there should be a creature in the world to whom a nestfull of eggs was not the utterly fascinating and precious and never-too-much-to-be-sat-upon object that it is to her."[9] Humans, likewise, want to do what we want—to be warm but not sweltering, to eat our fill, to love and be loved.

Innate action patterns.   Let us turn to some of the stereotyped behavior patterns we all accept as instinctive, to see why instinct is so often equated with fate. After that, we can work up to modules and programs that underlie complex human behavior, to see how instinct is fundamental even to the kind of learning we so often equate with freedom.

A mallard duck sitting on her first nest responds to a baby duckling's distress peeps. The peeping is like a key that turns her maternal lock for "lost infant." She prods the crying duckling into the nest with her duck bill. It quiets down; she looks at it. If it is a different duck species with the wrong-colored infant down, a key turns in the lock "intruder in my nest." She pecks it to death.[10] This oddly fragmented behavior is typical of highly fixed instincts. They are usually evoked by a few simple but salient stimuli and involve relatively few motor actions. That, of course, is why we often think of instinct as forcing behavior—as fate, not flexibility.[11]

The common factor in such instincts is that it is very important to get things right the first time. Many birds and small mammals have an innate fear of two round circles, if they are not too large or small and especially if they go from small to large as though looming nearer. You can send newborn chicks into immobile paralysis and newborn hog-nosed snakes into convulsions if you just show them two expanding circles. If they needed time to learn step-by-step that owl eyes are circles, they wouldn't get past the first step. We also have such innately triggered responses: orgasm (though we have some chance to learn about that) and various actions of mortal importance, such as the first sneeze of your life and the first time you are born or give birth.[12]

Your body, willy-nilly, starts to sneeze, goes into labor, opts to move from first- to second-stage contractions. A nurse told me once, "Don't you dare push!" "I'm not pushing!" I squawked, "It is!" You can consciously or unconsciously slow things down or speed them up as

Figure 13. Laughter is innate, common to all our species, and can be shared even when language fails. (After Bill Leimbach, in Morris, 1977)

you push, or suppress the contractions out of fear and nervousness, but the body's wisdom during labor does not rest on your conscious decisions.

Sneezing, orgasm, and childbirth obviously reach our conscious minds. We will come back to the question of animal consciousness later, but our own experience suggests we cannot presume an animal to be mindless just because its actions are triggered by innately recognized stimuli. Perhaps the naive mallard feels a surge of concern when the duckling peeps, and then a duck's version of disgust at finding an alien in her nest.[13]

Ritualization. "The appearance either bird presented to its mate had changed altogether in an instant of time. Before, they had been black and dark mottled brown; they saw each other now all brilliant white, with chestnut and black surrounding the face in a circle. In this position they stayed for a few seconds rocking gently from side to side upon the point of their breasts; it was an ecstatic motion, as if they were swaying to the music of a dance."[14] Thus Julian Huxley, in 1914, described the courtship of the great crested grebe, and launched a fundamental metaphor. He said the grebes' posturing to their mates was "ritualized." Ritualization in animals means exaggeration, stereotypy, and repetition. You recognize it by the same signs that you would use for human ritual, if you were suddenly parachuted into the monkey dance of Bali or High Mass at the Vatican.[15] Why have animal social signals evolved to be ritualized?

Natural selection has shaped both sender and receiver of ritualized signals wherever it matters to perceive and be perceived without ambiguity. We normally think of this evolution within one species, but it can also work between species. The neon blue scales and dipping approach of a cleaner wrasse signal to the moray eel that this fish will rid it of parasites. It is important to both (though much more to the wrasse) that the moray does not eat it instead.

Many of our own social signals are stereotyped across different human cultures and even among people born blind or deaf. We can presume them innately ritualized. Laughter and smiling, the eyebrow flash of greeting at a distance, the tense-mouthed frown of threat are universal human signals—not that different from the same faces in chimpanzees. Blind babies smile at the sound of a human voice and focus their sightless eyes at the place where the voice is. The smile and focusing do not have to be learned by copying an adult's smile or even by rewarding the baby for smiling. But visual copying is also highly innate. Babies as young as twenty minutes old may work to imitate an adult's expression. They open their mouths, frown, and even stick out their tongues in response to the antics of earnest psychologists.[16]

The evolution of ritualization is well understood. Darwin pointed out that social signals usually begin from "byproduct" actions, what he called "serviceable associated habits."[17] A textbook example, derived from Darwin, is the red visage of apoplectic rage. Still more dangerous is pale cold fury. A face going white, with blue beside the nostrils, may mean its owner is drawing a knife. The colors change as the nervous system shunts blood away from the internal organs to the exterior in preparation for fight or flight, then withdraws it even from the skin to concentrate in the muscles needed for immediate action. The colors are a byproduct of muscle readiness, but we respond to them as meaningful stimuli. People of European and Far Eastern descent do, that is. I have asked dark Malagasy friends about it: they also speak of cold, trembling fury. The basic functioning of the nervous system is more widespread in our species than its use as signals, though pale faces may have exaggerated the color changes as they evolved fair skin.

A second source for ritualized signals is intention movements. If a vervet monkey wants to approach another, it moves forward, only to be checked by incipient fear. In vervets and many other vertebrates, such threats and checks are ritualized as formal jerks or head bobs, often accompanied by an open-mouth threat face. Sometimes the first approach is provoked not by anger but desire. Watch any city pigeon bowing and cooing!

Why should perfectly good signals such as flushing, or intention movements of approach, need ritualization? Think of an imperfect telephone line—perhaps phoning home from Madagascar. You have just three options. First, shout: exaggerate the volume and bellow down the line. Second, keep it simple: "I'll come home Sunday!" not "The work has gone well, and it turns out I can change my flight, so I think it could be as soon as this Sunday." Third, repeat yourself until you get confirmation: "Fine, I'll meet you!" Only then can you risk more subtle endearments until the connection gives out.

Innate signals are often both exaggerated and simple. A male stickleback fish courts a crude fish model with a swollen belly as though it were a female gravid with eggs. He attacks a model with red-painted belly as though it were a rival male. In Niko Tinbergen's laboratory in Oxford, his experiments were disrupted by scarlet English mail vans passing on a distant road. Stickleback males raised their dorsal spines, dashed to the window side of their aquaria, and followed the mail van from one corner to the other, until the intruder fled out of sight of their territory. The fish weren't just stupid. They reacted even more strongly to real fish than to models and mail vans. However, the overwhelming evolutionary need is to get the first message simply and get it right.[18]

Many displays are not only simple and exaggerated but also given at a fixed, if exaggerated, intensity.[19] This again eliminates ambiguity. However, that means an animal cannot use volume, or gestural range, as a way to indicate the strength of its feeling. The volume is permanently set at "loud." The answer to that is repetition. Repetition drives home the message and also registers the creature's persistence. The vervet monkey threat-bobs over and over. The bowing city pigeon postures on and on while his female nonchalantly pecks crumbs.

When you combine this with the mates' determination to assess physical stamina or potential commitment, small wonder that animal courtship repeats itself most wonderfully, arriving at the hundred-eyed tail of the peacock or continuous somersaulting of birds of paradise to flaunt their creamy plumes.

Some animals may not actually have brain space or information-channel capacity to deal with a large number of such signals. As evolution refines new ones from the raw material of intention movements, it is possible that old ones become rare and unrecognizable, or watered down by overuse. The overuse could certainly apply to sexual signaling. To out-compete a rival, produce a "high-intensity" signal at ever lower motivation: declare undying passion when all you really want is a one-night stand. Recipients of such signals evolve to ignore them and pay more attention to a novel approach—which could eventually become ritualized in turn.[20]

The reasons why animal signals evolve to be simple, exaggerated, of fixed intensity, and repetitious are clear. Human rituals also convey important, and in one sense unequivocal, meaning, though extended to fill our human attention span. Human rituals take on their own momentum, even when their origins are forgotten. Perhaps messages in any noisy channel would converge on the same traits, like shouting down a Malagasy phone line. But just perhaps we also inherit a mental sense that ritualization is the right way to reach out to deaf and distant gods, who must be made to hear, understand, and in the end respond.

Complex Genetic Traits

There are two ways to approach the heritability of complexly varying traits. The first is comparing identical twins with fraternal twins. Identical twins stem from a single fertilized egg; they share all their genes. Fraternal twins come from two eggs; they share half their genes, on average, just like other full brothers and sisters. The question is this: If one twin shows a trait, such as rheumatoid arthritis or schizophrenia or high IQ, how often does the other twin show the same trait? If the concordance between identical twins is much higher than between fraternals, you can probably attribute part of that difference to heredity. Not all the difference, though. Identical twins look and act much alike from birth. Strangers confuse them. Their own similar responses provoke family and friends to treat them as much the same. In short, identicals share a more similar environment than fraternals do. The difference in concordance is not an absolute guide to heredity, just an approach to measurement.

But the differences are staggering. If one twin of a pair of brothers is alcoholic, 40 percent of identical twin brothers are also alcoholic, though only 20 percent of fraternal brothers share this trait. (Not so great a difference for female twins: just 30 percent against 20 percent.) For schizophrenia, the difference is 45 percent among identicals to 15 percent among fraternals; for autism, 55 percent to 20 percent; for reading disability, 60 percent to 40 percent. Many diseases show similar differences: 45 percent to 10 percent for rheumatoid arthritis; 80 percent to 25 percent for epilepsy. More abstruse traits give the same trend: shyness, memory span, vocational interests in adolescence, performance on IQ tests. Correlations on IQ tests reach 80 percent for identicals, only about 55 percent for fraternals. This has nothing to do with the fallibility of standard IQ testing in the face of social background, since the two twins being compared to each other grew up at the same time in the same family.[21]

A second approach is to compare adopted children with their biological and adoptive parents and siblings. Again, heredity plays a large role in all the same traits that appear in the twin studies. The substantial heritability of human behavior traits is hardly surprising to animal geneticists. You can easily breed animals for mental differences in just a few generations. Some strains of rats prefer to run mazes by feel; others attend to visual cues painted on the maze arms. Some strains of mice kill other males' offspring; some are benignly tolerant. Cocker spaniels are bred for docility and Border collies to herd sheep. Obviously we should expect genetic influences on our own personalities and intellectual abilities.

IQ correlations yield a figure: roughly 50 percent of the variation between people seems to be due to heredity, 50 percent to environment. Plus or minus 20 percent. In other words, somewhere between 30 and 70 percent of the differences seen are probably due to heredity. (This is about variation in a population. It does not mean that genes gave you or any other individual 50 percent of your intelligence and that culture added the rest on top.) I think this kind of statistic can almost be reduced to a qualitative statement: Yes, genes have much to do with intelligence; so does environment. Which is where most of us came in.[22]

The danger is that by trying to *quantify* it, we open up a political can of worms. So long as the agreed emphasis was wholly environmental (the SSSM straw man, and only partly true), the conclusion was to improve schools and day care and HeadStart programs, to support parents and families and maybe even good children's television. When the emphasis shifts to admitting a large hereditary component (how large does not matter), that old Bad Biological Bogey Man, "instinct is fate," looms up again. You try to protest: "Look, 30 to 70 percent of population variance is hardly individual fate, even if you accept all the assumptions behind those figures!" You still become vulnerable to the kind of politics that starts out: "Well, some people are just born stupid."

Genes for and genes against. I won't say much about genetic mechanisms. We do need to clarify what is meant by a *gene for* something. It usually means genes *against* something. A major gene effect that brings on recognizable changes is usually a disruption in some complex process. Many genetic disruptions result in mental deficiency: Down syndrome, phenylketonuria, the stuttering genetic repeat of "Fragile X," the commonest heritable cause of retardation in boys. We have no idea how many genes interact to build *up* the phenomenon that we call ordinary intelligence, let alone genius. The analogy is an arch. Pull out any stone and it all collapses. No one stone is the "gene for" the arch. The great exception is *SRY*, where the body has two complete evolved programs, ready to follow on and produce the sex differences of two different castes of people.

This is not to belittle the gains to be made by identifying major genes. For instance, Huntington's disease is a neurodegenerative disorder which destroys adults in middle age. It starts with tics and random waving gestures. Over a decade or more it progresses inexorably to mental disturbance and death, though patients remain agonizingly aware of their own decline. In 1993 it was tracked to a repeated fragment within just one gene on chromosome 4.[23]

More than twenty-five years of collaboration between teams of molecular biologists and the family of psychologist Nancy Wexler went into discovering the genetic anomaly. Three of Wexler's uncles died of Huntington's; then in 1968 her mother was diagnosed as beginning to succumb. A second family also contributed to the search. Eleven thousand descendants of a single woman with the mutation live around Lake Maracaibo in Venezuela. They have the highest incidence of Huntington's disease in the world. It was their blood samples and genetic pedigrees that finally led to pinpointing the defective gene. "We thought if we started right away, maybe we would figure out a cure in time for my mother. Then we thought maybe it would be in time for my sister and me," says Wexler. "Now, in a way, I feel like I have such a huge family of thousands of people at risk of Huntington's all around the world. We're hoping we will find a cure in time for all of them."[24]

For construction, as opposed to destruction, of complexity, you need many genes, with additive or interacting effects. Such conglomerates of genes underlie the hereditary component of quantitatively varying traits like adult height or IQ score. The possibilities are astronomical, and beyond one-gene effects or very short series we soon reach the limits of knowledge. This does not stop us from estimating the proportion of variability due to heredity. It simply means we cannot usually specify which genes, or how many of them, may be involved.

A "gene for" homosexuality? Dean Hamer and his colleagues announced in 1993 that they had found a small region on the end of the X chromosome that may influence male sexual orientation. Hamer's group found that gay men have a high proportion of gay brothers. "When we collected the family history," Hamer says, "we saw more gay relatives on the maternal side than the paternal side . . . in particular among the gay men's maternal uncles, and the sons of maternal aunts."[25] This suggests a hereditary factor, but it is far from proving one. The classic psychoanalytic explanation for male homosexuality was that it all stemmed from having a weak father and a domineering mother. Perhaps whole families of monstrous women dominated brothers and nephews, as well as husbands and sons? That would give the same kind of pedigree.

What the pedigrees did say was that *if* a male tendency to homosexuality is hereditary, the best place to look for the gene is on the X chromosome that mothers transmit to their sons. Hamer's team analyzed chromosomes of 40 pairs of gay brothers whose families showed the maternal inheritance pattern. Thirty-three of the pairs shared the same set of five marker genes at the very end of the X chromosome. That could mean that in those 33 pairs, a gene influencing their behavior lay in the neighborhood of the five markers.

Marie-Claire King, famous for discovering a gene for breast cancer, snapped a letter back to *Nature* the week after Hamer's group announced their findings. She pointed out that seven pairs of the brothers did not share the markers: what about them? The pedigrees were chosen, deliberately, to find a population of homosexuals that might plausibly share a single gene. The results tell us nothing about the frequency of that gene, or its associated markers, in the gay population at large, or among brothers of the subjects who are (or say they are) straight, or how many different alleles there are at this still-hypothetical locus, and what proportions occur in gay or straight men. Besides, how many other genes, not to mention environments, influence sexual orientation for men or women?[26]

King went on to say: "Sexuality—that is, being female—is now the world's most common reason for elective abortion based on a genetic trait. I would like to believe that, through education, we could pre-

vent the emergence of a comparable level of intolerance against homosexual males. But, like any fundamental advance in human understanding, modern genetics is vulnerable to misappropriation . . .
To identify variation and characterize genetic influences on it is to
recognize the significance and value of that variation, not to con-
demn it." Anne Fausto-Sterling and Evan Balaban added, "We urge
researchers to remember that widespread sexual behavior patterns
probably represent a multiplicity of pathways leading to a single end
point [homosexual orientation] that we in 20th-century Europe and
America have named as if it were a unique entity . . . Given the
increased frequency of hate crimes directed against homosexuals, it
is fair and literal to say that lives are at stake."[27] Further studies to
follow up on Hamer's original findings have been equivocal. A variety
of questionnaires of several thousand twins suggest that male homosexuality has around 50 percent heritability, like IQ. As with the IQ
estimates, the margin of error must be very large—even more so for a
subject where many people lie.[28]

How might such genes work? There is a tantalizing suggestion, not
from mammals but fruit flies. A single gene, if expressed in the anten-
nal lobes where a fly identifies scent, turns on the switch that causes
a male fruit fly to find another male's odor attractive, so that he
courts both males and females. No one suggests that human sexual
orientation depends on a single pheromone. However, the recognition of the opposite sexed partner is one of those simple, crucial
acts for reproduction that should have a relatively firm lock and key
mechanism in the behavior of our mammalian ancestors—at least
those that left any descendants.[29]

The gay community's reaction is mixed, of course: there are lots of
differing people in any community. Many are quoted as welcoming
Hamer's findings. In present American culture, a genetic basis for
male homosexuality means that the trait is nobody's decision, neither
the gay person's nor his family's. Gays can say they are simply born

different. It is not a very safe course. The idea of a genetic tendency was a rationale for Nazis killing homosexuals even before Jews.

Among American males, around 2 to 4 percent report being exclusively homosexual. Kinsey estimated that almost 40 percent of American males had had at least one homosexual contact to the point of orgasm; a more recent study lowers the estimate to 20 percent. If homosexuality has any genetic component at all, why is there so much of it? Why was it not edited out by natural selection?[30]

Several compensating advantages have been suggested. Perhaps gays benefit their immediate kin. A family with nonreproductive adults who are splendid uncles and aunts may do significantly better than a family without. Another explanation is that many gays do have children. In the recent past our society has so stigmatized homosexuality that some gays become defiantly so, when in a more tolerant culture they might be bisexual. Similarly, in a more tolerant culture, many people now living as straights would also be bisexual. Homosexual behavior occurs in most birds and mammals, including all our immediate primate relatives—but this is bisexuality, not pure homosexuality. Primate bisexuality may extend to females competing with troop mates for (nonestrus) female partners.[31] If pure homosexuality is one end of a continuum, possibly with multiple genetic as well as multiple environmental causes, there may also be multiple advantages to having some of those genes: for oneself, one's offspring, one's family, and even one's group. After all, according to the ancient Greeks, a soldier could hardly be expected to die for his fellow men if he was not also capable of loving them.

Is Instinct Right?

If instincts and moderately fixed genetic components of complex traits result from natural selection, we suppose that they were indeed right, in the sense of helpful, for a large number of our ancestors, over many generations. We have lived in cities for only about 5,000 years. We have made nature into farms for at most 10,000 years. Our instincts date from much longer ago. They were programmed into our hunter-gathering ancestors. This means that most of our instinctive biases were adaptive in the Stone Age. Though natural selection continues, there simply has not been time to change much to adapt to our new circumstances. It also means that the wonderful mix of selfishness and selflessness, of personal ambition and social responsibility that we see today has been around for a very long time.

To decide if a trait is right (in the sense of adaptive) today, we have to ask which trait and when. Take breastfeeding. It is convenient, clean, promotes mother–infant bonding, and has components of immune resistance and supremely adapted proteins that never appear in cow's milk or soy. In the Third World, where sterile water is often hard to come by, bottle-fed babies are twice as likely to die as breast-fed babies. On all counts, the natural course is better—unless a mother works away from home, or is carrying HIV, or really doesn't like it, or for some reason has difficulty producing enough milk. Despite the propaganda of La Leche League, not every new mother produces enough milk—just many more do than think they do under Western hospital procedures. Women are as variable about this as they are about other things reproductive, from fertility, to nausea during pregnancy, to preeclampsia, to difficulty during labor. In some cases it may be better for both mother and child to bottle-feed. Baby-feeding pitchers survive from ancient Egypt; women have long known there is a choice.[32]

Is the natural course better when it comes to the number of children to have? It would be natural for well-fed women in industrial societies to bear around twelve, by following our instincts. Or at least, our instinct to mate. We counter by following our instincts for self-preservation and a deep, deep parental instinct to do the best possible thing for the children we already have.

Am I fudging the issue, by talking about what is right in the sense of advantageous, not right in terms of an abstract moral code? The whole tenor of this book is that moral codes are also advantageous, by and large. Our ancestors evolved the capacity to fear and respect social norms, to test for cheaters, to value cooperation above selfishness in every single one of the world's societies. But this does not solve the problems of how to behave in any one case. It does not tell you what to feed this baby, or whether to have another. It does mean that we need not fear the knowledge that we are a chimera of body and mind, and that our minds are chimeras, constructed at each moment in time from all that is happening, and all that has happened before.

Promoting Differences

We have talked (a good deal) of the importance of variation among one's own offspring. Variation is so important it is even worth having sex to produce it. If instinctive, or hereditarily influenced, behavior predominates when it is important to get an action right in the face of environmental noise, how does behavior achieve variability?

One possibility is to program different hereditary instructions for different contingencies. If the approaching fish is rotund, court her; if slim and red, fight him. If the world is full of big male sunfish, become a female-mimicking sneak. There are of course limits to information capacity even in the genes, and limits to the number of situations so likely to occur that they can exert selection for the appropriate response.

Protean behavior.   Another possibility is to actually program randomness. The only unbeatable strategy in some games is truly random play. This is the sort of game that moths play with bats: a flying moth may start to loop and tumble erratically when it hears a bat's squeak. It is the sort of game that submarine captains played in

World War II: they sometimes rolled dice to determine their zigzagging approach to dangerous ships. Missile guidance systems have built-in randomized courses to foil tracking missiles. Random, and thus wholly unpredictable, behavior can be the best strategy in a very serious game.[33]

Genes channel learning. Some learning is sharply channeled by instinctive biases. Imprinting is circumscribed both in quality and in time. There is a sensitive period, generally in early infancy, when the organism is primed to accept and learn the stimuli that fit into pre-set slots. This is the story of the ducklings following waddling, quacking Konrad Lorenz.[34]

What about learning the rich variety of human culture? A mind is not a human mind without a culture around it. People may react against their culture and revolt against its constraints, or step outside to revel in cultures distant from their childhood's norms. But without any cultural models, our minds—even our brains—could not grow and develop. Even the retina needs visual input in early life to establish pattern recognition. (Keeping a kitten in a cylinder with only vertical stripes on its walls deprives the adult cat of the nerves that recognize horizontal lines.)

We reach out to learn faces and language and the meaning of social patterns. Without an answering environment, we do not grow. Fortunately, as with all biologically vital functions, we have enormous redundancy of ability: even gross childhood deprivation may not stop a person from piecing together a sense of her world. But our minds have evolved to incorporate a scaffolding of cultural models, as surely as a duck has evolved to learn that a duck is its mother.[35]

Learning channels genes. If culture persists in applying a selection pressure, the genes themselves can fall in line; this may explain our unruly head hair, which responded to millennia of grooming by losing its ability to groom itself. A well-studied, clear-cut case is tolerance for lactose, or milk sugar. Milk makes many people feel sick, because they lack the ability to produce an enzyme (lactase) that helps digest milk. Most Asians and many Africans lose tolerance for the nasty fluid soon after weaning. In cultures where adults consume dairy products, over 90 percent of the populace can digest milk, but typically less than 20 percent of people can digest it where there is no such tradition. This has nothing to do with overall race. African Masai drink milk curdled with blood; Tibetans in Asia float gobs of yellow yak butter in their tea. Masai and Tibetans have retained or evolved genes to deal with lactose.[36]

If culture can affect a single gene, like whether milk makes you queasy, can we expect there are differences in more subtle traits between people of different cultures? Bluntly, yes. If natural selection can affect skin and nose and the frequency of blood types, it would be extraordinary to find it had no effect on minds.

But we are not likely to find many differences in group means of mental traits. First, of course, most of us don't want to—but some scientists and statisticians are still bound to look. The second reason is technical: there is so much individual variation in complex traits like aspects of intelligence or altruism or conformity. The variation comes from gene–environment interaction—many, many genes, in many environments. If there is a difference of a fraction of a percent in some average value of a trait, it is trivial compared to the variability within each group, and it is masked by any systematic environmental bias. The third reason is much more fundamental, not a mere statistical problem. If culture selects for diversity *within* families and social groups, for all the reasons unthinking natural selection also promotes diversity, then the individual variation we see rests on a basis that will continue to confound future statistical refinements.

The final reason is historical. People have always moved from group to group, and cultures change from century to century. An environmental influence making much difference to the human genome generally needs to operate over millennia. Some races stayed long enough in northern climates to evolve pale skin to absorb vitamin D, avoiding rickets. Others moved south again, in India and South America, and re-evolved dark skin, fending off skin cancer, like the first ancestors of all of us in Africa. However, since humans created culture, we have never stayed in one style of life long enough for fishermen to evolve webs on their toes, or hunters to grow claws. We have apparently never even had time to evolve mental habits that would distinguish hunter from fisherman. Instead, we have, on a grand scale, evolved the need to learn the idiosyncratic habits of our tribe.

A situation that is static for many, many generations in humans or other animals makes it efficient to build in the response. Those animals that learn most rapidly are selected. Whatever genes help them learn become more and more frequent in the population, until the response itself becomes more and more likely, more and more innate. In contrast, if the environment is chaotic from generation to generation, the only possible response is trial and error. Everyone can learn, but they must make their own mistakes.

Channeled learning, rather than narrowly fixed instincts or random trial and error, is favored when the environment changes rapidly, but not too rapidly. Few if any cultures have survived unchanged long enough to build fixed traits of mind. Instead, societies favor the people flexible enough to learn the mores of their own place and time, to learn culture, not just a particular culture. There is interplay through all ages between the genes that make our minds ever more receptive and the likelihood that there will be one of the many variants of a human environment with the patience to teach.[37]

There was a child went forth every day,
And the first object he looked upon and received with wonder
    or pity or love or dread, that object he became,

And that object became part of him for the day or a certain part
   of the day . . . or for many years or stretching cycles of years.

The early lilacs became part of this child,
And grass, and white and red morningglories, and white and
   red clover, and the song of the phoebe-bird,
And the March-born lambs, and the sow's pink-faint litter, and
   the mare's foal, and the cow's calf, and the noisy brood of the
   barnyard or by the mire of the pondside . . . and the fish
   suspending themselves so curiously below there . . . and the
   beautiful curious liquid . . . and the water-plants with their
   graceful flat heads . . . all became part of him . . .

His own parents . . . he that had propelled the fatherstuff at
   night, and fathered him . . . and she that conceived him in
   her womb and birthed him . . . they gave this child more of
   themselves than that,
They gave him afterward every day . . . they and of them
   became part of him.[38]

13
Thinking

Consciousness is one of the endless mysteries. Do you know that I think? Perhaps this book is done by automatic writing, and I am actually channeling messages from an astral plane. Do you believe I see red as you see red and not as you see purple? Or are you called colorblind, and know that you cannot know what I see as red, just as neither of us knows what the ultraviolet heart of the petunia looks like to a bee?

We may never understand sentience. After all, philosophers have tried for at least 2,500 years, without much progress. Steven Pinker argues that it is outside the realm of anything our brain has evolved to understand, like visualizing objects in the fourth dimension. If our brain evolved to solve problems our ancestors needed to solve, how the brain works isn't one of them. It is like a Klein bottle, the analogue of the Möbius strip, the three-dimensional bottle with a twist in the fourth dimension so all the world is simultaneously inside and outside it. The brain in its own way contains the sky, but perhaps it is not able to contain itself.[1]

The "I" does not seem to be a single searchlight in the dark but a caucus of neural demons. (At this stage I wish the programmers had chosen to call their routines "angels.") You cannot pinpoint an arch-demon, archangel, or grinning homunculus who sits in your head and views the caucus's deliberations on a little screen. As the philosopher Daniel Dennett points out, such a scenario would leave you asking how the homunculus is conscious in turn, in infinite regress. Psychologists' and neurophysiologists' experiments imply there is no single center where all the components of consciousness are assembled at a single point in time. Dennett says, "One discharges fancy homunculi from one's scheme by organizing armies of idiots to do the work." Like it or not, the brain is a bureaucracy.[2]

Although we are baffled, at least for now, about *how* consciousness works, we are a little clearer about *why* it works. It is a decision-maker for those parts of the mind that need decisions, while many more get on well enough with their unconscious functions. The brain, with or without consciousness, continually deals with the question, "*Now* what do I do?" The sea squirt, a saclike relative of the earliest ancestors of vertebrates, starts life as a larva with a rudimentary nervous system, just sufficient to swim and then settle down on a rock. After settling, with no more decisions to make, it eats its brain (like an associate professor getting tenure). From this first step of anticipating what to do next in larval sea-squirt terms, the brain evolved little by little to our own complexity, as the first simple light-sensitive spot evolved into functioning eyes.[3]

Other Kinds of Minds

Dennett points out that we can easily imagine that robots are conscious. We do it all the time, for Frankenstein and R2D2 and HAL in *2001*. Any child does it for teddy bears and the Little Engine that Could. Ancient primate anthropomorphism makes crediting feelings to others a natural course. What we cannot imagine is *how* a robot could be conscious. We also find it hard to imagine that a hierarchy of functions, of lesser demons, add up to more than the sum of the parts as the system burgeons in complexity. And yet that is exactly how our own brains work. There seems to be no reason why brains of silicon, if ever they reach the complexity of human brains, might not think and feel much like brains made of meat.[4]

René Descartes. The test in this century is not robots but whether we credit evolved consciousness to others of Earth's creatures. Descartes is often blamed for cutting the world in two. The idea that consciousness is a purely human ability, not found among animals, and the related proposition that mind is a separate thing from body, has a long tradition in western thinking, often blamed on Descartes. It has been used to exalt the glory of the human soul, and it has been used to justify vivisection of screaming animals because "they don't really feel it."

Descartes' *Discourse on Method* in fact sees animals both as machines and as conscious beings. He attributed "animal spirits" to them—including emotions, simple thoughts, and the ability to dream. He drew the dividing line for human consciousness just about where we would put it today. For Descartes, only humans have metaphysical intelligence—that is, complex thoughts expressed in language.[5]

He even suggested a kind of Turing test. Alan Turing, one of the founders of computer science, proposed that you hold written conversations with a computer and a person, each in a separate room. If you cannot distinguish which one is human, you might as well call the computer alive. They did not have computers in the seventeenth century, but Pascal had already invented a calculating machine, and artificers made elaborate clockwork toys. Golden birds sang in fretwork cages; mannequins played toy organs. (They were forebears of Hoffmann's tale of Olympia, the singing doll, and the besotted young man who fell in love with her, or, in its recent version, William Gibson's novel of courtship with a computer-generated Japanese music video idol. Not to mention the Stepford wives.)[6] Descartes asked: could such a mannequin talk? He surmised that it could be programmed to produce a large number of phrases to appropriate stimuli: "Delighted to meet you," and "Ouch!" and even "I love you." But he concluded that if one moved on to probe for real conversation, the machine would never produce the combinations and innovations of human language.

There is another section of the *Discourse on Method* that bears consideration for style and gusto. Descartes gave several pages to a description of the human heart. He assumed his reader had not yet heard of Harvey's new discovery that blood circulates around the body and that what goes out the arteries is the same stuff that comes back through the veins. This solves the whole mystery of what the heart is doing! Descartes traced the circulatory system from the heart's right ventricle to the lungs and back to the left atrium, from the left ventricle to the body and back in turn to the right atrium. He got the circuits right but the mechanism wrong. He believed it worked by heat like a steam engine, not by pumping. The point, though, is his enthusiasm for this beautiful bodily machine.

Later centuries' take on Descartes has been, first, that mind is wholly separate from body and, second, that bodies are inferior dross compared to mind. It comes out differently to say that Descartes did indeed claim a separate essence of the human soul, and asked as objectively as he could what might be the special properties of the human mind—but that he was also fascinated by the body and recognized that a dog whining and twitching in its sleep may dream of the chase.

Donald Griffin. I now skip three centuries to Donald Griffin's *The Question of Animal Awareness* (1967) and *Animal Thinking* (1984). At the time Griffin wrote the first book, the accepted doctrine of animal behavior studies was a caricature of Cartesian thought. Any word that slipped out implying consciousness or intent in animals would be blue-penciled by editors and reviewers as "unscientific." Of course, there were rebels like Jane Goodall and the whole school of

Japanese primatologists. Indeed, anyone who dealt with mammals in the wild assumed animals were conscious: it was the obvious way to understand them. But scientists' censorship and self-censorship of such thoughts before they reached print rivaled anything imposed by Stalin.

In that climate, it was very brave of Griffin to reopen "the question of animal awareness"—to put mind and body back together. He got away with it largely because of his own towering reputation as a "hard" scientist. He had, after all, discovered ultrasonic echolocation in bats. He also got away with it because he attacked the foundation stone of the other school: parsimony.

Parsimony, or Occam's razor, is the principle that one should never accept a complex explanation if a simple one suffices. Simple tropisms, such as attraction to light or response to gravity, explain the directions that a seedling's shoots and roots take as they grow. No need to postulate that the plant "wants" to link the earth and the sun. But parsimony was being misused to say that acts where we ourselves are clearly consciously aware cannot be so in animals—exploration, innovation, even the emotions of love and hate. When actions seem homologous between humans and other animals, Griffin concluded that denying animal consciousness "is not only unparsimonious, it is egotistical."[7] He argued that attributing awareness to animals is the most efficient explanation one can give for the behaviors observed.

Mental Models

What reaches consciousness? What kinds of behavior reach consciousness, and what kinds do not? Most do not. We breathe, type on our keyboards, note that the cat is on the windowsill, sit upright, blink our eyes, all at the same time, without missing a beat in the conscious interior monologue about what word to type next. What is unconscious can be important (breathing). It may be relatively innately constrained (breathing, sitting down, blinking) or learned long ago (typing) or what is current but requires no decisions (the cat, unless she steps on the keyboard).

The key seems to be what requires a decision. What to say next, what to do in the future—and also often reruns of decisions not taken in the past. (Shall I ask her? Shall I let him? If only I had not forgotten the punchline! If only I had said to my mother before she died . . .) Actions like typing are conscious when they are being learned, but then they become unconscious routines.

Consciousness serves to *model* action. It is a way to choose between alternative courses without actually trying them out. It is a "selection arena" for acts and ideas. It is not only unnecessary but inefficient to bring into consciousness what requires no decision.

> The Centipede was happy quite,
> Until the Toad in fun,
> Said "Pray, which leg goes after which?"
> And worked her mind to such a pitch
> She lay distracted in the ditch
> Considering how to run.[8]

With the advent of PET scans, neurologists have begun to track consciousness. There are areas of the brain that light up when people consider problems but do not light when the problems run on automatic. Let me give just a single example, from the ancient central structures that deal with emotion. The amygdala is a region implicated in social responses in other mammals. Human volunteers were scanned while watching photos of faces. One was an angry face. The volunteers were trained by a burst of white noise to find it threatening and unpleasant—the kind of classical conditioning you can do to a rat, too. Later, in a test, if that face flashed by so quickly that it never reached consciousness and was then masked by a neutral, nonconditioned face, the volunteers' right amygdala still lit up with activity. If the threatening face stayed longer, so that the volunteers were aware of what they saw, the left amygdala activated. These were left-brained, right-handed people; social consciousness kicked in on the same side of the brain as language. The sum of this and many, many more experiments will trace out paths of consciousness in unnerving physical detail.[9]

What neurology has not yet done is to show when consciousness is necessary. Some injuries, in fact, block out skills from conscious awareness while leaving them still functioning. The most famous is "blindsight." In this rare condition, damage to the occipital lobe leads the person to think himself or herself totally blind. If asked to "guess" whether he is being shown a square or a triangle, a sheep or a table, he says it is ridiculous to guess at all since he cannot see, but then he guesses right![10]

What animal actions would we quickly assume were conscious in humans? Simple attention, to begin with. The cat on the windowsill watches a sparrow in the tree outside. She has no hope of catching it: it is forty feet away, not counting the problem of the glass pane. But her ears and eyes are focused, her head now follows its flight. I do not know what she thinks, if she has a category of "bird" or "prey" or just "small-scale movement." But it would be hard to deny that whatever awareness she is capable of is fixed on the sparrow. Of course, a relatively simple robot might do the same.

From attention, we can build up the scale. The cat eagerly explores a cupboard whose door is left ajar; the young baboon deceives his mother into attacking another female holding a root; the gorilla takes a straw from his bedding and delicately rakes in an orange segment from the next cage; the chimpanzee strides off, leading other animals away from the bananas he alone has seen. At some point in the list, we say in humans this behavior would have to be conscious and done with intent. Why invent another mechanism for animals?

Once you build a bridge to the most complex and innovative behaviors of our nearest relatives, there is no longer an impassable divide. Of course, we do not know qualitatively what is in others' awareness. It is often hard even to imagine what another person feels; it is logically impossible to feel what it is like to be a bat. Still, we can admit that the *bat* may have some awareness of what it is like to be a bat.[11]

Pretense and the Negative

With a little time out imagining bathood, this brings us to the whole question of imagination: dealing with what is not so. Gregory Bateson long ago sketched out the crucial importance of the negative. If a young baboon wrestles and play-bites with another, their play has all the components of fighting. In fact, it is physical training for dead-earnest adult fights. But play also has other signals: an open-mouthed play-face and a gamboling gait that say, "This is *not* real, *not* in earnest." Older, bigger juveniles play with younger ones, holding their strength in check. We all know the edge where play turns real. Someone loses control, someone gets hurt, someone lashes back with tears of rage in the nursery school yard. Baboons know the edge, too. If a young infant squeals "for real" during the game, an adult male breaks up the play group. [12]

Rough and tumble play is one of the simplest, clearest frames—the fight that is not a fight. But all our language and imagining depend on such frames. The word "chair" is not a chair. A thought or imagined image is not the thing it represents. That is part of what is meant by thinking.

Do any nonhumans seem capable of imagination? There are dreams, of course. Rapid-eye-movement sleep is common to all mammals. REM sleep goes with dreaming in humans and with suppression of nerve impulses to the muscles that would "act out" the dream. Little children spend more of their sleep time dreaming than adults do, small babies even more. Fetuses and premature babies have still more REM than newborns.

What does the unborn child dream of? We do not know. Perhaps it "looks at" mental images that serve to train and develop the nerves of the eye. Children and other mammals need to look from birth to develop their retinal connections to the brain. Perhaps the process starts in the womb. The content of those images is further off from our adult imagination than the thoughts (rabbits?) of Descartes' dreaming dog.

When a dream is going on, we believe it, except for that near-awake state when we struggle to escape from a nightmare by waking up. Dreams show that the brain can create images without detachment. But what about real imagination? Social deception among apes has a sliding scale from simple hiding up to elaborate misdirection. Are there examples of clear nonsocial pretense in our nonhuman relatives?

Yes, a few. Apparently any monkey can learn to respond to pictures and to treat them as representing real things. They can sort them into groups: all the animals here, all the humans there. The most famous case was the home-raised chimp Viki, who placed a photo of herself with the humans but put her chimpanzee father with the animals. Of course we do not really know her criteria. Perhaps she was sorting people-who-wear-clothes from people-who-go-naked, or people-like-my-human-family from people-I-do-not-wish-to-be-related-to.[13]

The capacity for recognizing pictures is not confined to primates. The most extensive experiment was Herrnstein's work with pigeons. Trained pigeons can sort thousands of photos, accurately, according to whether they contain any person (even indistinctly in the distance) or not, and whether they contain any tree or not. On the other hand, pigeons do not behave like the home-raised chimpanzee Lucy, who was given a copy of *Playgirl*. She turned pages and touched or scratched the male genitalia with her fingernail. Then on reaching the centerfold, she positioned herself carefully over the pictured genitalia, rubbed her clitoris on them, and dribbled urine accurately on the spot.[14]

Many primate experiments take the picture recognition for granted and use it as a stepping-stone to ask more subtle questions. Verena Dasser tested rhesus monkeys' concept of "mother–offspring" by showing pairs of photos of familiar animals. The monkeys were trained to respond "yes" if the pairs were mothers and offspring. The monkeys rapidly generalized to new pairs of portraits. They recognized relationship, not just age—mothers and adult offspring still got a yes while mothers and nonoffspring, adult or infant, still got a no. The result is interesting in itself, but the point here is that such a result depends in the first place on the monkeys accepting a picture as representing a real, known animal, even though it is clearly not the animal.[15]

At least, it boggles our own highly visual imagination to think that a monkey cannot distinguish a picture from another monkey. Perhaps if it were a representation in another modality, which is not so much our own specialty, we would not be so sure. Cheney and Seyfarth showed that monkeys recognize individual juveniles' distress calls and look at the mother who should be responding. Is the call a representation or just an attribute of the juvenile and its mother? Would we be even less sure if it were a smell? Perhaps even ringtailed lemurs associate individual scents not only with the individual but with its mother, since their whole society is built of mother–offspring dyads. Perhaps for the rhesus monkey a picture is merely an attribute, like the timbre of one's voice or a body odor?[16]

There are a few indisputable cases of representation, as always, in apes. Take dolls. Most young primate females like to play with and carry babies. In a few species young males do too. Among Barbary macaques, for instance, adult males often carry babies as pawns in their status games. In that species it is the male juveniles, not the female ones, who are most earnest "allomothers" who carry and hold tiny infants. So we discount infant carrying as pretense—the older juveniles are not necessarily playing mommies and daddies, just obeying some species-typical response to cuteness. Babying pets (Koko the gorilla with her fluffy kitten) could be an extension of the cute response.[17]

But any primate must know the difference between a cute live baby and a stick. There are two suggestive cases from the wild of chimpanzees treating sticks like a toy or a doll. One was an adolescent male wrestling with a branch, with play-face and play-panting. No more, perhaps than any kitten with a ball, but kittens are evolved as hunters, while chimps only wrestle with chimps.[18]

In the Kibale Forest of Uganda, Richard Wrangham watched 8-year old Kakama pick up and keep a small log for two days, hugging it, carrying it in every possible fashion, lying on his back in his nest and balancing it on his feet as mothers often do with their babies. He made a little day-nest and laid the log in it while he sat beside. He retrieved the log when it fell. Wrangham eventually lost the chimp's—and the log's—trail when he had to circumvent snorting bush-pigs. Wrangham writes: "Kakama was exactly the sort of youngster who might be expected to want a play partner most. He was an only child with a playful personality, his mother was relatively antisocial, and she was now pregnant . . . My intuition suggested a possibility that I was reluctant, as a professionally skeptical scientist, to accept on the basis of a single observation: that I had just watched a young male chimpanzee invent and then play with a doll . . . A doll! The concept was novel enough that I simply filed away my notes without saying much about it to anyone else, and left Uganda the following week. Four months later, two field assistants at Kibale, Elisha Karwani and Peter Tuhairwe, happened to be following Kabarole and Kakama. Neither Karwani nor Tuhairwe knew of my observation. Yet for three hours they watched Kakama carry a log—not the same one as before, surely—taking it with him wherever he fed. This time they saw him leave it. Once they were certain Kakama had disappeared, they collected it, brought it to camp, and stapled to it a label that described their own straight-forward interpretation of the object's meaning: 'Kakama's Toy Baby.'"[19]

In captivity there are a few more examples. Washoe, the first of the signing apes, had been regularly bathed. Sometime between the ages of one and a half and two years, she picked up her doll, filled the bathtub with water, dumped the doll in the tub, took it out, and dried it with a towel. In later repetitions she even soaped the doll. This is imitation, but it also must be a form of representation—indeed, of pretense.[20]

Far and away the most complex case, never disputed, was Viki's unseen pull-toy. Viki, a home-raised chimp who could pronounce four "words," was at the toddler stage when everything possible to drag along becomes a pull-toy. She suddenly began to trail one arm behind her as though dragging a toy on a string that did not exist, while looking behind her on the floor. Cathy Hayes, her psychologist foster mother, took notes but tried to resist interpretation. Viki repeated the game over and over. One day, the invisible "rope" apparently caught in the bathroom plumbing pipes. She fumbled, pulled, gave a jerk and was off again. Still later she sat on the potty, "fishing" the nonexistent object from the floor hand over hand.

Then, after several weeks, wrote Cathy Hayes, "It was one of those days when Viki loves me to distraction. She had pattered along in my shadow, [and] at every little crisis she called for 'Mama' . . . I was combing my hair before the bathroom mirror while Viki dragged the unseen pull toy around the toilet. I was scarcely noticing what had become commonplace, until she stopped once more at the knob and struggled with the invisible tangled rope. But this time she gave up after exerting very little effort. She sat down abruptly with her hands extended as if holding a taut cord. She looked up at my face in the mirror and then she called loudly 'Mama, Mama!'"[21]

Still unbelieving, Cathy Hayes went through an elaborate pantomime of untangling the rope, and handed it back to the little chimp. "Then I saw [her] expression—a look of sheer devotion—her whole face reflected the wonder in children's faces when they are astonished at a grown-up's escape into make-believe. But perhaps Viki's look was just a good hard stare."[22] A few days later Cathy Hayes decided to improve on the game. She invented a pull toy of her own, which went clackety-clackety on the floor and squush-squush on the carpets. Viki stared at the point on the floor where the imaginary rope would have met the imaginary toy, uttered a terrified "oo-oo-oo," leaped into Cathy's arms, and never played that game again.

Art

My mother, an artist, used to talk of the human urge to decorate things. The urge to improve on nature, to fix things up, seems to be very deep in us. By definition it is linked to the capacity to imagine what is not yet so.

Very occasionally animals also decorate things. Wolfgang Kohler, sequestered on Teneriffe with his colony of chimpanzees (much better use of one's life than fighting in World War I, which was then the main alternative), wrote down a case of "serious play." "On the playground a man has painted a wooden pole in white color. After the work is done he goes away leaving a pot of white paint and a beautiful brush. I observe the only chimpanzee who is present, hiding my face behind my hands as if I were not paying attention to him. The ape for a while gives much attention to me before approaching the brush and paint because he has learned that misuse of our things may have serious consequences. But very soon, encouraged by my attitude, he takes the brush, puts it into the pot of color and paints a big stone, which happens to be in the place, beautifully white. The whole time the ape behaved completely seriously."[23]

The more manipulative primates, in the pressure cooker of captivity, do take to decoration. Cebus monkeys and chimpanzees, orangutans and gorillas scribble or paint if given the tools. They even smear feces on the walls of their cages if deprived of tools. Cebus monkeys make "sculptures"—they will modify a ball of clay into a different shape. It is not part of the species' usual biology, like the constructed courtship bowers of bowerbirds. Only some individuals do it, even in captivity. Nothing like it is known from the wild, except the very occasional self-adornment of apes putting on a leaf like a hat, or, in a recent case, draping the skin of a red colobus like a necklace.[24]

Apes who paint work with extraordinary concentration. They have tantrums if anyone tries to stop them before they are finished. They know when a picture is finished, and rip it off to start a clean sheet.[25]

How much thought is in it? Apes counterbalance designs: given paper with a dot or blob on one side, they scribble on the other. Given an empty square or triangle, they fill in the space, or roughly trace the outline. They can play "connect the dots." It was Desmond Morris, himself an artist, who took chimp art seriously enough to ask if there are rules of spontaneous composition. His little chimp, Congo, preferentially drew either radiating fan patterns or circular sweeps. Easily explained, of course, as a simple motor activity—but just before Congo's early death, he began to paint circles with separate spots inside. In children, this is the final stage before they begin to draw recognizable faces. Children also say what their paintings mean at this stage, just before achieving representational art in grown-up eyes.[26]

Tantalizingly, some language-trained apes—the chimpanzee Moja in the Gardners' lab and the gorillas Koko and Michael in Patterson's—sometimes name their paintings with words. It is so hard to be sure what it means, and many chimp researchers remain skeptical. Trainers may have inadvertently coached their subjects to say a word of sorts after drawing. By definition, spontaneous naming is not subject to experimental rigor. The first time certainly involved a question-cue, though the answer was a surprise. "In April, 1965, Moja drew an atypical scribble with very few lines. Her companion, Tom Turney, put the chalk back in Moja's hand and signed to her to try more. She dropped the chalk and signed *All done.* The reply was unusual; so was the drawing. Turney signed *What that?* Moja placed the forefinger and thumb of her right hand to her lips, the sign for *bird.*" Moja extended her oeuvre, naming other pictures—but then, chimps are subtle social maneuverers, and she surely registered Turney's astonishment and delight.[27]

On the Gorilla Foundation's web page (www.gorilla.org) are images that Koko and Michael have "named," with tantalizing appropriateness, especially Michael's red-and-green "Pepper" drawn from life. A skeptic would still say they mostly show gorillas' love of bright poster paints. All but one, which is strokes of black on white, though Michael had the whole palette to choose from. That one he tagged "Apple chase," for his black-and-white dog, Apple. The painting's resemblance to the dog (whose photograph is shown on the web site) is astonishing. Richard Wrangham reports: "Francine Patterson used sign language to ask . . . Koko to paint pictures representing two emotions, love and hate. Patterson asked me to guess which painting was which. One presented a twirl of soft reds. The other was full of sharp angles painted in black. It was obvious. But it's hard to *prove* that Koko even understood the instructions, let alone that her paintings represent emotional states like those felt by humans."[28]

Sense of Self and Others

If a person recognizes herself in a mirror, she must know she has a self. Knowing oneself is much more than a game with mirrors, of course. But playing the game with mirrors is one clear, unequivocal proof. Responding "That, there, is me" means that you must have some concept of "me."[29]

Apes and humans have it; monkeys don't. Monkeys never learn to recognize themselves in mirrors. A mirror simply shows them an animal to appease or attack. A mirror left in the cage for a year shows a very boring animal. A mirror left with a mother and her juvenile shows the mother another female and juvenile—never ever a reflection of her own child.[30]

For apes, the reaction is similar for perhaps a week. Most, but not all, mature apes then reach a moment of truth. They rock, gesture, make faces, eyeing the reflection. They inspect body parts they have never seen—inside their mouths, their sexual swellings. In every way it is like that day in a human child's life, sometime around two years old, when she, too, discovers herself in the mirror.

The mirror test has become the benchmark for a sense of self. If you anesthetize apes, then paint a dot on an ear, or above a brow-ridge, when they regain consciousness and pass a mirror, they do a double take. They touch and groom the unfamiliar smear on their own heads. The test succeeds with most chimps and orangs, but not all gorillas. Koko the signing gorilla clearly passes formal tests (her forehead wiped with a washcloth either dipped in water or in pink paint). When I had the privilege of meeting her, she borrowed my lipstick, applied it to her upper lip and one brow-ridge, and swung over to inspect her face in the lens of the video camera. Bonobos treat closed-circuit television like a mirror, eventually starting sessions by making a quick, test gesture to see if the film is live or not.

Figure 14. Koko the gorilla regards herself in a mirror, having "made up"with chalk on her face, and touches her lip with her finger, her sign for "woman." (After Ron Cohn, in *National Geographic,* October 1978)

Their skills start late, in different individuals between three and six years old, whereas we start at two.[31]

If you could overcome the initial complexity of getting a computer to recognize objects at all, you might build in a fairly simple template to recognize itself. However, there is no way that wild apes could *evolve* such a template. They have been known to stare into pools of clear water, but not very long. It must be a new, and important, deduction to recognize the ape in the mirror is oneself.[32]

There is a corollary. If an ape thinks "That, there, is me," it is *thinking.* The unstressed word in Descartes' "I think, therefore I am" is the "I." I do not suggest that an ape thinks "I think," but it takes the first step if it can at least think "I."

Mindreading, or the theory that others have minds. What does it take to realize that other people also have selves? Yet again, the important question is a negative. What does it take to realize that others can have *different* knowledge, a different point of view from oneself? The positive prediction of others' actions from their behavior goes back a long way. Any squirrel can maneuver around another squirrel's actions. I wouldn't put that past a juvenile sea squirt. What does it take to play on or compensate for another's *ignorance*?

Monkeys occasionally deceive each other, but more often seem unaware of others' knowledge or lack of it. Monkeys' communication lacks the crucial aspect of language—an intent to communicate to other. Monkeys do not seem to have a theory of others' minds.[33] Children also do not imagine other people's ignorance until sometime around the age of four. The type test is "Maxi and the chocolate." You tell a child a story about little Maxi putting a piece of chocolate in a drawer to eat later. Maxi goes away. His mother sees the chocolate, takes a bit of it for cooking, and puts the rest into a cupboard. Where does Maxi look for his chocolate? Three-year-olds say that Maxi will look in the cupboard, because they know that's where it is. Five-year-olds say Maxi looks in the drawer. The five-year-old can make a second step: where Maxi will look if he needs his grandfather's help to open the drawer, and where he will send his greedy brother to look, if he wants to hoard the chocolate.[34]

Attributing false belief and effective deception develop through stages. Two-year-olds "pretend play," as in bathing a doll. Three-year-olds attempt deception ("I did not take the cookies!" they say with a face smeared with chocolate chips). Four-year-olds are progressively better at deception and misdirection, up to the five-year-old's mastery of others' beliefs as different from his own. Of course, even adults often assume we all share a viewpoint in many situations: "I can't believe you didn't know *that!*"[35]

The original effort to prove apes can understand intentions was showing videotapes to a smart, test-savvy female chimpanzee named Sarah. The videos showed a trainer with a problem. He stacked boxes to reach hanging bananas. He tried to play a record on a phonograph. (Technology dates; the conclusions don't.) Sarah chose photos to finish off: the tower of boxes with the trainer on top, or plugging in the unplugged phonograph. For videos of a trainer she disliked, she frustrated his intentions by choosing a photo of him on the floor with the boxes on top of him.[36]

Do primates understand others' attention, as well as intentions? Chimpanzees can follow the direction of another's gaze and will peer round an obstacle to see where another person (or chimp) is staring. Chimpanzees distinguish between a person who saw where food is put, and who is therefore a reliable informant, and a person who could not have seen, because he was out of the room or happened to have a paper bag over his head. Language-trained chimps can name objects when a trainer simply looks at the object without otherwise holding it or pointing at it.[37] Very young chimps, like young children, fail at all these tasks.[38]

Even more telling, chimpanzees understand to some extent the roles of others. The type test is Daniel Povinelli's cooperative game. Povinelli built a table-top with a screen, so only one player could see food placed to the left or the right and could point at it. The player opposite pulled one of two handles: if he chose the handle on the side with food, the handle moved springs that gave a bit of the food to each player. To get food every time, the players needed to cooperate: one had to point left or right to the side with the correct handle, and the other had to pull it. Macaque monkeys or chimpanzees were paired with humans in each role. Learning the task took several trials: nonhuman primates rarely point spontaneously and do not easily read others' pointing, but eventually they all learned the problem.

Then came the crucial trial: the box was simply turned around. For monkeys this presented a whole new situation, starting from scratch. The chimps, instead, responded at once. They simply shifted to the other person's role. Povinelli argued that instantly taking another's role means real understanding of another person's point of view. This is true empathy. It does not take a human mind to start the progress toward detailed knowledge of what is in others' minds.

Imitation and teaching. If psychologists skillfully define imitation as whatever humans do that other apes don't, it is by definition a skill unique to humans. Directed attention to the food that other apes are eating, or the objects in other apes' hands, is now merely called "facilitation." True imitation means *understanding* the process someone else demonstrates.[39] Such a skill rests on identifying with others, knowing that they know what you do not. Even more, true teaching means that you can recognize, and wish to help, another's ignorance. At the highest level, teaching and learning rest on the theory that other people have minds.

General "facilitation" is widespread. Most mammals learn which foods to eat by associating with older animals as they feed. Even chimpanzees do not actually imitate so often as we would like to think. Show a young ape how to pull candies into a cage, by turning a little rake over so that the bar scrapes the candies along. At an age when human children follow the demonstration exactly, the chimps merely gesture with the rake. As often as not they place the rake the wrong way up so the candies merely slip between the tines. Human-reared chimps do better. Close attention to a human demonstrator is as much a social as an intellectual process.[40]

In the wild, chimpanzees in Guinea learn how to open very hard nuts with stones. They learn by playing beside the adults. It goes on from there by personal trial and error. They start to bash nuts at three years old; the technique is fully operational at ten. Human children also don't open the hard nuts efficiently until ten years old. It may be that to get the knack of skill in aiming, and just enough strength so the thing does not fly into smithereens, both people and chimps need maturity and practice, not demonstration.[41]

There are a few clear cases of true imitation. Hand-reared apes can learn a general command for "do this, copy me." Viki, the chimpanzee with the imaginary pull-toy, was one; Chantek, an orangutan, was another. Given his cue for "Simon says do this," Chantek reliably imitated gestures: clapping, swinging, touching an ear. The anecdote that stays in mind, though, was the gesture he could not copy. His teacher jumped; Chantek offered one false move after another. At last his teacher realized that given their anatomy, orangutans cannot jump. Just then, Chantek devised his own solution: he put down his long, long arms and lifted his stubby legs off the floor with his arms as crutches.[42]

Biruté Galdikas's rehabilitant orangutans are the most elaborate ape imitators. No wonder Galdikas says she sometimes thought of them as furry red children! They unlatch doors, unscrew jars, paddle boats, brush teeth, light cigarettes. A female orang named Supinah washes clothes beside the washerwomen on the river. Given a piece of cloth, she soaps, scrubs, and rinses—a well-known routine, never taught to her. She once found the fuel shed unlocked. If anyone but the visiting psychologist Anne Russon had been there, Supinah would have been quickly shooed away, not just filmed. The orangutan went through the whole sequence of assembling wood, placing a hose to the kerosene tin for a siphon, and sucking on the hose, but fortunately failed to light her fire.[43]

Do other primates actively teach, as well as actively learn? Again, there are fine distinctions. Adults offer an opportunity to learn nest building, nut cracking, termite fishing every time they do it themselves, and they tolerate children poking in alongside each process with fascinated attention. Rather than erect a single step to something called teaching, we might accept a graded scale of sophistication.[44]

Simply tolerating a child's presence is *opportunity teaching*. Rewarding it with bits of nut that keep it interested (which adult chimps do) is *coaching*. So is whacking an annoying child—negative coaching. Any house cat or tiger teaches in this simple way when it brings home live prey for the kittens to practice killing.[45] The cases of intentional *demonstration teaching* are few indeed. Washoe, the first signing ape, actually molded her son's hands into the sign for "chair"— just as she had been taught herself. Koko, the signing gorilla, even molded her doll's hands into a sign. Just once, in the wild, according to Christophe Boesch, "a mother interrupted her activities to approach her juvenile daughter, who was having difficulty cracking nuts with an awkwardly shaped stone. The mother took the stone from her daughter and with slow and exaggerated motions demonstrated to her the most effective grip for it, monitoring her daughter's attention throughout. The mother then opened nuts with this hammer. The daughter watched the demonstration attentively, then, when the mother returned to her former activities, retrieved the hammer and painstakingly imitated the grip demonstrated. She had failed to open nuts with this hammer before the demonstration, but she succeeded afterwards, using the demonstrated technique."[46]

That is about all: three anecdotes. Note it takes a whole series of adverbs and adjectives to confront the skeptics: *awkwardly, slow, exaggerated, attentively, painstakingly*. Christophe Boesch, watching, knew what he was seeing, because of just those attributes. It is still a lonely example of teaching, only recognizable to the humane observer because humans do it all the time.

Protoculture

Many, many animals learn from one another. Ravens find food by following other ravens, and they fear predators that alarm other ravens. Rats trust food that other rats eat. Many animals learn their initial home range from traipsing after their parents. This is social learning. No need to call it culture.[47] Culture, instead, is both unconsciously absorbed from the cues of society and also deliberately taught and transmitted. Even other primates lack culture in the human sense.

But they do have what may be called protoculture. Japanese macaques learned to wash sweet potatoes in salt water and to wash the sand from wheat grains by copying other macaques. This may not have been deliberate imitation so much as social facilitation, but it was still the social climate that led to the learning. Chimpanzees use stick and straw probes for ant and termite fishing and leaf sponges to sop up fluids. They hurl sticks and rocks as weapons.

In each forest, tool types vary by social tradition.[48] In West Africa, chimpanzees crack nuts with hammer and anvil stones. East African chimpanzees do not. The same nuts grow in East Africa, but the chimps do not eat them. East African chimpanzees are a different subspecies from those in the west, so you might just about argue that they are mentally different, or perhaps that the nut trees in their environment differ. But chimpanzee populations in Gabon, in West Africa, do not crack nuts either. They have the nut trees. They have as many of them. Local people and other animals eat the nuts, which means they are nutritious and tasty. The chimps' diet is like other chimps' diets: they are not wolfing down nut substitutes. Stones and roots lie about, ready to use as hammers and anvils. The chimps seem as intelligent as any others: they modify sticks to fish for ants and dip up honey. There is only one reasonable explanation: the chimpanzees simply don't know how. They don't crack nuts like other chimps because it isn't their local tradition.[49]

Primate protoculture is still far from human teaching. It may be relatively hard to evolve from protoculture to deliberately fostered culture, so long as the only useful skills being demonstrated are things a child could learn on her own. If behaviors that really need demonstration are rare, there is little evolutionary pressure to teach, so the cycle is hard to set off.[50]

But perhaps we focus too much on teachers, too little on the child's rapt attention. We are the species that lives by culture. One of our strongest urges in childhood is to acquire that culture. Just as we are born asking for milk, warmth, and love, we have a profound and clamorous demand to be taught.

Herbert Simon, the cyber-theorist who offered the parable of Tempus and Hora to explain the importance of hierarchical organization, also wrote of humans' innate "docility." By this, he meant the property of being led. Children absorb what adults tell them and show them. Teenagers turn more to peer pressure, while adults intensify their deliberate efforts to initiate the adolescent into adult social norms. This is not culture divorced from biology, but culture rooted in a specieswide need to believe in others' example. The feed-forward process of individual docility in a bath of surrounding culture has got us where we are. We have become creatures who can think together.

Adaptive and Maladaptive Culture

On the whole, we presume that the culture around carries out the basic biological mandate to survive and reproduce—at least those cultures do which have survived. Sometimes a biological approach can even show underlying patterns of culture. One of the most fundamental kinship distinctions is inheritance patterns. Among many people of the world, a woman's brother, not her husband, traditionally passes his name and his possessions to her children. This seems odd if we take the long view that we descended from species with male-bonded groups and traveling women. Women in some societies keep such close touch with their own kin that their brothers are their primary male relations.

Not surprisingly, women have more, and more readily permitted,

303

partners in societies where a child's inheritance comes from the mother's brother. The calculation runs like this: a father is related by half his genes to the child, a maternal uncle by only a quarter. Therefore, in any one generation, if a man is only 50 percent certain of his wife's fidelity, he is at the threshold where he might as well leave his goods to his nephews. That is a very high uncertainty of fidelity. Surely whole societies rarely reach this average! However, if you follow the logic down generations, it takes infidelity by only one of his daughters-in-law to drop genetic relationship in the male line down to zero. Even if a man is pretty confident of his own wife's behavior, it is far safer over generations to stick to matrilineal inheritance, or else develop very strong sanctions on adultery.[51]

Another division between societies is that some parents offer a dowry with a marriageable girl, while some demand a brideprice. On the whole, brideprice correlates with fairly equal societies, or equal in terms of the woman's possibilities of inheritance. (Among the Tandroy of southern Madagascar, a bride usually costs something between the price of a cow and a bull.) Inheritance of both herds and social affiliation are strictly patrilineal, so a man is free to marry whomever he pleases, even a stranger from outside the tribe. However, he is gaining his wives' aid as well as their sexual favors and their children, so of course he has to pay for them.[52]

In contrast, dowry systems correlate with highly stratified societies, where a recognized wife stands to see her children inherit upper-class rank. Dowries escalate particularly in societies that maintain official monogamy. The bride's parents are buying her a status above other women of her class and securing the husband's goods and power for their grandchildren.[53]

Parents sometimes go much further than scraping together a dowry for their daughters. They may offer "claustration," physical proof that the girl is a virgin and dedicated only to pleasure and procreation for her high-status husband. Chinese foot binding was extreme claustration. It began, according to legend, with an emperor enamored of the tiny feet of one of his concubines, which he compared to crescent moons. This foot shape became essential for any woman with pretensions to an elite marriage. Mothers tightened the bandages around the feet of three-year-old girls until their daughters screamed with pain, and kept on tightening them until they produced crippled adults, just like themselves and their own mothers. In essence, this practice said to the prospective husband that a woman could never run away, could never act without her husband's permission, and would endure a life where each step is pain in order to be beautiful for him.[54]

It is in extremely stratified, extremely polygamous societies that parents so dedicate a daughter. If she succeeds in marrying up the social scale and bears a son, his many wives will bear her many grandsons—a benison of both reproduction and wealth for her descendants.[55] All this reflects in turn on the relative value parents allot to girls and boys. Way back when talking about the Seychelles warbler, I mentioned that a bird can skew its sex ratio of eggs according to which sex will help most in raising the family. Indian and Chinese parents who try not to have girls (through abortion, abandonment, or infanticide) if the dowry may be a catastrophic expense for the whole family, are only being rational like the Seychelles warbler. That is, they are both economically and biologically rational, within the constraints of their society.

Genital mutilation. The real extreme in female claustration is female genital mutilation as practiced throughout northern Africa upon some 80 million women a year. This ranges from cutting off the tip of the clitoris, to excision of the whole clitoris, to sewing up the vagina and leaving only a small hole for menstrual discharge. It was traditionally performed at menarche, often along with elaborate ceremonies and a sisterhood of bonds among the girls operated upon at the same time. Nowadays it happens at a range of ages from early toddlerhood up until school-leaving or just before marriage. In the most extreme form, the bridegroom is expected to force open the sewn-up slit with his penis, tearing the bride open. If the husband leaves on a trip, or while the woman is pregnant, she might be re-sewn, then opened again to deliver the child.[56]

Women inflict this upon other women, just as Chinese mothers bound their daughters' feet. Everyone knows that the wounds can become infected, leading to menstrual, urinary, and sexual difficul-ties, sometimes barrenness or death. The rationale is that women's appetites must be curbed so they will not dishonor their husbands and family. There is a subtext of cleanliness: to bring forth perfect children, the evil of sexual desire must be purged. In many places it is a simple fact that an intact woman will not find a husband.

Most people who now practice female genital mutilation are Mus-lims, but the Koran does not prescribe it, and Muslims outside North Africa do not do it. There are indications that female mutilation happened in pharaonic Egypt. It is an older tradition than Islam, rooted somewhere in the Sahara. How could such a custom persist when it attacks the very basis of fertility? Of course, if it is the only socially acceptable way to get a husband, people will do even this. But it does not buy special status, since nearly every girl in some countries undergoes it. It seems, to an outsider, as though a particu-larly nasty meme has infected a quarter of a continent, in spite of all biological sense.

It is not only African women who allow themselves to be mutilated. Male circumcision is a widespread practice. This is often reproduc-tively neutral. However, it is not necessarily neutral among the Tugan and many other Kenyan peoples, when it was traditionally performed on a whole age-set of pubertal youths by the oldest man of the tribe, a rusty razor blade in his shaking hand. It is not neutral among Australian aborigines, who prolong the cut as a "subincision" reaching almost to the base of the penis, opening the urethra, which presents a considerable challenge to fertilization.

Group suicide. Personal suicide may or may not be biologically irrational. The honeybee drone's suicidal monogamy, tearing out its genitalia to leave them guarding the female, or the red-bellied spider male as accomplice in its own cannibalization, are wholly functional suicides. You may object that human suicides are usually acts of despair, not of procreation, but you do not actually know whether the spider feels a spider-sized droplet of something like despair at the end.

We do need to ask what the effects are of a suicide. The person who feels that he or she would be better out of the world is, once in a while, right. His or her death may lift an emotional and economic burden from kin and let the others remake their lives. The fact that parents, children, or lover will bear the guilt must be set against the fact that sometimes the person who died was so ill that his or her survival would have been even worse for everyone else. There is no sign that we evolved the capacity for suicide per se, rather than as part of our general foresight and understanding of death. However, even suicide is sometimes, in the larger sense, adaptive.

Group suicide, in contrast, has absolutely no biological rationale. The cult at Waco or the drinkers of poison in Guyana killed not only parents but children. A larger-scale example occurred among the Xhosa tribe of South Africa in 1856. The belief arose that the spirits of the dead, speaking through the medium of a girl of the tribe, had promised that if all cattle and crops were destroyed, they would be replaced in abundance on a certain day, and the hated white men driven from the land. In obedience to their chief, the Xhosas destroyed their food supplies entirely. In the famine that followed, more than 60,000 are believed to have died.[57]

The only conclusion is that memes are now more powerful than "instincts" of personal self-preservation. The capacity to believe in cultural dictates, to have faith in leaders, to submerge one's own interests to the glory of the group, have been selected for their biological benefits. On the whole, our biological capacity for culture has got us where we are. However, anyone who tries to argue that *all* human behavior is biologically adaptive, rather than just a lot of human behavior, is plainly wrong.

14
Speaking Adaptively

Language is built up like genetic information. They are both hierarchical, they are composed of discrete elements, they are linear, and they depend on context at every level.[1]

Language is a multileveled construction. At the base, the smallest elements that can be called linguistic are phonemes—the separate sounds of speech. Strangely, phonemes seem to be discrete elements. There are a few thousand of them worldwide, many more than the elements of the genetic code that specify the twenty-two amino acids. We categorize phonemes as we hear them. If a sound is betwixt and between, we slot it into a known category. Play a computer-generated "pa" sound to a very small baby sucking on a bottle wired for recording. As the sound repeats, the baby habituates and ignores the noise. But now make the computer gradually shorten the unbreathed pause after the "p." Suddenly adults hear the noise as "ba." At the same moment, the baby notices a change and sucks harder. (We cannot claim that phoneme categories are uniquely human. Chinchillas apparently make the same distinctions a human baby does.)[2]

As early as six months, babies gradually cease to distinguish phonemes that do not pertain to the language they hear around them. English uses only forty, Khoisan or "Bushman" 141. Babies recognize all phonemes to start with, then they start to simplify their categories, grouping ones they do not need with the few they do. Strangely, adults who have learned phonemes at an early age can still hear the sounds, though they have long since forgotten the language. Hindi has two different T variants, indistinguishable to an English ear. Americans who spent their earliest years in India can still distinguish the two different Ts, though they may have no memory of hearing them.[3]

At the basis of language, then, are a discrete gamut of phonemes, innately recognized and then pruned to those apparently useful. The next level is the morphemes. These are the root meanings of words, which may be combined or modified to form yet more words. Of course morphemes differ in every language. At a still higher level comes grammar: the rules for stringing together, and/or modifying words, to take their place in sentences. All languages share the same deep grammatical structure. We all deal with agents, actions, objects, which may be coded only in ways permissible within a given language. It seems that this formal structure is more limited than the "real world" it reflects; there are forms of construction that no language permits. Again, grammar has an innate basis. There is even a known gene that specifically disrupts children's effortless ability to absorb their own languages' rules. If they want to speak correctly, those deprived people have to sit and memorize regular and irregular verbs, regular and irregular plurals in the grammar of their first language—dry-as-dust classroom language study.[4]

Grammar and the construction of sentences are not, of course, the highest level of language. Sentences in turn combine into narratives—layer on layer of them. We have strong opinions about the construction of narratives, too. Anyone recognizes a good story and can tell it from an unsatisfying story, even though tastes differ. We reject disconnected fragments that do not seem to add up to any story at all. (Very modern novels may pretend to dispense with plot; this amounts to a challenge to the reader to invent one.)

This amazing construction is a linear sequence, a string of phonemes, following one after the other like DNA bases strung along a chromosome. The novelist A. S. Byatt points out that "a novel, *Women in Love,* for instance, is made of a long thread of language, like knitting, thicker and thinner in patches."[5] Each element is also related to the next. Each exists in a context of sound and meaning. Dependence on context is equally true of genes.[6]

The sum total is that both language and genetics (and knitting) have the power of a combinatorial, discrete, linear system. Besides all that, language and genetics reproduce themselves or, better, are reproduced in interaction with their respective environments. These are the only two such systems we know.

The comparison is not new. Francis Bacon, often called the founder of the scientific method, remarked: "It is the common wonder of all men, how among so many millions of faces, there should be none alike; now, contrarily, I wonder as much, how there should be any. He that shall consider how many thousand several words have been carelessly and without study composed out of 24 letters; withal how many hundred lines there are to be drawn in the fabrick of one man, shall find this variety is necessary."[7]

Melodies of Motherese

Children have a strong innate program for learning language, but of course they learn it from others. Babies don't wait for birth to start learning. They know intonations of their mothers' own language, as distinct from other languages, even as newborns. Apparently they start listening in the womb.[8]

Other people help them. Primate vocal noise becomes human sentences even in *parents'* speech. People who talk to small babies raise the pitch of their fundamental frequency. Similar emotional contexts produce the same rises and falls of pitch in English, German, French, Italian, and Japanese.[9]

Do you wish to convey approval? One way is to slide your voice from moderate pitch to a high peak, and then down a glissando that may cover two octaves, whether you are mouthing "you CLEver boy!" or "BraVISSima!" Of course there is not just one variant, but the fluctuating pitch is recognizably child-oriented: "What a SMART little BOY you are!" "You are SO SMART!" Prohibition is discrete short syllables with little pitch change, as in "No/no/stop/that/now" or just "STOP it!" Comfort is repeated syllables, starting low but descending: "Po-or bo-oy," "Calme toi, Calme toi." These "melodies" probably derive from primate calls, and indeed resemble primate calls in conveying information without words. A new baby's ears respond to higher pitches than adults', and distinguish less. Infants will work to hear playbacks of such infant-directed speech, rather than more difficult-to-decode adult speech.[10]

These accessible intonations then support a peculiar form of talking to toddlers, dubbed "motherese." Motherese is repetitive, with simplified vocabulary, and absurdly grammatical. When speaking to a two-year-old, around 50 percent of statements will be repeated; over 99 percent of them will be grammatical. Children learn to speak whether or not they are immersed in motherese. Some cultures believe that it is not worth addressing toddlers who do not understand. Those children have enough innate framework to learn language like anybody else. But motherese provides a kind of Rosetta Stone to translate between the child's mental images and the complex grammar and vocabulary it will shortly master.[11]

What to me is the ultimate lullaby illustrates it all—the pitch melody of approval repeated fourteen times over, the prosodic repetition, the rallentando of the final line (and, of course, Odetta singing):

Speaking Adaptively

Hush little baby, don't say a word,
Momma's gonna buy you a mockingbird.
If that mockingbird don't sing,
Momma's gonna buy you a diamond ring.
If that diamond ring is glass,
Momma's gonna buy you a looking glass.
If that looking glass gets broke,
Momma's gonna buy you a billy goat.
If that billy don't eat clover,
Momma's gonna buy you a dog named Rover.
If that dog named Rover don't bark,
Momma's gonna buy you a horse and cart.
If that horse and cart fall down,
You'll still be the sweetest little baby in town.

Signing Apes

Language is a skill that makes us human. It is surely not the only skill. This generation, schooled on television, photojournalism, and the web, probably thinks in visual symbols much as previous generations turned to literacy. It is clear, though, that no other species thinks in language.

Is it an absolute Rubicon? People used to claim that an ape would never manage to use words, with their detachment of signified and signifier. Alan and Beatrix Gardner then taught ASL (American Sign Language of the Deaf) to the young chimpanzee named Washoe. Beatrix Gardner told how they originally plucked up courage to try. They watched a film clip of Viki, the home-raised chimp who invented the unseen pull-toy. Viki strained to produce her four words in a breathy whisper, sometimes closing her lips with her hands to make the consonants for "Mama" and "Papa." She simply did not have voluntary control of vocalization. Then, released from language drill in her high chair, she put a record on her child's phonograph. It did not start, so Viki went over to the wall to plug it in. Beatrix Gardner exclaimed, "An animal must be capable of *some* language, if it knows enough to check the plug!"[12]

Washoe eventually learned over 300 words. She generalized their use—*open* for boxes and cupboards—and even, creatively, asked Alan Gardner to retrieve a toy that fell down through a hole in her house-trailer wall. (She lived in a separate house-trailer of her own, on the University of Nevada campus.) She occasionally strung signs together, the famous one being "water bird" on seeing a swan. Skeptics suggested this was in fact the self-duping of doting parents. Then meticulous double-blind tests showed that Washoe named slides of objects reliably and generalized to new objects of a class, just as we use words. It was academic psychologists, not Washoe, who needed the double-blind tests.

More impressive to a layman was that Washoe would sometimes name objects to herself while looking at a picturebook, or she would play word games like a toddler's sing-song. One unforgettable film clip showed the ape by herself in a tree, signing *Dirty, dirty, dirty! Dirty, dirty, dirty!* In liberated modern days, this sign would be translated more literally: *Shit.* Washoe later extrapolated it as an insult to monkeys and people who threatened her. Viki had already shown that chimpanzees have only a tiny period of vocal "babbling," but signing apes play with signs.[13]

Other chimpanzees, two gorillas, and an orangutan followed Washoe's lead, in at least seven different laboratories around the nation. But Beatrix Gardner said she sometimes felt as though they were chasing a figure fleeing before them down a long dark tunnel. As soon as it was clear that apes could in some sense use words, linguists decided that words are not the criterion of language. They asked for grammar instead.[14]

David Premack embarked on the search for ape grammar. His star pupil, the adult chimp called Sarah, did indeed master simple word order to convey grammatical relations. Then came the debacle of Nim. Herbert Terrace, at Columbia University, tried to teach grammar to a young chimpanzee named Nim Chimpsky. Terrace installed rigorous controls to confound the skeptics. Nim did not make as much progress as Terrace hoped. Nim's bleak and institutional testing environment, and some sixty different caretakers over the infant's four short years, were enough to have discouraged many human children from learning. The crux came when Terrace looked hard at 3.5 hours of videotape of Nim's training. He saw that the teachers often inadvertently cued the very signs they were trying to produce. He did not see any evidence of grammar. Nim's longest recorded string of signs, to an orange held out of his reach, instantly shows how his begging differed from grammatical sentences: *"Give orange me give eat orange me eat orange give me eat orange give me you."*[15]

Terrace's book on Nim cut off much of the funding for ape language research. Others, notably Lynn Miles working with the orangutan Chantek, countered that most of their own subjects' signing was spontaneous and that there was much more evidence for grammar. Francine Patterson and Roger Fouts struggled to keep their apes alive and fed and to slowly push forward their records. Still, several of the early cohorts of language-trained apes wound up in medical research, signing from their cages to technicians who never understood.[16]

The next breakthrough came when Sue Savage-Rumbaugh turned to bonobos. Savage-Rumbaugh and her husband, Duane Rumbaugh, had already worked with common chimpanzees, including Washoe. They doubted that even the "words" of signing chimps were quite like human words. True, they could be produced appropriately, but were they *understood*? Savage-Rumbaugh trained two chimpanzees called Sherman and Austin to communicate using signs. These were symbols on computer keys. The computer could record just what was "said." The two apes faced problems to solve jointly. Either Sherman or Austin could see food in a puzzle box that could be opened with a particular tool: a stick for poking, a key, a hoe. The other chimp had the toolkit. The first one named what he needed. The second passed over the named tool, like a nurse passing forceps and sponges to a surgeon. If they got it right, both Sherman and Austin shared the reward.[17]

They mastered the game, but it did not come easily. They knew all the separate signs. They used their signs, like other apes, to beg from humans, or solve the naming tasks that seemed to delight human psychologist foster parents. Language as chimp-to-chimp communication was a difficult step for common chimpanzees. Savage-Rumbaugh became more than ever convinced that comprehension, not production, is the true criterion of language.

She then tried teaching the computer language to an adult bonobo named Matata. Matata, wild-caught as an adult, did not learn. Matata came back into estrus and returned to the main bonobo group for breeding. Her 1½-year-old toddler, Kanzi, stayed behind in the lab, with the keyboard. On the very first day, with no instruction, Kanzi started to use words: "I was hesitant to believe what I was seeing. Not only was Kanzi using the keyboard as a means of communication, but he also knew what the symbols meant—in spite of the fact that his mother had never learned them. For example, one of the first things he did that morning was to activate 'apple,' then 'chase.' He then picked up an apple, looked at me, and ran away with a play grin on his face. Several times he hit food keys, and when I took him to the refrigerator, he selected those foods he'd indicated on the keyboard. Kanzi was using specific lexigrams to request and name items, *and* to announce his intention—all important symbol skills that we had not recognized Kanzi possessed. How could this be? We had spent two years systematically trying to teach Matata a small number of symbols, with meager success. Kanzi appeared to know all the things we had attempted to teach Matata, yet we had not even been attending to him—other than to keep him entertained. Could he simply have picked up his understanding through social exposure, as children do? It seemed impossible."[18] Savage-Rumbaugh threw away the rule book. She decided to teach this ape as children learn, with symbols used in context. Kanzi galloped ahead with signing. Eventually he and his younger siblings showed they could understand spoken words as well as signed ones.

When I first met Kanzi, I hazarded a question, too. There was a six-foot-long construction of plastic pipes in his cage. I asked what they called it—it wasn't a test object with a name, just a toy. Very sheepishly, I said, "Kanzi, can you climb on your toy?" all the while thinking, "I've cued him with my eyes, then by asking a name for that thing, and then by looking at it again. If he goes over and sits on it, all *that* will show is that he is polite enough to pay attention to me." Kanzi went over. He up-ended the toy to balance on one end, then climbed up it, hand over hand, till it (and he) fell over. "Climb" meant *climb* to him, like climbing a tree. He had answered my verbal request, not my cued intention.

Kanzi also deals with grammar. Savage-Rumbaugh, in a mask, behind a screen, or even over earphones, gives him novel spoken commands: "Kanzi, can you put the pine needles in the refrigerator?" and using grammatical word order: "Kanzi, can you make the [toy] doggie bite the [toy] snake? or "Kanzi, can you put the Perrier water in the jelly?" (not vice versa). Kanzi copes.[19]

He can even deal with subordinate clauses, so long as they come after the noun they modify: "Kanzi, can you get the pine needles that are in the refrigerator?" In the last case his hand goes out toward a pine branch lying on the floor, then as the sentence continues he swings around to retrieve the other pine branch he had put in the fridge. What he cannot do, or does with difficulty, is retain what he hears as a sequence of separate ideas: "Kanzi, can you go to the refrigerator / and get the pine needles?"[20]

Figure 15.   Kanzi the bonobo, in a playroom full of different objects, responds to Sue Savage-Rumbaugh's novel spoken command: "Kanzi, can you take the vacuum cleaner outside?" Kanzi himself communicates through pointing at symbols for words on the folding board. (After Bass, 1995)

Kanzi and his family actually imitate some human words: yogurt, melon. (Kanzi does have a family; Matata returned, and he has grown up with mother and sibs as well as humans.) I once concentrated on hours of videotapes, made by the Japanese station NHK, to choose some for reuse. I suddenly heard the bonobo voice say "M-and-M," as Kanzi reached for candies. I exclaimed out loud, but Savage-Rumbaugh remarked dryly: "You have heard that sound in that context perhaps twenty times over the past two days. It is just that you've finally got your ear in."

Do bonobos, or any other ape, naturally use something like language? Savage-Rumbaugh reasoned that their most likely need for representative symbols might be stating an intention to travel to a given place or eat a certain food. Bonobo groups fission and fusion through the day but often rejoin at night. It would be very useful to be able to fix a rendezvous. She taught Kanzi symbols for different sites in the woods. He would announce his intention to travel there, and then go, leading his keepers.

Is there any indication that wild bonobos do the same? No, there is not. But would we know if they did? Their voices are unlike ours: soprano whistles and strings of vowels, unbroken by consonants. They do make long strings of such sounds. Bonobos' soft, whistling notes are as different from the emotional repertoire of screeches and squeals as human conversation is from our own screams, sobs, and moans. But then, birds' subsong is also very similar to the bonobos' "conversation." Even if we succeed in separating out components on a tape recording, how would we agree on a referent? Just suppose a bonobo actually did announce it was going to the ginger patch, but went there by way of succulent vegetation in the swamp, would we deduce the sign for ginger?

The best we can say is that we have no indication that bonobos' linguistic skill has any use in nature. But it might be wise to keep an open mind.

The Word That Is Not What It Says

"In the days before Babel, before God punished the human race for its presumption in raising its winding structure towards Heaven by dividing its tongues, by setting confusion amongst its speech—in the days before Babel, the occult tradition went, words had been things and things had been words, they had been one, as a man and his shadow perhaps are one, or a man's mind and his brain."[21] Today, the word *chair* is not a chair. The word *cow* is not the word *vache* nor the word *Kuhe,* but they all have the same relationship to a real cow.

Sassure, Bertrand Russell, and whole libraries full of academic books of the twentieth century wrestle with the nature of word—whose essence is that it merely *signifies* what it says.

And yet we recognize, even thrill, to the magic words of the occult tradition evoked by novelist A. S. Byatt—the magic words which are continuous with what they name. We recognize them. We know them from childhood, when so much was magic. We know them from adulthood in the forbidden names of excrement, sexual intercourse, and God. When those names pepper speech as the commonest of adjectives, they degenerate into words, or even less, place-holding sounds. Names which still have an emotional link so strong that they summon the power of the thing itself are not words; they are invocations.

At some stage in the evolution of language a few vocal sounds became detached enough from their connotations to become tools of the mind. They moved from invocation to metaphor, and across the permeable boundary toward analogy. Again we return to the fundamental importance of Bateson's negative: the conscious decision what route not to take, the imagination to pretend what is *not* so. Words confer the linguistic power to name what is not here, not even to be invoked, but merely a thought to discuss.[22]

Suppose the bonobo, or some long-ago ancestor of our own, did have a noise meaning "the ginger patch." If it could only say that noise when desiring ginger, or intending to travel to eat it, that was only half way to being a word. Useful, yes: it could tell the others where to meet, it could influence them to go where it wanted by conjuring up the spicy taste of ginger. Some such noises led our own ancestors through the first steps of communicative speech. They could even argue (as primates do without words) where to go next—one faction starting toward the ginger patch as they named it, others the swamp, with numbers and social dominance deciding the troop course.

One day (or one millennium), our ancestor said, "No, not the swamp, the ginger patch," naming the alternative which did *not* matter. She did not shout; she was not registering fear or disgust. No statement worth making is wholly detached from emotion. She may well have felt antagonism to the swamp faction of the troop, but the swamp itself was merely a negative. And that was a word—a word freed from magic to become an idea of its own. A new space was born, the space of the human mind, home to ideas, or memes.

I nearly wrote: "a space which encompasses the unreal as well as the real." But that is not what I mean. The mindset of magic words that are continuous with things is not particularly real. It can be the partial view of the small, self-centered child, who may be mistaken, who may see cause and effect where there is none, whose world is painted in the colors of his own desires. It is not unreal, either: the greatest poets find words that bring the smarting of tears. We never outgrow that emotional realm of language (and would not want to), but both the scientist exploring the spaces of nature and the novelist creating other worlds of imagination began their journey from that first step of detachment: the word that is not what it says.

15

Are Babies Human?

Human babies are born as embryos. If our brains had the same proportion to adult brains as the newborns of other apes do, we would require a twenty-month gestation. Imagine giving birth to a baby like our one-year-olds! No, best not to imagine it in any detail.

Humans evolved two incompatible demands: we stood up on our hind legs, and we grew very large adult brains. The solution has been to give birth to infants at an ever earlier stage of their lives, with the brain only partly formed. The result is that our newborns have almost a whole year of social behavior, of learning and communicating with their mothers and other caretakers, at a stage of brain growth when other apes are still tucked safely in the womb.[1]

Primates as a whole are precocial. That means they are born with eyes open, ears functioning, and furry enough to help maintain their own body temperature. Primates are not the extreme of precociality: a baby gnu can run after its mother an hour after birth. It may not be able to outrun a cheetah, but it had better be able to try. A baby elephant may have to march five miles on the day it is born to keep up with his mother's female family. Primate babies do not do forced marches, but they cling onto their mothers' fur while she bounds through the treetops.

Our near relatives the great apes make a few allowances. Mother chimpanzees and gorillas support infants under a month old with one hand when they are moving around. Those mothers can put up with walking three-legged because they have few predators, they spend most of their time on the ground, and they do not leap wildly even when clambering in trees. The babies are still precocial—open-eyed, furry, and clinging.

The opposite style of development is altricial. Newborn mice or rats are like pink, naked, squirming worms with black eyeballs encased in a layer of translucent skin. Even a brand new kitten's eyes are sealed closed. Altricial young are nest young, kept safe at home. They grow at phenomenal speed from a semi-embryonic start.

We are primates. We were geared originally to bearing breast-young, not nest-young. When we ran into the conflict between brain and bipedalism, our human lineage jury-rigged a way to become secondarily altricial. We now give birth to helpless, half-formed primate babies, with profound consequences for both child and caretakers.

Birth

We reached this solution step by step, like most other changes during evolution. The first, great change between our lineage and the other apes was standing up. Three and a half million years ago, our ancestors the australopithecines—Lucy and her kin—had brains no larger than modern chimpanzees, but they were almost as efficient as modern humans in striding across the savanna.

A quadrupedal animal supports its abdominal organs by connective tissue hung from the backbone. Its pelvis translates the pressure of the backbone down and out to hip and hind legs. The pelvis's central opening can be long in the vertical dimension, from backbone toward the ground, without any problems. Once you stand upright, however, the pelvis takes over holding up the abdominal organs: stomach and gut and even the weight of a growing embryo.

Australopithecines' pelves became more like the wide, basin-shaped supporting structure that we inherit today. The central opening for the birth canal also changed shape. Its largest dimension was not front to back but side to side.

Perhaps this pelvis shape presented a problem even for Lucy's kin, with their small brains and presumably small-brained babies. They could solve it by elaborating a birth process known today even in a few monkeys: the child, instead of slipping out directly, turns and twists in the course of being born. Most apes and monkeys are born facing forward. Infant squirrel monkeys aid their own birth by pulling on the mother's belly fur as soon as the arms emerge. A modern human fetus in normal birth presentation begins to enter the birth canal facing sideways to the mother, the head aligned with the widest dimension of the upper part of the birth canal—the same shape as Lucy's. Then the child twists, as the head reaches the lower part of the birth canal, which in modern humans is widest front to back. The child is now facing back toward the mother's hips. Meanwhile, the shoulders, above, are entering sideways. The child twists again as it emerges, to get the shoulders through the bottom of the canal.[2]

During a normal birth, if a human mother pulls on the child to help, as many primate mothers do, she could actually pull backwards on its neck and spine and perhaps harm it. If the fetus has an unusual presentation instead, facing forward or feet first, the birth through the doubly constricted canal might be fatal to both child and mother. Other primates also have fatal complications during childbirth, but not so often as we do.

Enter the midwife. An experienced female could ease the child gently in the direction it was emerging and hand it to the mother when the dangerous moment of birth had passed. Aid during birth may even have begun as mothers became bipedal, way back in the days of Lucy the australopithecine—not to deliver the baby's small head but its wide shoulders. Midwifery is a candidate for at least the second oldest profession.[3]

Problems increased dramatically when the human brain grew larger. Twisting and turning during birth would not solve that. When adult brains mushroomed in *Homo erectus* and eventually *Homo sapiens,* the solution was to give birth to infants with brains at an ever smaller fraction of the adult size—an external, newborn embryo. That development made still more behavioral demands on the midwife, to cuddle and warm and clean the child if the mother was not yet able to, sometimes even to blow in its nostrils or slap it to start it breathing. Soon enough the mother would need to pull herself together to deal with an infant who claims not just an occasional hand in support but total care for years on end.

Birth as behavior.   Birth is not just a mechanical process. It is an interaction between child, mother, and caretakers. It is not clear what sets off normal labor. It seems to originate, however, both with the fetus and the mother. The child itself has at least a share in the decision when it is ready to be born, producing oxytocin from its pituitary gland that helps trigger the contractions of labor. The whole process involves a complex of hormones to soften and "ripen" the cervix and prime the muscles with receptors for the oxytocin.[4]

Normal human births are statistically more likely to happen at night. The largest study was 601,222 births in the United States and Europe between 1848 and 1960, but even a few hundred will show the effect. Most labor occurs at night, ending in a birth peak between 1 and 6 AM, highest in the hour from 3 to 4 AM. Births with unusual complications show no such difference. Of course, births induced in hospital for the doctor's convenience peak in the daytime.[5]

Unaided labor is shorter for nocturnal births than daytime births by about two hours on average. I first heard that from a midwife who was ready to deliver our third child at home in the cosy English system. She informed my husband that if I started labor at night he could telephone her. Otherwise he should drive to her house and look at the blackboard in her window, which would let him know her schedule of visits for the day. (This was before cell phones.) He expostulated, while I remarked that her plan would suit me fine if I were a rhesus monkey, but being a human, I was not happy about it. "Oh," she assured us, "if it is a daytime labor, you will have plenty of time."[6]

Nocturnal labor is the rule in other primates. There are exceptions, of course. Species that are active by night often give birth in the day, as do a few others, for their own reasons. It is obvious that if a mother must move around with a diurnal primate troop and fend off the eager attentions of too many others who wish to groom the newborn, she does much better to have her baby at night, in peace. If evolving humans foraged alone or in small bands as chimpanzees and human gatherers do today, it would be even more important to give birth in the safety of the camp at night. Indeed, a somewhat longer daytime labor could be an advantage. It allows time to go home to camp or at least to find a midwife, as in England in the 1960s.

Of course, cultures differ. !Kung San mothers traditionally gave birth all alone, away from camp in the African bush. In contrast, a friend who had an interesting tropical disease in early pregnancy told me that seven different doctors dropped by in a Washington, D.C., hospital to watch her give birth. (Her daughter was disappointingly healthy.) Most people in most places just want a midwife.[7]

Is labor shorter when the mother feels safe and comfortable? There are a few suggestions from other primates. Gertrude van Wagenen, who pioneered captive breeding of rhesus monkeys, remarked that one could not avoid the feeling that the rhesus quietly held back labor until the caretakers went home. The only measured study I know is the "weekend effect." Chimpanzees that live in laboratory colonies are more likely to give birth on the weekend, when they do not have the disturbance of routine colony activities and care. A very few medical texts admit that human labors are shorter when women are confident, or even that the entry of the obstetrician can stop contractions temporarily, as the woman reacts like a chimpanzee confronting the laboratory keeper. This is sometimes called the "white-coat effect," and it is commonly seen when a doctor walks into the office to take a patient's blood pressure—just the sight of the physician causes blood pressure to rise. Of course, human mothers, ideally, could just say they prefer to feel safe and comfortable, without waiting for statistics.[8]

The first stage of labor is horizontal muscular contractions, which squeeze the fluid-filled amniotic sac rather than the baby and helps to open the cervix, though labor may also start with the "breaking of the waters." The second stage is longitudinal pushing, to actually expel the baby. Some mothers are favored with naturally produced opiates or analgesics that let them doze in the intervals between contractions. It is not known what the child feels, but it has a remarkable series of buffers that let it endure the strain. These range from the softness of the head bones and their junctions, which mold under the pressure, to the physiological ability to withstand oxygen deprivation up to the moment of birth. One may wonder whether there are also means for it not to feel pain, or as much pain as an adult or child would feel, while its body undergoes the most extreme human physical transformation up until death itself.

When the neonate is born, it begins at once to communicate with the humans around—an embryo ape, perhaps, but a human child.

The Newborn

Newborn infants turn their heads to a touch on the cheek and suck if something is put in their mouths. They are ahead of the game in this:

mothers will not have any fluid in their breasts for several hours, but the newborn is ready. Suckling or even licking the nipple stimulates the mother to release oxytocin, which causes the uterus to contract, expel the placenta, and reduce bleeding. Oxytocin is often injected now by doctors, but in the past, infants' sucking may have saved some mothers' lives.[9]

Newborns grasp with their hands and make stepping motions with their feet. On a father's hairy chest, where little hands can grip onto fur, the clinging may actually turn into crawling along. The grip of a newborn may support its own weight. (I suspect, though, that the tale that Spartans hung their babies off rooftops by their hands, to see if they could hold on hard enough to survive, was a calumny put about by Athenians.) Newborns go quiet or jiggle into motion in the presence of human voices, depending on their state of arousal. If the infant's head drops backward, or you slap the mattress, its arms flail convulsively toward the midline and its hands clutch whatever they find. This "Moro Reflex" was used for years by doctors testing infants' coordination before people realized that the Moro is the ancient primate gesture of grabbing hard onto the mother when she starts into motion. Our babies still brace themselves for bounding through the treetops.

Helen Blauvelt called the newborn's repertoire "the reflexes of love."[10] They are innate responses to innately recognized trigger stimuli, but so adorable. Reflexes communicate the child's needs to the mother—clinging for comfort and warmth, mumbling of the mouth to suck. The child does more. As we saw already, infants as young as twenty minutes old can copy facial expression. They open their eyes wide and stick out their tiny tongues in some innately primed echo of the face they see. They also learn quickly, at least in the realm of smell and sound. Babies a week old can distinguish their mother's milk from another's, as well as her special voice, which they may even learn in the womb.[11]

Maternal bonding. Meanwhile, the mother goes through her own changes. Many, though not all, women feel a rush of euphoria an hour or so after giving birth. This is, of course, what our culture tells us to feel, so it is hardly an untutored response. It also seems quite likely that there is a natural burst of endorphins. That would help a primate mother accept this new stranger clinging to her body, not brush it off as a wet and sticky intruder.

Much has been made of mother–infant bonding in the two hours after birth—the normal time to first meet one's infant, in the hormonal state when both are primed for the meeting. However, it is obvious that this is not the only time. Mother–infant bonding is far too important for survival to leave it to a single moment with no backup. We are well aware that a mother can love a child she first sees hours after a difficult birth, or first holds a month after a premature birth, or adopts when it is years old. So can fathers. In wild primates, even siblings and aunts have adopted orphans. What is natural is to have a huge redundancy of mechanisms for survival wherever possible, and baby primates, including humans, need love to survive.

Breastfeeding. They also need milk. A few hours after the birth, the mother begins to produce colostrum. This is a yellowish clear fluid filled with antibodies, which helps to protect the infant until its own immune system begins to function. Many cultures throughout the world dictate that a mother should express and discard the colostrum because it looks like inferior milk. Knowing its protective function for the child, and its role in establishing breastfeeding for the mother, it is amazing that such an unhealthy belief is so widespread. Present-day western culture has some bizarre beliefs surrounding childbirth, but other people do too.[12]

After about two days, real milk comes in. This is also high in protective chemicals, especially lactoferrin, which guards against infant diarrhea, the chief killer of the world's children. If a mother is exposed to bacteria and viruses, she produces antibodies, like any other adult. These find their way through her bloodstream and into the milk. Her child receives antibodies to whatever is in its environment "for free," without having to mobilize its own immature defenses to fight off the germs. A breastfed four-month-old gets about half a gram of antibodies a day in its milk.[13]

The nutritional content of human milk reflects our primate heritage. It is relatively watery and high in protein and sugar. That is milk designed for a child who will be carried, with frequent feeding from an ever-present breast. Cows' milk is higher in fatty acids. The highest fat content occurs in milk from animals that leave their young alone for long periods in a nest, like rabbits or tree shrews, which turn up only every day or so for a few minutes of squirting milk under pressure into their young. The most fat of all is in animals that not only leave their young but get them cold and wet, like seals and otters. We have primate milk, but it could be special even among primates, geared to growing that demanding infant brain.[14]

Lactation is one of the most exquisitely adjusted biological feedback systems that we know. The hungrier the baby is, the more it sucks, and the more milk the breasts produce. Giving bottles between breastfeedings fills up the child, so it does not fully empty the breasts at the next feed. The breasts take the cue that less milk is needed. The foremilk, which is left in the breast since the last feed, is let down first. Freshly produced hindmilk follows. Hindmilk is much richer in fat. Full, or frequent, feeds give enough fat to satisfy the child. Bottles between not only tell the child to suck less but leave it sucking only watery foremilk. Then it wakes up early, hungry, and cranky. This can convince the mother she does not produce enough milk. UNICEF calculates that *at least* 97 percent of mothers are capable of giving enough milk. Three percent is still a big number, big enough to mean it happens to someone you know—or to yourself. Humans vary, in childbirth and milk production as in everything else. We are fortunate to have a choice.[15]

However, it is worth knowing that some hospitals in the United States, and many abroad, regularly offer bottle feeds when the baby is still in the hospital. This can start the negative cycle, with less chance of establishing adequate breastfeeding. UNICEF has started certifying hospitals as "baby-friendly" if they do not offer bottles unless the mother specifically asks. In Third World conditions, water itself is not sterile. Wood for fires to boil it must be bought or carried in on the mother's head. Poverty tempts caregivers to skimp on the amount of milk powder when making up formula. In poorer countries, bottlefed babies are twice as likely to die as breastfed babies.[16]

As with maternal bonding, of course, breastfeeding entails both redundancy and trade-offs. Millions of successful, healthy, intelligent children have been bottle-reared. Clay nursing bottles survive from ancient Egypt. There were surely earlier ones if we could recognize them, probably as early as the first domesticated sheep or cow. Bottle-feeding was imperative if the mother sickened or if she died in childbirth or of puerperal fever, and nowadays if she carries HIV. Still, most bottlefeeding and wet nursing have been the mothers' choice.[17]

In Europe from the Late Middle Ages and through into the nineteenth century, women who could afford it gave their children to wet nurses to rear. They wished to keep taut, girlish figures, they wished to avoid the fatigue and constant demands of breastfeeding, and they wished to conceive again sooner, without lactation-induced delay. Medical manuals uniformly inveighed against the practice, lecturing women that they should feed their own children rather than put them aside for the mother's convenience or vanity. Women continued, and continue, to make their own choices.[18]

The rhythm of the dance. Babies only a day old wriggle their limbs and orient their faces in tune with tape-recorded female speech, whether in English or Chinese. They do not do it to disconnected vowel or tapping sounds. They are primed already to social rhythms, probably even in the womb.[19]

Most human babies are carried for much of the day against an adult's body. All primates are, too, except for some species of prosimians, who build nests or park an infant immobile on a branch. Parked babies have a suite of protective devices: almost no smell, almost no crying. If a predator approaches a parked baby potto in the rainforest of Cameroon, it just lets go of its branch and falls like a stone to the leaf-litter, where it remains totally motionless, unless called by its mother's own voice. It's not hard to recognize that we did not evolve as pottos.[20]

Carried babies respond to their caregivers' motion. They rock to sleep at a steady walking pace, they wake and cling if the pace changes. They give little wiggles to indicate that they will defecate or urinate, which lets a mother primate hold them off to the side. I am not recommending this—diapers are easier. Unless, of course, you are poor and live in a warm climate, where the old primate system seems to work fine.

Babies in most of the world's families sleep in rhythm with their parents. Meredith Small tells in detail how sleeping babies surface through the layers of sleep and drop back down in synchrony with their parents' sleep. This lets mothers breastfeed without actually waking up very far. It apparently also helps protect against sudden infant death syndrome—the unexplained dying of infants in their cribs. The lighter sleep of a child who is co-sleeping, and the rhythms of nearing consciousness in the night, help remind it to breathe.[21]

So are babies human? Of course they are—sort of. But go back to the Social Science Straw Man: Aren't humans supposed to be creatures of culture? Aren't we the species that learns what to do and how to run our lives? By that definition, a baby is an animal. It isn't even much of an animal. It has instincts, but many, many of the instincts are geared to coaxing some caretaker into attending to its needs. It wants feeding and warmth and co-sleeping, which any primate mother will do. It also seems to want attention and voices and language. It snuggles and sucks and sleeps and wakes as a social being. A baby is an amphibian between the world of biology and the world of adult humankind. It is fortunate that parents also have their responding instincts, for a more objective view might echo Vickram Seth:

> How ugly babies are! How heedless
> Of all else than their bulging selves—
> Like sumo wrestlers, plush with needless
> Kneadable flesh—like mutant elves
> Plump and vindictively nocturnal,
> With lungs determined and infernal
> (A pity that the blubbering blobs
> Come unequipped with volume knobs),
> And so intrinsically conservative,
> A change of breast will make them squall
> With no restraint or qualm at all.
> Some think them cuddly, cute and curvative.
> Keep them I say. Good luck to you;
> No doubt you used to be one too.[22]

The First Year

What happens in that first year, when the child is still physically helpless, except for social skills that induce others to help it? Its motor control is weak, but its mind is already active. Babies go through a process of imprinting like those mallard ducklings who followed Lorenz through the grassy field.

A duckling responds, as a tiny infant, to very general stimuli. It will try to approach any object that is not too large, moves, and makes a repetitive noise like quack, quack, quack. Then the duckling learns in detail the characteristics of its own particular mother, so that it can follow without mixing her up with other mallard mothers. Humans, for all our brain power, have a hard time telling one duck from another, but a duckling manages. Then comes a third stage: the duckling grows frightened of strangers who are not its mother. Again, this matters. The wrong mallard is likely to give them a good swift peck; the wrong species would lead them far astray—and some strangers are foxes.

Humans under a year cannot follow, of course. They lure their caretakers to approach them instead. They have two main signals. Bawling irritates the caretaker and motivates him or her to stop the unbearable noise. The other signal—the infant's smile—delights the caretaker into cuddling. We have seen that babies respond to faces, and recognize them from birth in some sense. Six weeks or so later, the baby rather suddenly develops the full display of focusing its eyes on a face, then giving that wide, beaming, deliberate-looking gummy smile. In fact, at that early stage the child is willing to smile at two black eye-spots drawn on a piece of paper—as general an image as the owl-eyes that drive infant hog-nosed snakes into convulsions. The smile originated in early primate evolution from a fear response, so the instinct to smile in the presence of eye-spots could have some tinge of its origins in social appeasement.[23]

As the baby matures, it demands that faces be more and more facelike. Soon it wants a head, then a nose and mouth to go with the eyes. It probably recognizes its mother (remember, it already knows her voice and smell), but at three and four months it beams at strangers with equal readiness. By five or six months, it is clearly more responsive to the people it knows.

Then comes what is sometimes called eight-month anxiety. This is highly variable; some babies remain socially extrovert throughout. Many, though, go through a period when they look solemnly back and forth between their own mother's face (or the faces of other caretakers they know) and the well-meaning friend who comes up to admire the baby. Then, having made its comparison, the child buries its face in its caregiver's neck and bursts into tears. The friend perforce looks away and retreats with hurt apologies.

This sequence could have been very important in a small band of hunter-gatherers—our lifestyle up until a mere 10,000 years ago. Babies were carried all the time when tiny, but a three-month baby is heavy. It needs some powerful signal that it can be held and cuddled; hence the social smile. It is not likely to meet many strangers at that stage, and if it does meet one that is ill-disposed, there is not much to be done, like the duckling that meets a fox. When the child starts crawling and walking, though, between eight and fourteen months of age, it can barge in on the more irascible members of the tribe. That is when it needs a fear of strangers, just as mallard ducks do. The sequence is exactly the same.[24]

Autism.   The genetic anomaly of autism seems to involve a fundamental inability to mindread; autists do not seem to understand that other people have minds. The majority of autistic children are, or soon become, deeply mentally retarded, but about 25 percent fall within the normal range of general intelligence. A few have spectacular special gifts in mathematics, music, or art—or, like the autist played by Dustin Hoffman in *Rain Man,* can remember the play of all the cards in the deck.

People with autism find it hard to deal with social situations. In fact, the disorder may be even more specific. Autistic children interpret a mechanical picture story as well or better than normal children—for instance, a man kicking a stone into the sea. They can read behavior: one child grabs another's ice cream cone and the victim cries. Where they fall down is in interpreting expectations, intentions, beliefs. A child puts down her teddy bear and turns to pick a flower. A second steals the teddy. When the first turns back, she is surprised the teddy is gone. Her surprise baffles the autist. Apparently, autists do not tell lies, having no conception that it is possible to hide the truth.[25] Such small tests do not tell how it feels not to empathize with others. Perhaps no one can tell, except autists who have reached out to an alien world. Jim Sinclair writes, "I really didn't know there were other people until I was seven years old. I suddenly realized there were people. But not like you do. I still have to remind myself that there are people."[26]

Dr. Temple Grandin describes her childhood sensory anguish: "Sudden noises hurt my ears like a dentist's drill hitting a nerve." Many autists seem deaf, because in desperation to screen out painful and confusing noise, they retreat from all noise. They also show aversion to many kinds of touch: "I often misbehaved in church, because the petticoats itched and scratched. Sunday clothes felt different from everyday clothes . . . Most people adapt to the feeling of different types of clothing in a few minutes. Even now, I avoid wearing new types of underwear. It takes me three or four days to fully adapt to new ones . . . Animals placed in an environment that severely restricts sensory input develop many autistic symptoms such as stereotyped behavior, hyperactivity, and self-mutilation. Why would an autistic and a lion in a barren concrete zoo cage have some of the same symptoms? From my own experience I would like to suggest a possible answer. Since incoming auditory and tactile stimulation often overwhelmed me, I may have created a self-imposed sensory restriction by withdrawing from input that was too intense. Mother told me that when I was a baby, I stiffened and pulled away. By pulling away, I did not receive the comforting tactile input that is necessary for normal development."[27]

Grandin reviews animal studies which show that such early deprivation permanently alters brain structure. Children artificially deprived of social contact in the gulag of Romanian orphanages may similarly (but fortunately not always) react with permanent deficit. Grandin also reports that the cerebellar deficits that underlie autism have left her with no sense of rhythm. She cannot clap in time to music. After all her self-training, she still interrupts others' talking. She never had the chance as a baby to move in time with adult speech, or engage in a give-and-take of "conversational" babbling, hugs and glances. Her mistiming, as well as sensory overacuteness, may also have isolated her during the crucial first years when babies' instincts are primed to learn that other people exist. Even now, with a Ph.D., a university post, and extraordinary talent as a visual designer, Grandin says most of her social contacts are with autists or with the livestock she studies. As for other people, she makes a conscious effort to understand them like "an anthropologist on Mars."[28]

Objects in space. Is imprinting on one's caretakers just a subdivision of more general comprehension of objects in space? Not really. Children do sometimes smile at a moving object, like a bright mobile, but it is nowhere near so specific as the social smile. Understanding objects goes on in parallel to the social development of imprinting.

Children have some concept of objects even in the first two weeks of life. New babies cannot reach out and grasp in a coordinated fashion, so early psychologists supposed they had no mental conception of objects either. Then it turned out that a newborn baby presented with a hologram image of a cube flails more toward the side where the cube seems to be. Much more telling, if the baby's hand goes right through the space where it "perceives" the hologram, it is likely to cry. An infant watching videos, while sucking on a bottle wired for recording, can answer more subtle detailed questions. If a ball rolls behind a black screen and comes out unchanged, the child is unsurprised. If the ball comes out smaller or a new color or reverses direction or turns into two balls, the child is startled enough to suck harder for a moment. In other words, even new babies come equipped with some of the same expectations as adults about the behavior of objects.[29]

At around four months, babies progress in the course of a single day from random flailing to a directed, ballistic reach that is quite likely to connect. This becomes the great new game. But as Jean Piaget noted in his diaries of his own children's development, the babies are not far along with cause and effect. They repeat successful action as though to magically produce the result again. If pulling a string jiggles a mobile, they will tug the string even when the mobile is visibly disconnected. Little Laurent Piaget swung his hand while holding a cigarette box, then swung it again after dropping the box, as though the box would reappear. At this age, an object that disappears for more than a few seconds seems to go out of existence. The child does not look for it, as though she has no concept of its continuation.[30]

This bald statement needs qualifying, and has been much qualified. Tests with videos show that children have indeed a concept of object permanence. Baby monkeys totter round behind a barrier long before they lift a cloth to find concealed objects.[31] Four-month-old babies cannot yet crawl, so they do not have the monkeys' option. Whatever it means, they do not actually hunt for hidden objects until about seven or eight months. Then (again rather suddenly) babies do indeed lift cloths and boxes to reveal a hidden toy. But they make a strange mistake. If they find a toy in one hiding place and then, in full view, you put the toy in a second hiding place, they go back to the first! They show every sign of disappointment when the toy is not there. Piaget's interpretation was that the child now knows the object continues to exist in time but cannot track its trajectory in space. There are other interpretations, such as that the child again repeats a "magic" act: lift that cloth to make the toy reappear. It worked once, so keep trying the same move.[32]

Adults occasionally take the same approach. (I know my glasses *ought* to be here, even though I looked five minutes ago and they weren't.) Whatever the theoretical gloss, children do not manually track an object through multiple hidey-holes until they reach about one year old.

Production and comprehension. One theme links all these

Figure 16. Donald the boy and Gua the chimpanzee were brought up together. Donald, at 11 months, is just finishing his "embryonic" year. His motor control resembles Gua's shortly after her birth. No question who gets the hat. (After Kellogg and Kellogg, 1933)

realms, social, mechanical, and linguistic: a baby seems to understand stimuli long before it can do much about them. Socially, the newborn responds to and recognizes people for weeks before developing a magical smile that draws adults to cuddle it, and long, long before it is crawling or walking at ten months or a year, with the first wobbly means of reaching others by itself. A baby cries at waving its hand through a hologram months before it can aim that hand at a silver rattle or a rubber frog. It knows sounds, rhythms, even words before its lips and larynx can squeak out a single word. It looks as though that extra "embryonic" year leaves us at the end with only a little more motor control than a newborn ape. What we have gained is a year of maturation surrounded by the world of objects and language and other people—an extra year to begin understanding the world.

And yet, how distant is the one-year-old's mind, "innocent of past or future." The poet Pamela Wilkie writes for her grandson:

> We seem so close
> like a pair of skaters round our pond.
> From your one year
> To all of mine,
> no distance really.
>
> But now you must go home—
> thousands of miles.
> How can I smile so wide
> and wave enough
> for such a leaving?
>
> Worlds apart, my words
> and your perceptions.
> Innocent of past or future
> you cannot know
> that you are going to forget me,
> your infinite, engrossing Now
> rises like valley mist between us—
>
> you must travel further
> than tomorrow,
> you have to navigate all the years it takes
> to grow and discover what I mean
> by my secret tears at the gate.[33]

One to Five

A one-year-old child begins to walk, and most begin to talk. They build up from one to two words to sentences. They have a feel for linguistic structure. They use even more regular grammatical forms than those they hear around them. Children say "he goed," "he seed," extrapolating from regular verb forms to the irregular ones. There is also profound language hunger. Toddlers demand talking, they demand listening, they play with words, and learn as much as one new word every hour of the day. The experts argue about how specific is the innate module for learning grammar, whether it is purely about language or overlaps more generally with the logic of understanding actors and agents in the world around. Either way, it is something that a child works at—a felt need, like being allowed to walk or cuddle.[34]

We have already seen that older great apes make the long step to self-recognition. Late in the second year, the human child toddles past a mirror, stops, touches it, tests it, and begins to make funny faces. Mirror recognition goes with a set of other behaviors: the beginnings of embarrassment and guilt.[35]

Socially the child under two is still attached and clingy. It may play well with others and may cope well with having been long in childcare. However, it is not yet actively trying to separate from its main caretakers. Freud called this the "oral stage," from birth to two years old. His language can still be somewhat shocking, with its equation of virtually all pleasure with sex. However, his picture of the infant's joy from sucking a nipple or thumb (or even big toe) and from the contact of skin to skin still rings true. Freud emphasizes the continuity between child and caretaker at this stage. The child has little need to make distinctions between itself and what it loves.[36]

All this changes as the child discovers that it has, or is, a self. Enter the terrible twos. The two-year-old is often ready to withhold whatever it can. If its battles are not over potty training, they are over sitting still in the high chair, or eating what is offered, or wearing tights on a cold day. The child now talks, often starting with 'No!" Not all children rebel, of course. Some sail through with perfect charm and poise. Others push the conflict to the point of tantrums.

It would be interesting to know if the terrible twos have anything to do with a stage of weaning conflict in our earlier ancestors. Modern humans give no clear indication. In many cultures people wean at about two and a half, sometimes sending the baby to a grandmother or aunt to get some relief. The gamut of weaning, though, goes from birth to about five years old, with little indication of any biological norm.

The year from three to four brings a disjunction between two of the great mental steps we pointed out in apes. The younger child has a sense of self and of its distinction from the rest of the world, as shown by mirror recognition, by the subtle signals of guilt and embarrassment, and of course by the flood of stories about "Me, I." What the child lacks is a clear picture of others' beliefs, knowledge, or capacities as different from its own. It knows about other people's existence and their connection to itself, of course, and has known this since birth, but not until around four does the child imagine that other people have their own knowledge and ignorance. Then it starts to "mindread." A suite of further social skills appears around the same time.

Freud called this the Oedipal stage. Most children masturbate (as they have before) and are clearly orienting to the genitals. They are also much more aware of the gender of those around. Little girls begin to flirt with adult men, their father and others. Less markedly, little boys become much more butch in their play. I have wondered how much this is a very general increase in social and bodily consciousness. It was at that age that my own children began to spontaneously report whether a friend who came over to play was black or white. (I do not know of studies on this, or if such studies would show anything about children in general, or only the immediate culture.)

Of course, much much more happens at this age—stories and games, nursery school and adventures and imagination. The fact that adult apes have the mechanical and social skills of a human four-year-old does not mean that apes will ever listen to a fairy tale.

Five to Puberty

Later childhood is the second great peculiarity in the life history of humans. The first is our early birth, with quarter-formed brain. However, the years from one to five are not so very different from the prepubertal years of other apes. Apes are slower to reach the stages of self- and other-recognition, around three and six, respectively, as compared to two and four. Their emotional independence is not much different at all. They deeply depend on their mother up through weaning at four or five, and remain largely with her, though playing with others, until females reach menarche at about eight. Males approach physical puberty then, too, but are not socially mature until around twelve, since they cannot compete until fully grown with the mature males for their place in the band.

We humans have stretched out the apes' three-year post-weaning period into six or seven years of childhood, when we play with our peer group and in most cultures help with chores around the home or farm or hunting camp. Freud saw this as the time we become human. Later childhood is when we develop an abstract consciousness of right and wrong that transcends the immediate personalities involved. We learn the moral rule of our tribe.[37]

We are only just beginning to ask about the role of the juvenile period in other primates. Is it partly a way to exist in a different ecological niche from the adults, feeding on branches too thin for adults' weight, so there will be some food left for the young? In humans, at any rate, we use it for learning. Oddly, Freud may have actually underestimated the sexual interest of children between the ages of five and ten, by drawing on his experience in straitlaced Austria instead of, for instance, the children's sex-play huts of the

!Kung San. However, he was right in pointing to this as the time when we shift to rule-based learning of our culture—and it seems to be the time that we take longest, compared to the growth of our kin.[38]

Puberty. The next great reshaping of anatomy comes with puberty. The sequence of action, as always, is steered by evolution. In boys, the genitals begin to take on adult size and even activity before or in the first stages of the spurt in growth. For girls, some physical growth usually comes first, then menstruation, then the first fertile, ovulatory cycles. This makes reproductive sense. A precocious boy who finds a willing woman might actually father a child. A girl who conceives when her body is not grown risks death in childbirth.

In the hunter-gathering life where we evolved, a girl might need fat stores as well as adequate bone size to keep herself and her baby from starvation. In Chapter 4 we talked about the evolution of fatty breasts and buttocks as signals of desirability. Rose Frisch launched a much-disputed theory that menarche does not come until a girl's body reaches a critical proportion of fat. One suggestive finding in support of this hypothesis is that leptin, a hormone which controls the set-point for appetite, can also trigger female reproductive maturity in mice. In other words, the signal that you are fat enough to cut down on eating may also help register when you are fat enough to be fertile.[39]

The age of menarche has been falling for as long as we have records. The oldest survey is a population of Swedish girls in the early nineteenth century. They did not menstruate, on average, until eighteen or nineteen. Slightly later records from Britain give the average at sixteen, but with a difference of a year by social class. The age fell decade by decade in Sweden and Britain and is now down to about eleven and a half years in developed countries. So far as anyone can tell, the reason is entirely nutritional. We are better fed now. The alternatives are clear in anorexics who starve themselves and in athletes who are in fine muscular condition but burn up their body fat.

In both groups, menses often start but then stop for the duration of training or starving.[40]

The mood changes of puberty are obvious in our culture. Are they biologically based? Is it possible to train the young to keep the docile tenor of childhood, even while their bodies are being remolded? In the 1920s people pointed to Margaret Mead's *Coming of Age in Samoa* as the perfect counter-example. Mead's Samoan confidants claimed to happily take lovers, avoid pregnancy by adolescent sterility rather than conscious choice, and at last settle into marriage with no trauma at all. Derek Freeman's onslaught on Mead's view implies that the story was largely wishful thinking; but Bradd Shore, another expert on Samoa, suggests that the teenage informants were talking about real behavior, not the uptight cultural norms that Freeman cites—perhaps with some wishful thinking on their own part as well as Mead's.

Perhaps adolescence is, everywhere, a time of potential hormonally provoked chaos. Of course, for some peoples, and people, initiation ceremonies with bodily deprivation, scarification, hallucinatory brain-washing, late-night homework, and examinations successfully channel adolescent energy into forms approved by adults.[41]

> Fourteen-year-old, why must you giggle and dote,
> Fourteen-year-old, why are you such a goat?
> I'm fourteen years old, that is the reason,
> I giggle and dote in season.[42]

Babies' Gender

All through infancy, childhood, and adolescence, society teaches us gender roles. If a pregnant woman has amniocentesis, the "gendering" of her infant starts even before birth. "Oh, you are expecting a little girl! How lovely!" "Oh, a little boy! You must be so proud!" Other primates care about gender, too. Mothers and troopmates inspect and finger newborns' genitalia and treat babies differently by sex.[43]

In a justly famous experiment, mothers coming to an office were approached by another mother in the waiting room. The stranger just *had* to go to the toilet—would the waiting mother mind holding her six-month baby, little Beth, or Adam, for five minutes? The waiting mother of course agreed. She took the baby and offered it toys from the assortment spread about. Beth in her pink dress was likely to be given a doll. Adam in his blue overalls got the train. Beth was held close and cuddly, Adam farther away. Then Beth's or Adam's mother came back. The helpful mother went in with her own child for her interview. Among other things she was asked if she thought little boys and girls should be treated differently. (This was in the feminist seventies.) All the mothers said, "Oh, no, children should be treated just the same, to develop as they please."[44] The joke is that Beth and Adam were the same (male) baby.

In our own culture everything from decorations of the baby's crib on up to teachers' treatment in high school reflects adults' gender expectations. Cultures differ. Gilda Morelli studied forest-living pygmies in Zaire, and the Bantu households with whom the pygmies often live. Pygmy hunter-gatherers made no difference in the amount or kind of requests for help they directed toward small boys and girls. Among the Bantu, settled agriculturalists, girls were getting demands for help in chores eight times more often than the boys—at the age of two and a half![45]

In short, the development of adult gender is a hugely interactive process. It is as good an example as any we have of the interplay of learning and instinct. To claim too much for either flies in the face of common sense and common experience. The only people who really have much right to testify as individuals are intersexuals and transsexuals—people who, for one reason or another, were subjected to biological sex and cultural conditioning at cross purposes. For the rest, culture mainly reinforces—or over-reinforces—an initial bent.

It does seem odd that feminists ever hoped for androgyny. There are urgent reasons to do away with prejudice that belittles the worth of one sex or the other, but it is not likely that the two sexes will ever be just the same. Maybe we don't need to be. Maybe individual freedom includes the right to be comfortable with one's own personal combination of sex *and* gender.

We have traced the development of human beings from a blob of fertilized egg to a creature capable of surviving the turmoil of adolescence. We followed the growth of mind from the reflexes of love, through knowledge of self and others, to the exalted roles of teacher and student. Throughout, at every step, it is amazing that such organization exists in any organism. Of all the most beautiful and most wonderful forms that have been, and are being, evolved, we ourselves are surely filled with glory and wonder.

IV
The Age of Humanity

Figure 17. It is highly speculative when midwifery and cooperative care for older children began, but both are older than our own species. Here, an imagined birth in *Homo erectus,* whose brain was already larger than those of apes, with a probable need for more "embryonic" newborns and a long childhood. Most births were at night, as in modern apes; humans show the same tendency. (Body form after J. Gurche, *National Geographic,* February 1997)

16
Feet First, Brains Later

Now let us turn again from this story of individuals to the story of evolution—evolution as it merges with history. Our transition to human status has the elements of a good fairy story. That does not mean it is a false figment of the imagination. It means that it has narrative structure.[1]

Classic fairy tales go like this. The hero is a frog, an ugly duckling, a widow's boy, a drudge to haughty stepsisters, an orphan raised by gnomes. His father's or her mother's death forces the hero to begin a lonesome journey. Trials reveal both inner strength and vulnerability: she must sort the grain from the ashes; the gnomes scheme to betray him. A magical donor endows him with a gift of power—a magic sword, a beanstalk, a pumpkin coach. With the gift, and true bravery and sincerity, he kills a volcano-tempered dragon, brings the giant crashing from the sky, and she vanquishes sisters who would cut off their own toes and heels to walk bipedally in glass slippers and marry the prince. Our hero and heroine emerge transformed and triumphant. They then live happily ever after, though in adult versions they may succumb to hubris and meet a tragic end.[2]

Evolution by natural selection lends itself to this format, whether you think of humanized apes or a lungfish crawling out to start life on land. Only individuals who are both gifted and lucky survive the challenges of any generation. Each really new species springs from fit and lucky individuals who are physically transformed away from all their relatives, as they occupy a new niche in a new landscape. The story of trial and transformation is a form that fits. Of course, it resonates better when we are the heroes.

When humankind is the hero, the narrative is not fairy tale but myth. Human consciousness becomes a terrible gift which challenges the power that gave it. We rise armed with Promethean fire, the taste of the forbidden fruit of knowledge fresh in our mouths. That moment of transformation lurks even in the cautious catalogue of human fossils. Evolution's story, as outlined by the theory of natural selection and the shards of bone in the rocks, is still a tale of the gaining of knowledge, and the power that has come from knowledge. It is the transformation of the nonhuman to the human, and perhaps the human into some future Other.[3]

Australopithecines

Darwin (as usual) got the narrative of human evolution about right. His version was that we left our long-term home in the African forests to begin our lonesome journey into the savanna, which then selected us to walk on our hind legs. The challenges of savanna carnivores and competition within our own species led to brain growth. Big brains endowed us with the gift of group morality, which let us cooperate against carnivores but also make war on one another, using weapons and tools. From there, group support and transmitted material culture led on to the triumphs—or complexities—of civilization.[4]

However, Darwin's version was often challenged or eclipsed—especially by the believers in Piltdown Man. Piltdown Man was actually several fragments of skullcap and jawbone found separately around 1912 by Charles Dawson, a local solicitor, in the gravel of a stream bed in Sussex. I live, at times, in Lewes, near Piltdown Village. When

I first arrived, the Lewes town historian was still sore at Dawson for buying a house in the name of the Historical Society, then living in it himself—his reputation was apparently not as high locally as it became in the world of anthropology.

Leading anatomists of the day pronounced Piltdown the key to human evolution, the crucial missing link. Piltdown suggested that humanity emerged brain first. Its skullcap had an almost modern aspect, while its lower jaw remained remarkably apish. This seemed eminently right and proper, a tribute to large brains and intelligence as the defining trait of emerging humanity, especially since the first human was so clearly an Englishman.

Not until 1950 did Kenneth Oakley, using fluorine dating techniques, finally establish that the skull looked human because it was, in fact, a medieval human. The apish jaw belonged to a modern orangutan. Someone had broken off the region where they should have fit together, then stained the fossils with iron and expertly filed down the ape teeth to sufficiently human form to pass.[5]

Why did it convince people for so long? Well, there may have been some higher level skullduggery than Dawson's, and eminent careers were staked upon it. Why was it finally tested and found out? Because by then, it was a major embarrassment.

Twelve years after Piltdown, two crates of fossils from the Taung limestone quarry were delivered to the home of Raymond Dart, an anatomist teaching in Johannesburg. He was, in fact, about to host a friend's wedding party, struggling into a stiff-winged collar and mothballed morning coat for his role as best man. He could not resist peeking into the crates. He tore off his collar and wrenched open one box, then the other.

As soon as I removed the lid, a thrill of excitement shot through me. On the very top of the rock heap was what was undoubtedly an endocranial cast or mold of the interior of the skull. Had it been only the fossilized brain cast of any species of ape it would have been ranked as a great discovery, for such a thing had never before been reported. But I knew at a glance that what lay in my hands was no ordinary anthropoidal brain. Here in lime-consolidated sand was the replica of a brain three times as large as that of a baboon . . . The most impressive feature of this endocast . . . was the marked distance separating two well defined and unmistakable furrows at the back of its outer surface . . . [so much] expansion had occurred between the [furrows] that they were separated by a distance *three times as great* as any existing endocast of a living ape's skull, whether chimpanzee or gorilla. So . . . [I knew] instantly that the creature whose skull could give a cast of this sort must have been at least three times as intelligent as any living ape. I stood in the shade holding the brain as greedily as any miser hugs his gold, my mind racing ahead . . . Darwin's largely discredited theory that man's early progenitors probably lived in Africa came back to me. Was I to be the instrument by which his "missing link" was found?[6]

Then the distraught bridegroom tugged at his sleeve: "My God, Ray, you've got to finish dressing immediately—or I'll have to find another best man!" Dart told the story again at the age of ninety at the great 1984 "Ancestors" conference in the American Museum of Natural History in New York. There were assembled the world's leading paleontologists and untold precious fossils, including the Taung child's brain and face and jaw, displayed to the public under bulletproof glass, more rare than the rarest of diamonds. Dart thought such a brain "laid down the foundations of that discriminative knowledge of the appearance, feeling and sounds of things that was a necessary milestone in the acquisition of articulate speech."

The connection to the spinal cord lay beneath the skull, not behind it as in apes. The creature stood upright! After seventy-three days, working with tools that included his wife's sharpened knitting needles, he freed the skull's face from a second block of matrix. It was a juvenile, its milk teeth still in its jaw. It resembled a modern human six-year-old. The child was actually even younger, for australopithecines developed more rapidly than we do. Dart named it the "Southern Ape from Africa": *Australopithecus africanus*.[7]

The Taung child is still controversial. Was its brain actually startlingly different from that of other apes? The human anthropologist Ralph Holloway supports Dart's view that the furrowing of the brain suggests a clear reorganization from apes, even so early. Dean Falk, another expert who specializes in brain anatomy, argues that the cast has been misoriented, and its brain was wholly apish. This is one of the many jigsaw pieces in the argument over when, and how gradually, began the transition to modern humanity.[8]

Just possibly two holes in the top of the skull are traces of the child's last moments. They fit the talons of the African martial eagle, which still seizes young baboons and similarly punctures their skulls. Perhaps an eagle carried its victim to its nest in a tree above a limestone sinkhole, ate it, and dropped it with other prey. Water dripped down into the sinkhole cave, gently fossilizing the assemblage of bones of the eagle's offerings, replacing the soft brain with eternal stone.[9] Two and a half million years later, the first of only five endocranial casts ever found in Africa was spared by the limestone quarry dynamite and sent to one of the three or four men in the world capable of recognizing its significance.[10]

As more and more *Australopithecus* fossils emerged, a shocking conclusion emerged with them. In fact, they had brains no larger than modern apes—only about 400 to 500 cubic centimeters (cc). But they walked almost like ourselves—perhaps somewhat less efficiently, perhaps even more so. Their flaring pelvic bones relayed less pressure than we now put on the head of the femur—the fragile hip joint that surgeons replace with ball joints of stainless steel.[11]

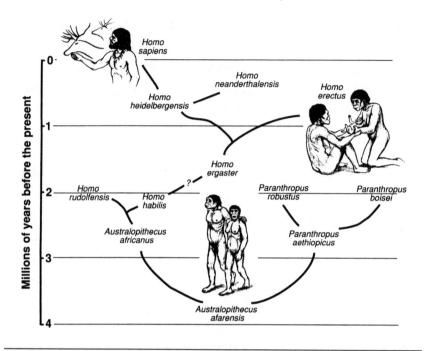

Figure 18. The route to modern humans was a bush of many species, not a straight-line progress from one to the next. Each date, name, and branch point is still being argued, but the increase in brain and behavioral complexity is clear. (After Tattersall, 1995)

A climax came one evening in 1976, when Andrew Hill, a paleontologist, tried to dodge a lump of elephant dung tossed at him by David Western, an ecologist. Hill fell over and landed on his knees, his nose near the floor of a dry river bed, in the low cross-light of sunset. He noticed odd dimples in the rock. They were fossil raindrop impacts 3.6 million years old. Hill and Western told Mary Leakey, chief of that site near Laetoli in Tanzania. As she uncovered more of that dimpled layer, they found the meandering tracks of animals from hares to rhinoceri—and then, miraculously preserved, the footprints of australopithecines.[12]

Two—or maybe three—creatures had walked across a new fall of volcanic ash, gluey with rain. One was larger, one smaller. The two trackways were only 25 centimeters apart, almost too close to be simultaneous, but the creatures matched their stride to each other's. Whoever they were, they walked together hand in hand, or more likely with one's arm around the other's shoulders, as couples do today when bigger and smaller promenade. The larger one's feet slipped a little in the soft ash—or else a third followed behind, stepping in the big one's trail of footprints. Their feet fell as ours do, heel first, then rolling forward to the ball of the foot, with a pushoff from the big toe. Then the volcano erupted again, spewing out a new ash-fall that preserved the footprints for the next 3.6 million years.[13]

Was Lucy a male?   Australopithecines are the earliest ancestors we know after the split from the apes. Lucy, the nearly complete skeleton found in Ethiopia by Don Johanson, dates from 3.2 million years. Lucy was a tiny creature, not four feet high, weighing about seventy pounds. Before her came other species, still known only from fragments of jaw and teeth and bits of shinbone, dating from around 4.0 to 4.4 million years ago. After Lucy comes Son of Lucy, half a million years later and half again as large as his "mother." Later australopithecines clearly split into many species. The Taung child and other small, gracile fossils lived 2.5 million years ago. A separate, large-bodied lineage with several branches lasted up to a mere million years ago. The various species of australopithecines had a long run for their money. A bipedal run.[14]

Was Lucy actually female? Johanson's team admits they cannot tell from the pelvis. They would not even expect to tell. Australopithecine brains were ape-sized, only around 400–500 cc; therefore females did not need enlarged birth canals. Lucy is judged female just because she/he was petite. Johanson and many other archaeologists believe there was just one species in any one place 3 million years ago, with big males and small females, like gorillas. It could be, though, that there were at least two species, one big and one small. There is a reasonable amount of variation in the material, enough to allow for males and females, though with nearly equal-sized sexes within each species. The argument over the number of species has been rumbling along since the very first assignment of a name to the Ethiopian finds.[15]

The choice could shift all our ideas of australopithecine behavior. If the males were half again as large as females, it implies a mating system in which a male mates with many females and competes with his rivals through muscle power—as solitary orangutans, harem-forming gorillas, and multimale savanna baboons all do. The upshot is he can physically thrash any female who opposes him. In contrast, if there were two different species, each with same-sized sexes, that implies a much more equal mating ratio between the sexes. It does not tell for sure what kind of society they formed, though probably something like monogamy.

Guess which version appeals to anthropologists who are also feminists: males equal-sized with females or hulking heavyweights? Now come back to those footprints at Laetoli. Sculptor John Holmes has reconstructed their makers with uncanny immediacy for the American Museum of Natural History in New York. A big male has his arm protectively around the shoulders of a tiny female. She has, apparently, just caught sight of you, the viewer, through the glass of the diorama, her eyes wide and her mouth opening in alarm. Perhaps she and the male, who is also apprehensive, are reacting to the blasted ruins of the landscape that was their home, blanketed with grayish-white ash that will hold the trail of their footprints for all future generations.[16]

"The expulsion from Eden!" objects paleontologist Adrienne Zihlman. She contests the big male/small female hypothesis: the footprint makers could have been a large-species parent and teenage offspring. To Zihlman, the protective gesture of the male is outrageously sexist. Actually, it is John Holmes's convincing artistry that most infuriates her, because he makes us believe what we see. Zihlman protests: "The 'Second Sex' continues to be portrayed as the handmaidens of society's male players, who, if we are to accept current theories, were out there innovating and making hominid evolution happen. What is particularly ironic is that this back eddy runs counter to the increasingly powerful current of research demonstrating the centrality of women and female primates in social life and the evolutionary process. But the concept of women in evolution remains encased in the glassed-in Old Testament diorama held down by a Paleolithic glass ceiling."[17] There is no resolution at the moment for Lucy and her/his ilk. The fossils remain enigmatic.

The australopithecine lineage starts us out on our journey. Around six million years ago the common ancestor of chimp, bonobo, and australopithecine lived as it always had, foraging in dense tropical forest. There are apish fossils from 6 million years ago, and then a gap, which has dramatically shrunk with the discovery of *Australopithecus* (or *Ardipithecus*) *ramidus* at 4.4 million years ago. *A. ramidus* consists of teeth and a bit of skull base connected by intellectual ectoplasm to fragments of arm bone. The skull apparently sat on top of the neck, not jutting out in front—*ramidus* was starting to stand upright. Then, at 4.2 million years, the somewhat different *A. anamensis* has bequeathed us an undoubtedly upright shinbone.[18]

A world-scale change began around 5 million years ago. The Himalayas were rising. (They still are.) A land bridge through Panama closed off the tropical link between the Atlantic and Pacific oceans. World climate grew cooler and drier. In Africa, savannas spread at the expense of forest.

Many species evolved to explore the new habitat. Savanna apes ventured out from the forest among many other animals. At three to four feet high, they were dwarfed by the large carnivores and elephants and huge baboons that, when they too reared up on their hind legs, could look an ape in the eye. The savanna ape was setting out on a dangerous journey indeed.[19]

Why did they walk erect? Probably for many simultaneous reasons. Bipedalism is more efficient than ape-type quadrupedalism for covering long distances; the simplest answer is that it was a good way to get around. For modern apes, bipedalism is most useful when they feed. It is a good way for a heavy animal to balance on a branch when picking fruit from the twigs of a branch above. Lucy still had shortish legs, swiveling shoulder joints, curved fingers and toes which also had some capacity to curl downward. S/he still climbed trees to pick fruit and take refuge from carnivores. For that matter, people still do. The founding vice-chancellor of the University of Zambia told me of the night when he and his friends lost their way trekking across the Luangwa Park from their rural homes to boarding school. They all slept in one tree. He was the smallest, so he got the lowest branch. He tied himself on with his belt, terrified of falling in his sleep. Lion prides roared to one another north and south and west, but the little boy kept telling himself that lions usually don't climb trees. Then he heard a leopard cough.[20]

Walking erect let the man-apes scan above the long grass for predators. Many primates, even ringtailed lemurs, stand up on their hind legs to gain a long view. Walking erect might also have kept that growing brain cooler, catching less sun and allowing better radiation of heat, especially if the skin was already losing its fur coat. That in turn could be crucial for brain growth.[21]

Hero stories. Of course bipedalism freed the hands for doing other things, such as using tools and weapons. Australopithecines' canine teeth were much smaller than apes', relative to body size. Perhaps males threatened one another with their hands, and what they held in their hands, instead. There is no evidence of shaped tools, but they could have flung stones or flailed with branches like modern chimpanzees. The quarterstaff then is the oldest hand-held weapon.[22]

In the 1960s and 70s arose the vision of "Man the Hunter." It was clear all along that most human hunter-gatherers eat far more vegetables than meat, as do all other primates. But hunting was a splendid "gift" for the emerging hero. He could cooperate with others of his band. He could develop foresight to plan ambushes. He could sharpen his wooden spear and fling it with masculine precision. After the hunt, he could march back bipedally with an antelope haunch over his shoulder, to share with admiring females, who, without him, merely gnawed on nuts and roots. Various authors have since tried to demote Man the Hunter into a mere scavenger, but even so, he had to drive large carnivores from their kills—quite as heroic, and still a male game.[23]

The most fundamental hero story of all, though, goes right back to Darwin and the outrageous statement that "man [could] have become as superior in mental endowment to woman, as the peacock is in ornamental plumage to the peahen."[24] If big australopithecine males were under intense sexual selection to gain little Lucy-like females, and if intelligence mattered for success, then sexual selection would act on a harem-master more acutely than on his females.

Heroine stories. Then, in 1981, two revisionist scenarios appeared almost simultaneously: Adrienne Zihlman's *Women as Shapers of the Human Adaptation* and Owen Lovejoy's argument that the "unique sexual and reproductive behavior of [wo]man may be the sine qua non of human origin." Both of them began by saying humans stand apart from all other primates in the degree of our food sharing. Chimpanzees do share meat. Chimps, tamarins, and other primates share other food that is hard for the young to catch and extract, such as fast-moving insects or hard-shelled nuts. But we share much, much more often. Food sharing underlies both the cohesiveness of our society and our eventual division of labor.[25]

Zihlman pointed out that people, like all other primates, depend on gathered plant foods. Bipedalism would let people carry vegetables as well as meat to one another. The first tools, then, were containers made by women—carrying vessels to transport food like the big gourds people still use to hold gathered seeds or water. Little children might be carried in a kaross, the flexible leather sling that is the all-purpose tool of !Kung bushwomen. Before inventing the kaross, it would have been a huge help to walk along with at least one hand free to hold the baby. A bipedal woman with both hands free could carry a stick in her other hand. Not a spear—a digging stick would be much more useful than spears.

Males in Zihlman's scenario would learn the skills of gathering from their mothers. As they, like their mothers, began to share food, they would primarily share with mother and siblings, then with the wider group, including potential mates. A widely polygamous system within smallish communities would mean that any child might be the male's offspring, so he would have an incentive to share out fairly evenly among the clan. Of course he could go off hunting with the boys, but this played a relatively minor economic role, perhaps a trivial one in the evolution of either tools or intelligence, compared to the roles of females. Zihlman's gift was still food sharing, but food provided by an ever more intelligent mother.[26]

Owen Lovejoy started from the same set of facts. Here were bipedal, small-brained creatures with small canines. He argued that the crucial constraint was the long period of infant dependency and mothers' need to carry and care for vulnerable children. Modern apes have a five- to eight-year interval between births, and mature at about age eight. If human females did not reach menarche and ovulation until still later, and the children were even slower growing, how could they raise enough young to persist? He concluded that some sort of improved child care must have emerged to increase survival and shorten birth intervals. He put this change early, even before our heroine came down from the trees.[27]

One way for the female to gain food might be to divide up foraging space. The females could feed in smaller ranges; the males could range more widely. They would not at first seek different diets; they would just have spatial separation. Why would a male go the extra distance? Because it reduced competition with his own monogamous mate and offspring. Given monogamy, there would be strong selection for bipedality and carrying so that he could bring home food. In other words, the male would become like paired birds and carnivores, only instead of a crop full of seeds or a stomach full of meat to regurgitate, he carried veggies home by hand.[28]

Meanwhile, the female progressively concealed her ovulation so as not to inflame all available males into brief bouts of competition. But she was ready to mate with her own man whenever he came home. Both female and male acquired permanent attractiveness— big breasts, big penis—which continually advertised both sexiness and reassurance to the mate, as in mated birds that join in "triumph displays" of greeting whenever they have been apart. For Lovejoy, division of labor, concealed ovulation, and the nuclear family thus predated even bipedalism. He saw the gift that led toward humanity as monogamy.

Feminists were outraged. The idea of hulking polygamous males was horrid, but this kind of monogamy seemed even worse. Though Lovejoy's scenario started with the needs of females and young, he relegated females to staying home and washing the dishes several million years before dishes were invented. All the innovations— bipedal ranging, hunting, provisioning—came from males, while the females merely became sexier to keep hubby in line. This did not sit well in the consciousness-raising 70s.

An australopithecine scenario. Lovejoy's initial premise that human females need help with children is familiar enough. I also made the point back in the "human ape" scenario that closed Chapter 8.

The difference is that I called it child care, and said a mother could get help from friends or kin, not just from a monogamous male.

So what is my scenario for australopithecines? I have no expertise in the fossils. Fossil experts are well and truly divided; it may be decades before they reach consensus. I do note that we are moving to recognize more and more coexisting species at each stage of human evolution. Our lineage at each stage was an evolutionary bush, not a evolutionary palm tree with a spindly trunk yearning upward toward eventual humanity. So a priori, we should expect that more than one australopithecine species roamed the savanna in the time span of 4–3 million years ago.[29]

However, on behavioral grounds, I would opt for big males and small females and a medium level of promiscuity, rather than Lovejoy's monogamy. If the savanna apes lived in a chimplike band of related males, jointly controlling access to the band females but mating with relative freedom, then you would expect males to be larger, as in modern chimpanzees and bonobos. It is the gang-war behavior of chimpanzee and human males that most influences me. I see the patrolling platoon of males defending their females as the deciding factor that carried through from our common progenitor. This is not simply because of relationship. Peaceful bonobos in their rainforest are just as close relatives as common chimps. However, australopithecines ate open-country grasses, and possibly meat, as well as small, tough fruits and leaves, though their teeth do not have the scratches that would come from grit-covered roots. Their diet suggests a fission–fusion society, like savanna and forest-edge chimps today—a hint they were also prone to chimp-type warfare.[30]

If australopithecines lived in fission–fusion, male-bonded groups like chimps, the females may well have mated promiscuously. The males would have been at least as fatherly as male chimps, tolerating and sometimes playing with the young and letting hero-worshiping bigger boys tag along. As males were related to one another, all children would be somewhat related to every male, while females' behavior would mean that almost any male could be a father.

One possibility is that concealed ovulation was first useful for attracting many mates, to tempt them to share food with any female who might in turn share her own favors but without the time limits of visible estrus. Males do hunt more than females, in chimps, baboons, and every human group. Concealed ovulation would give a female shared meat and other food, as well as the chance to confuse paternity, avoid infanticide, and have several goes at picking the finest fathers' genes for her children. This, after all, is a ploy of both modern chimpanzees and bonobos, and a possibility for many individual modern men and women, though few whole human societies are set up as "group marriages."[31]

Only later, long after the evolution of both concealed ovulation and bipedality, would more monogamous bonds lead to private sex and a father's paternal obligations to his own beloved young. This, in turn, could invoke severe conflict both between the mates and within the female's own psyche over how much she was committed to stick with her first or official mate and how much she was still free to roam. Perhaps, the tendency to love one mate at a time did not come in until the evolution of *Homo*. One story for the australopithecine diorama in the American Museum that has not, I think, been seriously proposed is that the little female was in estrus and waggling her bottom at somebody who followed after, where the first had been.[32]

Homo the Conqueror

About 2.6 million years ago, somebody began to bash pebbles systematically on one side, to make an edge. Alongside the pebble tools lie skulls thinner and rounder than australopithecines'—those of *Homo habilis*, the skillful or handy man. A suite of further fossils follow with brains of 600–750 cc. Like the australopithecines, this was a bush of many populations and probably several species, not a straight line leading to ourselves as a kind of goal.

A million years later, a tall boy died in a marsh in the Turkana district of Kenya. He was buried so quickly in the mud that neither rescuers nor scavengers could reach him. His skeleton remains almost intact. He stood about 5 foot 4 inches tall, though if he had lived to adulthood, he would have reached just over 6 feet. His limbs were long and slender, like the people who live in the arid climate of Turkana today. His teeth were like those of a modern child between nine and eleven years old, but his long bones had already started closing, like our thirteen-year-olds.

In other words, his bodily growth and stature were ahead of his teeth. It seems he had kept growing through childhood, as modern chimps do, instead of growing slowly through late childhood, followed by an adolescent growth spurt, as modern humans do. On his evidence, our lengthy juvenile period had not yet evolved. In modern humans the spinal nerves serving the thorax are enlarged, to give the elaborate breath control that underlies speech. The Turkana boy's thoracic spine is much narrower. Perhaps this means he did not talk, certainly not in our long melodic phrases.[33]

There followed a species worthy of fame: *Homo erectus*, appearing just over a million years ago. The brains of these creatures mushroomed to some 750–1250 cc. They began making tools to a set patterns: leaf-shaped hand-axes, chipped skillfully on both sides and both edges. And then they set out on an epic journey that took them from Africa as far as Java and Beijing. In a cave near Beijing they launched the tradition that "everything is first invented in China": she and he domesticated fire. By that time they must have had a longer childhood and the later human pattern of childhood dependence.

Homo erectus had over a million years to glory in these feats. The species lasted, apparently, right up to 53,000 to 27,000 years ago on Java, overlapping in time with *sapiens* and several species in between. In all that time, their tools did not change any more rapidly than bones might evolve. The finest sites, like Olorgesailie in Kenya, are littered with hand-axes. Some hand-axes are huge, demanded great skill to fashion, and are very beautiful to our eyes. But there was little difference between the work of a flint knapper across the span of Africa or at tens of thousands of years' separation. The pace of change in space and time was still occurring at the speed of biology, not culture.[34]

Their brains were larger, so their life cycles also must have changed toward the modern pattern of embryonic babies and slow later childhood. When that happened, children and mothers needed higher energy food and much more help with child care. The modern human brain consumes 22 percent of our basal metabolic energy, a chimp's brain only 8 percent. It is arguable that meat eating, whether from hunting or scavenging, became much more important to a society that had to provide for hungry brains. Perhaps this was when men began provisioning their own wives in earnest, to provide meat for the milk for their children's brains.

By this stage we can see a strong advantage for economic interdependence and, with it, forms of monogamy, or at least bonding to one's own wife or wives. The paleontologist Alan Walker calls *Homo erectus* "the velociraptor of its day. If you could look into its eyes, you wouldn't want to. It might appear to be human, but you wouldn't connect. You'd be prey." Maeve Leakey (another paleontologist Leakey, wife of Richard, who is in turn the son of Mary and Louis) counters: "Anyone who has spent just a few seconds with a chimp or a gorilla knows that these creatures relate closely to us. Certainly *erectus,* with a brain capacity greater than living apes', would relate even more closely to us." *Erectus,* to me, is even more of a mystery than the australopithecines: a hugely successful species, almost human, widespread, skilled, perhaps with conjugal sharing and love, but still somehow lacking imagination.[35]

Homo Sapiens and a Couple of Other People

Archaic skulls that are not *erectus* appeared around 500,000 years ago in Europe and England. Bigger brained, yet again, within the modern range of 1,000–2,000 cc. One approach is to say that anyone who was not *erectus* is *sapiens.* Another is to say they do not fit well with modern *sapiens* and deserve a species of their own, *Homo heidelbergensis.*[36] These people began making tools not only from the core of a stone but from large flakes carefully split off from a prepared core. The remains of a *heidelbergensis* shelter, with postholes braced by a ring of stones around an oval room, date from 300,000 to 400,000 years ago at Terra Ammata in southern France, along the ancient shore of the Mediterranean (an eye for real estate, even then?). Wooden spears 400,000 years old, dug up from a German coal mine, have points shaped from the hardest part of spruce trunks and balance at a third of the way from the point, like a modern javelin. It is clear that our predecessors worked with patience, skill, and foresight, and they were hunting big game. From about 300,000 years ago, brains grew rapidly larger. By then they absolutely must have changed toward modern childhood growth and, as we shall shortly see, to tongue-wagging.

The tool types of these new people became more varied only gradually. Different sorts of pointed and bladed flints proliferated up to the cave homes of the next species, or perhaps subspecies, the people we call Neanderthals. The Neanderthals' caves and camps are littered with a real toolkit of punches and knives and scrapers. Neanderthals successfully lived at the edge of the Ice Age glaciers. They make good fictional villains, but they were actually tough, intelligent, and technologically advanced.[37]

There is huge controversy whether Neanderthals left any sign of symbols or art. They wore necklaces of punched bone, but this practice developed late and may have been borrowed from still more advanced people. Neanderthals may even have gone deep into the French caves and built structures there: a four-sided collection of stalagmites dates to almost 50,000 years BP (before the present). This means its makers used fire not just for warmth but to explore and build underground. It seems sure that the Neanderthals buried their dead, flexed in fetal posture. They placed no grave goods with the bodies, though it is possible that one burial was strewn with flowers.[38]

Meanwhile, the first completely modern humans appeared. The earliest known skulls of true *H. sapiens* go back more than 100,000 years; finds stretch from South Africa to the Sudan. I took a South African friend to the Ancestors Exhibit in 1984. He burst out laughing until the hall of the American Museum rang with his voice. Hordes of amazed schoolchildren stared at this madman who howled with laughter at a skull. At last he wiped his eyes and recovered enough to point at the fossil from Border Cave, dated to 100,000 BP. He murmured, "To think the Boers claim they were the first human beings to settle in South Africa!"

Mitochondrial Eve Meets Y-Chromosome Adam, or Not?

In 1987 Rebecca Cann and her collaborators suggested that modern humans all descend from a single woman who lived in Africa between 200,000 and 100,000 years ago. She may not have been *sapiens*: maybe her great-great-grandchildren evolved into *sapiens*. But her lineage of mitochondria crowded out all others, by chance or by genocide or simply through successful patterns of parenting.[39]

The mitochondria are the oxygen-using, energy-producing cellular organelles whose asexual clones are passed down only in the egg, and whose intolerance of other clones may be a principal reason the sperm is so small and contributes little or no mitochondrial DNA to offspring. This means that mitochondria become an almost perfect tool for deciphering relationships purely in the female line. Sperm may occasionally contribute some of their own mitochondria occasionally—perhaps once in 500 generations—or at least some recombined bits of DNA. This does not change the basic argument that mitochondria tell us maternal inheritance, but it could double Eve's estimated age, back to the time that *heidelbergensis* built that villa on the Mediterranean.[40]

If one woman's female descendants survived to become all of us, their success probably did not relate to having particularly good mitochondrial genes. Many of the variants seen are "neutral," having no chemical effect. Perhaps that first woman had other good genes, or perhaps her immediate descendants had daughters while their friends had only sons—or perhaps they were just lucky.

Mitochondrial Eve stepped into a huge controversy. The technical questions concern mutation rates, neutrality or selection on the genes chosen for study, the parsimony of taxonomic trees, and, now, the contribution of paternal mitochondria. Rebecca Cann, however, felt the strident protests went far beyond "scientific objectivity." "It is my perception that when the discussion of diversity gets too close to us, taking on a female face and a black skin, the intensity increases dramatically. Academic platitudes about the need for free and open debate about diversity fail. For example, as a result of speaking about women's genes and human origins for public television, I became bombarded with hate mail from both creationists and various political groups. One family member admonished my parents for my behavior, asking them why I would tell perfect strangers that my ancestors were black when all the family picture albums showed the real truth! After participating in a heated forum on human evolution at a large museum, I was taken aside by a trustee and told that I should learn to respond to hostile questions in a more feminine manner."[41] The major view opposing the Eve hypothesis proposed that *sapiens* emerged as a diffuse entity from *erectus* ancestors on several continents simultaneously. This huge species would have retained the capacity to interbreed because of scattered migrants between its far-flung populations. This is not how animal species normally form. Animal species come from small populations in genetic isolation from one another, who develop their own particular peculiarities. The argument was that people were special, with both a very broad habitat niche and itchy feet. On this view, the subpopulations on each continent would have slightly different characteristics, inherited from differing *erectus* populations. In short, races would predate modern *Homo sapiens*.

Note that this version says nothing about the relative abilities of the races. You can admire or despise other people whether you believe their ancestors diverged from your own 30,000 years ago (the Eve hypothesis) or 300,000 years ago (the *erectus* hypothesis). However, the *erectus* hypothesis, with its earlier split, kept picking up an emotional nuance that other people—whoever they are—may not be quite human, if racial traits predated some definition of "real" humanity. When African Eve suggested instead that we are all relatively recently related, the emotional backlash came not simply from senders of hate mail but also from within the ivory tower.[42]

Mitochondrial Eve has now been joined by Y-chromosome Adam, though they could have missed meeting by 100,000 to 300,000 years. The technical controversies continue to blaze, but at least one likely scenario is emerging.[43]

First, our species descends from a small group of people who lived before—perhaps long before—the transition to *sapiens*. Much of the success may have been happenstance, as some last names persist just because their parents have children of the sex that transmits their name. Many other people's nuclear genes can be represented still—but not their mitochondria or Y chromosomes. It is also likely that Eve's and Adam's lineages had a selective advantage, at least for their respective sexes.

Second, descendants of the two progenitors went through a "genetic bottleneck," a population of 10,000 or perhaps only 1,000 people. This group interbred, which allowed them all to acquire similar mitochondria and Y chromosomes. It also allowed them to express recessive variants of their genes that might be suppressed in a larger population; the result would be more rapid evolution. After this period of interbreeding and rapid evolution, their tribe spread to eventually occupy the world.[44]

Third, the bottleneck people lived in Africa. If Eve and Adam never met, they each gave birth to families who eventually met. Then came the bottleneck. Then, after the bottleneck, descendants of the family evolved into *sapiens*. The timing is in the right range for ancestors of early modern skulls like the one from Border Cave. From there, their progeny spread out, acquiring new mutations. Modern Africans have a much wider range of variation in both mitochondria and Y chromosomes, and other DNA, than do Asians, Europeans, Native Americans. Africans have been here longer; the rest are parvenus. Each stock that migrated to a new continent began from a small subsample of the variants that were accumulating within Africa.[45]

Women, it seems, migrated more than men. The variants of Y chromosomes tend to be found in much smaller geographical areas; the average variance of mitochondrial DNA is more widespread. What that means is that mixing of genes from place to place did not come mainly from the sorties of the men, the Genghis Khans and their hordes sweeping across continents and leaving their offspring among conquered peoples. Most men stay put, or come home when ready to settle down. Some 70 percent of human societies are classified as patrilocal. It was, and is, women who regularly move when they marry (or are taken in raids). I would argue that this has been going on since women shared adventurous ancestors with female gorillas and chimpanzees.[46]

There is a final conundrum. Those first *sapiens* were genetically and anatomically distinct from their contemporaries, but they were not apparently culturally distinct. They made the same kinds of tools as the Heidelbergers and the Neanderthals, for tens of thousands of further years.[47]

The Great Leap Sideways

Between 40,000 and 30,000 years ago, something important happened. It has been called the Great Leap Forward. People invented art.[48] Not just art, but apparently religion. And calendars. And fashion. And fancy burials. And visible signs of status. These people were the Cro-Magnons, the first true *sapiens* of Europe, who invented society as we know it.

Beads made from coral and belemnites, the fossilized shells of an extinct giant squid, date from 36,000 years ago in the Don Valley of Russia. (Belemnite beads look like translucent yellow amber, and some have natural striations that make them ripple in the light.) Before 30,000 years ago, French Cro-Magnons also carved beads of mammoth ivory and soapstone.[49]

Three burials at Sungir, near Moscow, remain intact from 28,000 BP because the mourners burned out graves deep in the permafrost. According to the anthropologist Randall White, the bones and ornaments are still charred on the back where the corpses were laid on burning cinders. They were possibly all from one family—a mature man, a boy about thirteen, a girl about eight. The man wore six polished mammoth-ivory bracelets on each biceps and six on each forearm, with alternate bangles painted in red and black. From his neck hung a pendant of schist stone, painted red with a black central dot. His burial held 2,936 ivory beads like smoothed oblongs, the boy's 4,903, the girl's 5,274. Some were strung as necklaces; some were sewn onto clothing and caps. The beads are carefully shaped, so that when you shake a necklace, each successive bead settles at right angles to the one before. White estimates that each bead took about an hour to make. By the boy lay an ivory carving of a mammoth, a massive ivory lance almost 9 feet long, and a polished human femur packed with red ochre. He wore a belt of 200 fox canine teeth. Foxes have just four canines; that represents a minimum of fifty foxes. The girl's grave held reindeer antler batons and small pierced ivory discs and lances. Were these corpses from a family of proud, aesthetically inclined aristocrats, or were they communal sacrifices in time of crisis? Whatever the reason, their people buried with them well over 13,000 hours' worth of skilled labor.[50]

This new wave of people coming out of Africa lived for a time in caves in Israel, before eventually taking over the Neanderthals' European stronghold. The Cro-Magnons lived beside the Neanderthals for tens of thousands of years before demolishing them little by little, but recent molecular evidence suggests that the two races hardly interbred. This failure of interbreeding would seem to qualify them as biologically separate species, not just separate cultures. There are still many questions, but the present opinion is that though Neanderthals hung on in Spain long after *sapiens* came, we have not inherited their genes. Truly modern humans had arrived, genocide and all.[51]

The Great Leap Forward has two somewhat confusing meanings. One is a break within Europe from Neanderthal culture. Though Neanderthals sometimes buried their dead, they left almost no sign

of the kind of art, music (vulture-bone flutes), status, and ritual that
the Cro-Magnons did, and nothing like the continuity of Cro-Magnon artifacts. There was apparently a real dichotomy between Neanderthal and Cro-Magnon mentality.[52] The other meaning is an abrupt
break within the *sapiens* lineage. Was it really an overnight transition
to modern human imagination? What could have so reshaped society
and perhaps the brain itself? The only physical suggestion I know is
Jerome Dobson's idea that the Neanderthals suffered from iodine
deficiency, giving them a skeletal form like modern cretins and a
lower intellect. If the Cro-Magnons had mutant genes that increased
their iodine uptake, they might have indeed achieved a forward step
in intelligence rather suddenly.[53] The more common argument is that
there was an equally sudden transition to symbolic language.

I find that hard to believe. The innate roots of language are deep
and complex. The same deep structures of language are shared by all
the descendants of the human diaspora.[54] The genes that underlie it
must have been in place among early *sapiens,* before the Africans
trekked to Australia, Asia, and Europe. The people before the Great
Leap Forward had the biological bases of grammar, symbol, metaphor, and lullabies—or else their descendants converged on an im-
plausibly similar set of language capacities.[55]

The apparently sudden burst of European culture could then be
the trace of travel and conquest by immigrants who flourished in a
new home. Travel they did. Early *sapiens* spread not only to Europe
but right across Asia, and even crossed the sea to settle in Australia.
The oldest Aborigine sites are much disputed. They may date to
60,000 years ago, or even more, which is a major challenge, since
their progenitors must have left Africa well beforehand and used
boats for some sea crossings. By 40,000 to 30,000 years ago in Australia, people were certainly there, slaughtering rhinoceros-sized
marsupials and seven-foot lizards and incising lines onto rock walls.

They, like Europeans of the same date, had ritual burial: the skeleton of Mungo Man was daubed in red ochre and personal decoration: a necklace of pierced cone shells survives from 32,000 years ago.[56]

If the Americas were settled as early as 30,000 years ago (highly controversial), then those people came over the Bering land bridge from a similarly early Asian wave of *Homo sapiens sapiens* migration. As all present-day humans share the same capacities for language and ritual and did so over 30,000 years ago in both Europe and Australia, it seems biologically reasonable (but far from proved) to suppose that modern intellect had a long, slow run-up within Africa. Then the explosion of culture would be a great leap, to be sure, but a great leap sideways from Africa into Europe, Asia, Australia, and eventually the New World.[57]

There is a problem with that story, too. *H. sapiens* lived beside Neanderthals in the Middle East for some 60,000 years, from 100,000 BP to 40,000 BP, before the Great Leap. They seem to have shared much the same technology in stone working. If they did have a slow growth of modern tool-making capacities, they were not showing it. There are tiny fragments of deliberate, very early art. A flint engraved with semicircular lines from Israel dates to 54,000 years ago. A small bit of volcanic tuff, also from Israel, has natural lumps for "head," "arms," and "bust," with apparently human-incised grooves to enhance the neck and arms—some time before 223,000 years ago. There are not many clues to go on from the rest of Africa itself. Africa is short on permafrost to preserve delicate artifacts. There are only a few stone flakes, some 90,000-year-old carved harpoons in Zaire, and possibly still older pierced cone-shells and engraved ostrich shells with traces of red ochre. Perhaps we shall find more, or perhaps anatomically modern people in Africa or the Middle East did not see much need to make such things, until the cultural changes of the Great Leap.[58]

If the Great Leap did happen wholly within a mere 10,000 years, shifting from prehuman to human artifacts, it moved at the speed of culture, not biology. Any of us in the twentieth century knows the cultural changes that can happen in a generation. Perhaps the Great Leap was a growth in social structure and in demography. If it was an advance in language, it was likely not language in a vacuum—the sudden birth of poets relegated to garrets in the cliff—but a more public use of symbols. More people, living in larger groups, would have needed to reshape the rules of life and warfare. They could develop specialization of labor and hierarchies within communities. They might well have felt the need to distinguish themselves from other ethnicities, including the Neanderthals—a central function of both dialect and body ornament today. If so, the mind's biological evolution was largely accomplished, and the baton passed to society. One can imagine the horror of that Russian male if his teenage son had grown up to spurn fox canines in favor of body piercing with bird-bone safety pins.[59]

They were us. When European cave art was rediscovered in the last century, people saw it as a bridge to the modern mind, the hunting magic of the "savage." Now our timescale has increased so much, it seems a modern phenomenon. They were not "primitives"; they were us. We can deduce that by then people had achieved elaborate mental images of the world around them and the significance of human life. They attributed cause and effect. They speculated about unseen causes to link otherwise inexplicable events. They feared death. I'll bet that they fermented the wild grapes of Bordeaux to produce altered states of consciousness.

Is this too much to build on a few thousand artifacts—delicate sewing needles, fish hooks, beads and bracelets, and burials and food-freezer pits dug into permafrost? It would be, perhaps, without the art of the cave walls. People who ventured deep in the earth, carrying stone lamps of burning bear grease, hollow blow-brushes and powdered pigments, were creating an order of the imagination.

They drew linear scratches that might be female pubic triangles and torsos, and cartoons of mammoths with curly trunks. There is the circle of adolescent heel prints around the clay bisons of the Tuc d'Adoubert. There are the blow-painted silhouettes of hands, and of hands with one finger-joint missing—amputated, or just symbolically dedicated through art to the powers of the cave? To us, the caves are magic. We are engulfed by the darkness, amazed by the paintings. How could those who made them, and those who were led deep into the earth to see by the light of a flickering lamp, not have been awed?[60]

What to us are the grandest murals, the horse and reindeer friezes at Lascaux and the painted bulls of Altamira, are not "primitive" art. They do not share the abstraction, repetition, flatness of surface of tribal people trying to hold a magic world at bay, under some sort of visual control. They are instead confidently classic, like the horses of the Parthenon. They breathe realism. The shaded bosses of their shoulders bulge out with both the natural contours of the rock and the trompe l'oeil of air-brushed paint. They are caught in action—the male reindeer licking his female's forehead, the leaping cow, the bellowing bull. A galloping boar plays with stop-action imagery, its four pairs of legs in successive positions. Other engravings not only have many legs but several heads lowered, outstretched, flung upward. As Randall White tells, by firelight, under the sensory deprivation of a deep cavern and the sensory heightening of ritual, the animals would begin to prance before your eyes.[61]

This is not to ascribe one simple, shamanistic interpretation to the high art of the Paleolithic—any more than the Michelangelo ceiling of the Sistine Chapel is only an expression of Catholic faith. We are talking here about more than two thirds of the entire history of art—a period from 36,000 to 10,000 years ago. Those people lived from long before to long after the last great glacial advance, which peaked at 18,000 BP. The murals of Lascaux and Altamira date from as "recently" as 20,000 and 17,000 years ago, Grotte Chauvet from 32,000. We may think of them as something like the flourishing of the Renaissance, but it could not represent anything like a continuous culture. How would we interpret historic Western art, if there survived out of all the ages a dozen cycladic statuettes from 1000 BC, Alexander Calder's wire "Circus," and, miraculously, all alone without any context, the Sistine Chapel?[62]

And what of those naked "Venus" figurines, mostly clustered at a horizon around 26,000 BP? Were they mother-goddesses, or portable porn, or a bit of both? This is furiously contentious. Older traditions assumed they were made by men, for male conjuring of fertility or just male delectation. A feminist view opened up the idea that they were female cult figures made by and for women. Some are fat, many are thin—and many are pregnant. The series from the Grimaldi cave, as late as 14,000 BP, include statuettes with such potbellied protrusions you would think they're expecting twins. Possibly male-inspired fertility symbols tended toward nubile (though sometimes well-fleshed) playgirls, but at least the Grimaldi series look like female amulets for success and survival in childbirth.[63]

The one sure conclusion is that those people were us. They made symbolic art to speak to one another. It still speaks, even though we can never fully understand.

Who Spoke First?

The most recent of the Rubicons offered for humanity's uniqueness is language. It is convincing. Learning language is a powerful, innate urge of small children. It is structured in our brains, both phonemically and grammatically. Specific genetic defects can affect people's ability to learn specific kinds of nouns or to pick up grammar without having to memorize conscious rules. Apes do have some language capacity, but only under the kind of full immersion instruction we bestow on our own children. We have powerful needs to get every child to talk; so much the better if we can extend that skill to apes and parrots.

There is a huge emotional investment in arguing whether language came early or late in human evolution, fast or slow. At one extreme, some say that the Great Leap Forward was itself the result of inventing symbolic language, that it came with a whoosh, fast and late, only 40,000–30,000 years ago. At the other extreme, people say that australopithecines, with ape-sized brains but a new kind of brain organization, could have thought, and talked, at least as well as a modern bonobo.

The most exciting bit of fossil evidence comes from measuring the hypoglossal canal—the hole in the base of the skull that carries the motor nerve from brain to tongue. Our own canals average about 1.8 times the cross-section of apes'. The canals of 2-million-year-old australopithecines from South Africa (or they may have been among the earliest *Homo habilis*) are no larger than apes'. In contrast, 400,000 to 300,000-year-old *Homo heidelbergensis*, from that archaic period between *erectus* and modern humans, had canals as large as we do today. So did Neanderthals. That seems to mean that premoderns and Neanderthals waggled their tongues as much as we do. (These are recent findings, already challenged, and sure to promote a good deal of further tongue-wagging.)[64]

It does fit with a couple of other fossils. The tall Turkana Boy, 1.6 million years ago, did not apparently have the spinal nerves to control the breathing of melodious speech. At the other end of the timescale, a Neanderthal hyoid, a forked bone at the base of the tongue, looks just like that of modern humans. This is a surprise. For decades people have thought that Neanderthal anatomy precluded varied vowels, and thus speech-as-we-know-it. The discovery of an actual hyoid perhaps modifies that conclusion. [65]

In fact, I never quite saw the logic of saying you need varied vowels, or a modern naso-pharynx, for language. Our modern arrangement has dropped the Adam's apple and enlarged the back of nose and throat, giving us a kind of flexible organ-pipe for producing speech, at a very high cost in increased likelihood of choking on gobbets of food. This, to me, indicates that the ancestors of people with such an organ had already been speaking for a very long time, giving the selection pressure for varied articulation. Our pharynx shape could not have preceded speech but must have evolved along with it, after it got started and became increasingly important.

I think that the primary need is mental organization and voluntary control. Even if, like bonobos, you had no consonants (much more crucial than vowels), you could vary pitch and a few nasal vowels as bonobos do. In the unlikely event that a human of modern mentality were struck dumb, left with only a flute to make bonobo whistles, she would work out how to use those whistles as a linguistic code. Indeed the signings of the deaf, which are codified gesture instead of any sound, are full languages, residing in the language areas of the brain.

In short, my own predilection is to say that such an extremely complicated function as speech probably has an extremely long evolutionary history. It did not appear in an afternoon, like the trunk of the elephant's child in the Just So story, but over millions of years, like a real elephant's trunk. I find it difficult to think we could have reached modern brain size about 250,000 years ago without the main tool of the human brain also being more or less in place.[66]

What were the selective mechanisms for language? To make tools, making clear the finer points of demonstration and apprenticeship? To hunt? "The wildebeest are behind that hill; you go this way and I'll go that way and we'll trap them." Or good old gossip? "Simply shocking, my dear, how Quog's mate makes eyes at Hampton. Does she see a power shift coming or is she just stirring it up again?"[67]

I do like the gossip theory, championed by the primatologist Robin Dunbar. As social group size increased, we desperately needed a means for contact that was more efficient than grooming. Grooming is essentially a one-to-one relationship. The only way to enlarge it is to groom one friend while being groomed by another. Grooming male chimpanzees sit one behind the other like children playing train—a way to cement male bonding. But intimate talking can be done within a group of four different people at once. (With more people, even the talking group becomes unwieldy and breaks up into smaller conversations.) This fits Dunbar's calculations that neocortex size correlates with the size of primate social groups.[68]

It is instructive to just listen to the content of human conversations. Most talk is small talk, social noise, stories, not technical teaching or strategic planning. Most talk is about people. Of course, in our culture, some people are public. Talk about sports stars, pop singers, and presidential peccadilloes is still gossip.

Language was sexually selected. Seductive speeches and convincing promises always gain mates, in this lineage where both sexes are adventurous and both are choosy. It was politically selected. Oratory and clever arguments give personal power, which in turn translates into survival and reproduction. It was selected in child care. Parents who talked to their children launched their young with the gift of speech. Neither sexual nor parental language nor gossip excludes the use of "shop talk" to deal with the world, but they provide a ratcheting forward of success within the social group to speed evolution.

There may well have been bursts of more rapid evolution of the brain, concomitant with the periods of transformation we can make out in fossil brain size. Social language at those periods would have also been under accelerated pressure, to keep one's social comrades interested. The Great Leap, whether forward or sideways, was surely such a time.

Psychologists make it clear that, on average, little girls speak earlier and more fluently than boys and that their slight linguistic edge persists into adult life. This is as slight an edge as the much touted male edge in mathematico-spatial constructions, but it is an advantage in a hugely important skill claimed for the traditionally disadvantaged sex. I would like to think it true, and true because of women's role in the hugely important nexus of human social relations.[69]

17

Demanding Control

The makers of cave paintings envisaged cause and effect, categories, and myths to explain the unseen. From our point of view, they were almost powerless, with only the lightest of footprints on the realm of nature. But to the extent that they used their wits to survive and to understand the world around, they were already taking control of their lives.

Are modern hunter-gatherers inheritors of the same way of life, endlessly repeated? Today, they live where agriculture is impossible: deserts, ice floes, the depths of the least hospitable jungles. Astonishingly, they can live off marginal land where nothing else succeeds. Not only do hunter-gatherers make a living in extreme human habitats, they do it with less work than anyone else.[1] It may take only two hours a day to ensure the day's food.

Tell that to two-income American parents dragging themselves home to pick up the kids from day care, cook dinner or phone the take-out, pay the bills, and try to keep their tempers until they doze off in front of the television. Of course, as a hunter-gatherer, if famine comes, you must be prepared to walk a long way to the next source of food or water and, indeed, accept that some of your group will die en route. Death and emigration keep population density low enough for the rest to survive.[2]

Such people have few possessions. They are usually nomads. They make camp and build a shelter on the spot from woven thorn bushes, or enormous rain-shedding leaves, or blocks of compacted snow. They unpack the weapons of the hunt and empty their luggage cases, which then double as food-carrying skins or baskets. Of course there is inequality of skills. Some are expert weavers or orators; a few become medicinemen or wisewomen. But there is relatively little scope for inequality of material wealth.

Such people have myths of creation, rites of passage, shamans and trances and healing. They have a huge store of environmental knowledge, of medicines, foods, house construction. In this they do throw light on the distant past. You can be human with few possessions, but not with an empty mind.

On the whole, hunter-gatherers are monogamous. This is the monogamy of a sparse life and equal status. Few men can buy or coerce more than one wife, and few wives need more than one husband. Each needs someone, though. Husband and wife complement each other with their separate roles.[3] This does not necessarily imply sexual fidelity, however. In a few tribes, notably among the Inuit, sharing a wife with a visitor was accepted hospitality. This baffled but delighted early European arrivals. At one point the women of Thule, in Greenland, confronted a dilemma. A rare ship reached their port, full of sailors who would give them presents for their company. Unfortunately, all the husbands were away hunting. The women eventually decided to take the initiative, go to the sailors, and ask their husbands to beat them on returning from the hunt. The husbands would thus save face for having not given permission. It was plainly better than being beaten in earnest for passing up the chance of gifts, as well as seeming inhospitable.[4]

Lusty and independent women have many lovers, and vice versa— though with the usual share of jealousy by the current official spouse. Sexual jealousy is a common reason that camps of neighbors break up—among the San, the Efe pygmies of the Congo, and virtually anyone else. The saving grace is that such problems can be settled at least temporarily by one faction moving away. People do not have to live together in smoldering fury like villagers tied to a village.

Nisa of the !Kung San speaks for the continuity between being a woman of her tribe and having feelings familiar to the rest of us in all our tribes:

I would never leave Bo. If I found another man I liked, I would have him as an unimportant one, as a lover. Even if another man came and wanted to marry me, I would sleep with him when Bo was away and I would probably let myself like him, but I would never leave Bo. I have no need to do that.

My husband and I argue a lot. We complain and yell at each other, all the time. But it was like that from the beginning. Sometimes we argue about my not working when he asks me to. He tells me to do something and I refuse. He yells at me and then we start fighting. He says, "Aren't you going to get any water today?" When I refuse to go, he says, "Tell me, what do women do? When a man marries a woman, he tells her what work she should do: get water, bring home firewood, and set the blankets down for sleeping. That's what a woman's jobs are. Now what kind of a woman are you that you won't make the bed, fetch water, and bring home firewood?"

But I say, "Don't the other men help their wives? When their wives go out for firewood or water, don't their husbands go with them and help bring home wood and help fill the containers with water? What's telling you that you don't have to help me?" And we'll continue to fight like that . . . He yells at me, "Nisa, you're really bad! . . . I'll leave you and find another woman, that's what I'll do!"

I say, "Fine! Go! Find yourself another wife . . . I wouldn't care. Do you think I'd feel bad? I wouldn't. I'd have lovers and they'd help me with everything. Do you think you're the only one who's a man?" . . . Even yesterday we had a fight . . After, we made up and loved each other again. Everyone said, "Eh, Nisa and Bo do very well. After they fight, they like each other again." We just continue to love each other and sit beside each other. We fight and we love each other; we argue and we love each other. That's how we live.[5]

Rich hunter-gatherers. The Spartan lifestyle of contemporary hunter-gatherers was not universal in the past. Before the Neolithic invention of agriculture, rich lands as well as poor offered their game and fruits to be taken. Before white men conquered the American and Canadian western coast, the Haida and Kwakiutl lived largely on venison and the huge summer run of salmon. Haida believed that human souls returned as salmon, and vice versa; the sea god gave endless bounty from a world that was the alter ego and mirror image of the human world of the land. Northwest coast tribes created some of the world's boldest art. Their totem poles, spirit masks, carved and painted war canoes bespeak a past of noble achievement—and are now being recreated by Canadian artists who revive their ancestors' traditions.[6]

These coastal tribes were the potlatch people. A chief showed his status by inviting his friends and rivals to a feast and showering them with gifts. Then he might go on to destroy things—breaking ancient inscribed copper plaques, burning jars of oil from the oil-fish (I wonder how it smelled?), perhaps igniting a fine canoe, or at an extreme burning the whole house down. Guests were supposed to show their own fortitude and manliness by not moving back from the fire as the fish-oil jars were poured on. I am not sure when etiquette allowed them to move if the house itself was ablaze.[7]

At the beginning of this century Canadian authorities felt they must clamp down on this wasteful, primitive, heathen procedure. They forbade potlatches. The tribespeople then discovered it was permitted to hold white men's festivals called birthday parties. During the decades of the ban, there have never been such birthday parties. Potlatches (and birthdays) fulfilled an essential social role of redistributing wealth, reinforcing bonds, and establishing status.

The point here is that wealth came from the wild: salmon, deer, oil-fish, furs for wear and barter, old-growth fir and spruce for massive house beams and totem poles. These were not nomads, trekking from camp to camp with their gear on their backs or packed into a dog sled. These tribes had elaborate art, organization, and conspicuous consumption. That family buried near Moscow with 13,000 hours of beadwork, not to mention the mammoth-ivory bracelets and fox-tooth belt, could have come from such a wealthy group, though they lived 28,000 years ago.

For millennia, forest people have known that agriculture is an alternative and yet have chosen to cling to the riches of their own lifestyle, though coerced by their neighbors and excluded from the valley land. A Tamil epic from South India tells that a king rested on the bank of the Periyar River, lulled by the sweetly murmured songs of the forest bees: "There appeared before him the hill-folk like vanquished kings, laden with tribute. They came carrying on their heads such presents as the white tusks of elephants, whisks of deer hair, pots of honey, chips of sandalwood, loads of anjana and beautiful aritara, cardamom stalks, pepper stalks, luxuriant kavalai, ripe coconuts, delicious mangos, garlands of green paccilai leaves, jack fruits, flowery creepers, rich bundles of areca nut from luxuriant palms, fawns of the musk deer, harmless little mongooses."[8] Such tribal hunter-gatherers have bartered forest products to their agricultural neighbors for at least two thousand years. Records in ancient Tamil tally the export of forest resin and incense from the Malabar Coast to the Roman Empire, and tigers and elephants destined for the Coliseum.[9]

Expelled from Eden?   Why are the forest people like vanquished kings, while agriculture has invaded the planet? Is agriculture self-evidently a system that, once seen, must be loved? No. Hunter-gatherers were pushed into agriculture by increasing population pressure on the land. In Europe, the tall, strong skeletons of the Ice Age hunters are followed by Middle Stone Age people, stunted in comparison, as the great herds were hunted out or followed the ice sheet northward. Southern European diet shifted from mammoths and reindeer toward shellfish and frogs' legs. Still the population grew. Their own success meant that eventually too many people lived together to forage in an unenriched landscape. In the Levant, at the eastern end of the Mediterranean, cycles of population growth and climate change fostered the planting of grain, which in turn increased the numbers of people on the land.[10]

Romantics sometimes imagine that hunter-gatherers once lived in tune with nature. Did their traditions tell them how to maintain sustainable yields? Most evidence suggests that they were and are much like the rest of us. When the game grew scarce, they increased hunting pressure. In spite of many useful taboos, their ultimate solution was to exhaust the food in one area and move on to another. No-man's land on the fringe of a hostile neighboring tribe sheltered more game. (There is a nice parallel with off-limits military test grounds today, where wildlife flourishes among the unexploded shells.) Moving one's base often led to overt warfare with neighbors but was well worth it. A factor in accepting agriculture was devising ways to keep the peace, so you and your neighbors could invest your labor in land instead of in warfare, and you would have a little more assurance that standing crops and storage bins would not be raided.

Agriculture: Colonizing the Biosphere

The Neolithic revolution began about 10,000 years ago. "Neolithic" simply means "new stone age." It is usually diagnosed by the start of broken crockery. It was also the biggest change in human economics that there had been until our own times. Neolithic humans began to shape the biosphere to their own ends.

In the perspective of evolutionary time, the industrial revolution and the twentieth-century world are simply later phases of the Neolithic. Once you start breeding sheep, it only requires a few technical advances to start cloning them. The real change was realizing that our species could shape its environment, not just placate it with myth and ritual. The Neolithic began the day someone deliberately planted a seed in the ground, instead of merely defecating it like any other primate. Or maybe the day some child pleaded with a parent to keep the fuzzy wolf puppy for a toy, instead of eating it right away.

It is only about 10,000 years, 500 generations, since the Neolithic began in the Fertile Crescent and along the Yangtze River, and later in Mexico. With it, we slowly turned from nomadic to village life, based around fields, and commenced to weave baskets and shape clay pots to hold the produce. Pots are a sign of the Neolithic in fossil layers, though baked clay figurines date from tens of thousands of years earlier. Most people in the world today still live in villages—a lifestyle invented in the Neolithic. The game song of Didinga girls in Uganda might have been sung any time in the last ten thousand years:

> We mold a pot as our mothers did
> The pot, where is the pot?
> The pot, it is here.
> We mold the pot as our mothers did.
>
> First, the base of the pot.
> Strip by strip and layer by layer,

Supple fingers Kneading the clay,
Long fingers molding the clay,
Stiff thumbs shaping the clay,
Layer by layer and strip by strip,
We build up the pot of our mother.

We build up the pot of our mother,
Strip by strip and layer by layer.
Its belly swells like the paunch of a hyena,
Of a hyena which has eaten a whole sheep.
Its belly swells like a mother of twins.
It is a beautiful pot, the pot of our mother,
It swells like a mother of twins.[11]

Enriching the forest.   The Amazon is not pristine. Almost no place is, except, till recently, the Antarctic. People cut little pockmarks, burn a few trees, plant a variety of plants for a couple of years, and then move on to make another little pockmark. Meanwhile, they leave the old garden filled with banana palms and other useful trees as the clearing reverts to jungle. People do not forget the site. For them the jungle is a productive home.

"Slash and burn" people are responsible today for a huge share of forest destruction. "Slash and burn" is their pejorative name. "Shifting cultivators" or "swidden farmers" are more politically correct terms. Of course they do cut trees to make the fields, but they do not see themselves as destroyers. Their ancestral traditions make them caretakers of the forest. When they exhaust the game, the whole village moves, but they leave the forest behind enriched, where the game will also return. In a cycle of decades, the tribe brings back their new village to a spot near the old one.

In many ways the life of early shifting cultivators changed little from the hunter-gathering life that preceded it. People had possessions that were harder to move, including those breakable clay pots.

Polygamy was more possible, as was a somewhat greater division of status. Lands were still often held by the clan, not the individual, like clan lands for hunting and gathering. (This raises Cain today when forest people try to claim their rights: there is no individual to name if they seek a legal deed to their ancestral domain.)

The long fallow period of forest regrowth accomplishes many agricultural tasks. The forest fertilizes; it gradually accumulates carbon, nitrogen, and minerals that enrich the ground when the trees are burned. It reconstructs the soil, aerating it with burrowing tree roots. It weeds; the herbaceous growth that competes with crops dies out under trees. It clears out crop pests; fungi and insects that concentrate on crops disappear in the decades between clearing cycles. To be sure, peccaries and baboons come out of the forest to raid planted crops, but it is easier to stand guard to kill one meaty mammal than a hundred thousand tiny insects.

> Oh baboon
> I greet you, possessor of hard-skinned swollen buttocks
> Whom the hunter pursues and in the process besmears his
>   smock with earth.
> Animal from whose hands the hunter has not received a
>   wife, yet who receives self-prostration homage from the
>   hunter like a father-in-law . . .
> He who, after raiding a farm, returns to his perch, his cheek
>   pouches hang down like a Dahomean's pocket.
> Possessor of eyes shy like a bride's, seeing the farmers' wives
>   on their husband's farms.
> Bulky fellow on the igba tree, uncle to the red patas monkey.
> Gentleman on the tree-top, whose fine figure intoxicates
>   him like liquor.
> Ladoogi whose mouth is protuberant and longish like a
>   grinning rod.

Whose jaws are like wooden spoons and whose chest looks
    as if it has a wooden bar in it.
Whose eyes are deep-set as it goes a-raiding farms . . .
He whom his mother gazed and gazed upon and burst out
    weeping
Saying her child's handsomeness would be the ruin of him.

Singing:
Stout and noisy,
A baboon I saw on my forest farm, as it was munching away.
Stout it was, munching away.[12]

Neighboring people often think swidden cultivators destructive, primitive, ignorant, and above all lazy. In Madagascar, Java, and Peru, settled farmers scornfully look down on bush people. This is exacerbated by differing social systems. Sustainable shifting cultivation needs a basis of large land area. Anyone with ambition, and the agreement of the clan, can clear a new field. That translates into lenient customs of marriage and divorce. If you need not, and indeed cannot, marry for land, then there is freedom to choose partners. This provokes horrified (and jealous) disdain from people with stricter rules.[13]

Shifting cultivators have known about their neighbors' fields and plows since their invention. Like hunter-gatherers, they *choose* not to change. In rich ground, they can sometimes afford the luxury of leisure, art, and architecture. Mayan temples rose from an economic base of swidden agriculture, and so did the stones of Stonehenge and Avebury. It was not until the second millennium BC that traces of bounded, settled fields appeared in England, well after the raising of Stonehenge III, the colossal megaliths we gawk at today.[14]

Whenever people have the chance, they return to swidden agriculture. At the end of the Middle Ages, in the ghastly fourteenth century, Europe's population was so reduced by war and plague that forests grew back and wolves roamed to the gates of Paris.[15] Many of the laborers were dead, but there was enough forest to reinvent shifting cultivation with its lower demands of labor. In the Amazon, European and Japanese immigrants may also give in to surrounding opportunities and copy local methods of shifting cultivation. Why ever give up a system where the forest does most of the work? The answer again: population pressure.[16]

Settling down when the forest goes. As the population density increases, the long fallow period shortens. The forty or fifty years that the forest takes to reach full height (though not for centuries full species richness) reduces to ten or twenty years. Farmers then fell little saplings whose burned wood scarcely fertilizes the fields. Soon five years: now there are only twisted bushes to burn, with gnarled roots to dig out one by one, while weed seeds persist in the soil of the reused fields. As pressure on the land increases, the time goes down to two years or one. Now fallow fields are grass meadows.[17]

Grass is a whole new challenge. Grass roots mat in a tangled interlace below the ground surface. Firing grass produces a "green bite" of new shoots if you are pasturing cattle, but it is no help for planting crops. Confronted by grass, you plow.

A plow is a complex tool, best made with an iron blade. You must keep draft animals to pull it if you are not strong as an ox yourself. You spend the days of spring turning over the sods, trudging back and forth to trace the heavy furrows. The alternative, if you grow paddy rice, is to drive vertically into the field with a long-handled, narrow-bladed hoe or shovel. Clod by clod you dig and turn the earth. The blade is narrow because no man could lift wide shovelfuls of rice-field clay.

When you have spaded the surface, you let in water, turn cattle loose and chase them through the mud. You whoop and holler and dress up in leafy branches like a spring spirit, accompanied by all your real and honorary male kin. This is fun, though exhausting. At

Figure 19. Women replant rice seedlings in a Madagascar paddy field, while a man herds zebu cattle to trample the next field for planting. With settled agriculture, people and animals do all the work that the forest would do "for free."

the end of the day the field is a slurry of muddy ooze, well mixed with the dung of frightened cattle, and every man of the family is caked with the same mixture. Then the men go off and wash, ready for the feast the women have prepared. It will be the women's turn soon enough—the backbreaking task of replanting shoots of rice from the green chartreuse "nursery fields" and weeding the paddy through the summer, while the children scare away birds.[18]

In short, everything the forest used to do is now delivered by human or animal labor: aerating soil, fertilizing, weeding, fighting pests. The Malagasy have a proverb: "The rice of the first harvest is not for the lazy man." No wonder settled people despise the soft life of shifting cultivators!

Involuted societies. Rice can absorb an almost infinite amount of human labor. In addition to planting, there is the effort of building canals and dikes to bring water to the fields, terracing the fields on hills and mountainsides, and the further work of harvest. Population and rice output both doubled in Java at the beginning of this century, long before chemical fertilizers, let alone the green revolution. As population rose, there was more labor to grow more rice. There the cycle culminates with a harvesting tool more like a razor blade than a sickle, to cut each grain-filled head and save the maximum possible rice straw for other uses.[19]

In this farming system, improved or improvable land is critical. By all means marry your cousin, or the son of your neighbor. In Madagascar such a suitable marriage is called "sewing up the fields," as you join your lands edge to edge. Don't imagine divorce will be easy, given the whole family's stake in the land. As for extra laborers, this system allows all shadings of family, quasi-family, sharecropper, and serfs. To each is measured out their allotted share of the harvest.[20] This is an "involuted society." The agricultural system demands more and more human input. Human relations are elaborated to match.[21]

And even at this stage, people begin to be nostalgic for "unspoiled" nature. The romance of the wild, the longing for our past, is no mere invention of Europeans. Around the time of the birth of Christ, classical South Indian Tamil poetry pictured the mountains as the home of love. The hero is the hunter-prince who pursues a wounded elephant down to the plains. The lowland girl who runs away to join him speaks to her confidante:

> *Bless you, friend. Listen.*
> Sweeter than milk
> mixed with honey from our gardens

is the leftover water in his land,
low in the waterholes,
covered with leaves
and muddied by animals.[22]

Eating fossil forests.   Small-holding, settled, traditional farmers produce by far the most output from a given patch of land. What now succeeds them is large-scale, industrial farming. This drops down to less output per acre, though more than in shifting cultivation, and even less human labor per acre than in shifting cultivation.

Slashing and burning fossil forests takes up the slack. Energy builds the tractor and the combine harvester. Some of that energy comes from water and wind power but most is from coal or oil. Diesel oil powers the machines themselves. Oil-based fertilizers replace forest ash or cow-dung fertilizer. Oil-based pesticides and herbicides replace the weeding women and bird-scaring boys, or the forest regrowth that used to kill off pests.

Meanwhile, industrial science creates new crops to live in an industrial environment. American hybrid corn was one of the first large-scale enterprises—corn that cannot seed itself but must be mixed anew by commercial suppliers in each generation. Green revolution wheat and rice were developed originally from mutant dwarf strains. The reason these miracle grains give so much higher yield is that their allocation of effort is skewed—less to the dwarfish growth of stem and leaf, more to the head full of seeds. Green revolution stalks are just stiff enough not to fall and lodge against each other, or collapse flat on the ground, but only if the farmer grows them right. They need cosseting. They want fertilizer, and weeding, and water, or, for rice, exquisitely accurate adjustment of water height. Their high yield derives precisely from surrendering the vegetative toughness that would buffer them in the uncontrolled environment of the natural world. Their only habitat is the slashed and burned fossil forest of the Oil Age. Genetically engineered crops take us many steps further, under the control of companies, not farmers.

The green revolution originated from a conscious effort by scientists and foundations to feed the growing population of the world. The development of oil-based agriculture is a part of the whole structure of the industrial revolution. Population growth has both driven the industrial revolution and been its result. Population has climbed from 1.5 to 6 billion in this century and is now expected to level off at 10 billion in the next, but food supply per capita has climbed even faster. So far, the fossil forests are capable of nourishing us well.

The upshot is our present impact on the globe. Over 40 percent of the world's terrestrial primary production, which means both the natural and agricultural increase in organic matter, is used or wasted by humans and our domestic ecosystems. That is, 40 percent of each year's output of all green things is diverted toward our own species— naturally, the easiest 40 percent for us to reach and use, but there is still some out there to gobble up. The end of the Neolithic revolution, with its ever-accelerating increase in human power over the natural environment, has not quite arrived.

Bigger Bodies

During the last three hundred years, the average body mass of humans in developed countries has increased by over 50 percent, and our average longevity by over 100 percent. This extraordinary fact has nothing to do with natural selection of genes for bigger bodies; it has moved far too fast for that. It is a change—call it a technophysiological change—that has come about because we are better nourished.[23] It also has not much to do with the variation in individuals' genetic potential. Some people will always be shorter than others: natural selection both allows for differences and selects for them. This story is about the rising *average* of population height.

It began around 1700, when agriculture was being brutally rationalized but also when trade between regions relieved some of the seasonal and yearly gluts and dearths of one's own homestead. It goes with the almost vertical rise in world population since then. The economist Robert Fogel argues that it reflects a whole suite of advantages. People with greater height, and appropriately larger body mass for height, live longer and suffer fewer illnesses in old age. Also they can work more: they do need more calories for basal metabolism, but, if fed, give proportionately higher physical output.

Most of the data concern men, because nations measure soldiers. This means there are sets of statistics to compare the veterans of the Union Army who lived until 1910 with veterans of World War II at the same age. There are also sets that track Europeans through the centuries. Adolescent males in Britain have grown taller each fifty years, just as the age of female menarche has been falling. Currently the tallest population in the world are the Dutch, who average 181 cm (just over 5′11″). In 1860, however, Dutch males averaged only 164 cm (under 5′5″). Other European countries had similar rises, as did the United States, to its present male average of 177 cm (under 5′10″). Fogel calculates that if the populations of Britain or France in the eighteenth and much of the nineteenth centuries had reached modern stature, they would have eaten up their whole countries' food supplies just for basal metabolism, with nothing left for work.

People who are children of immigrants to the United States frequently look down on the tops of their parents' heads. (Not all the growth is achieved in one generation: it takes several for the effects of previous maternal nutrition to wear off.) We laugh at the child-sized suits of armor for great medieval warriors. Modern men who live in quaint English Tudor houses develop a peculiar bobbing gait, to duck under every door lintel, and still they keep on cracking their heads. But seeing Fogel's figures makes us realize that something very peculiar is going on. For the first time since the strapping skeletons from Cro-Magnon days, we are living up to our genetic potential.

Reproductive Technology: Changing Our Children

Shaping our own bodies, feeding them as well as we are able, and painting and piercing ourselves for beauty dates at least from the Great Leap. Shaping our children is scarcely a decade old.

Parents have always wanted the best for their children: the best food, schools, tennis and ballet lessons, or the best technique with a fish spear, the best initiation dream. The geneticist Lee Silver warns us that parents cannot be stopped from trying to get the best genes. If a government bans the technology, the rich will simply fly to fertility clinics in the Cayman Islands and combine conception with a holiday at the beach. (That particular combination has been done before, of course, but not through hi-tech.)[24] As I write, an advertisement offering $50,000 for a batch of eggs is circulating on Ivy League campuses, specifying a donor who is tall, blond, and with SAT scores above 1400. Never mind what simplistic notion of genetics the prospective parents must have, or even if the advertising firm previously hopes to corral a stable of possible donors—there is a market.

Genetic triage has already started. Parents at risk for Tay Sachs, a hideous disease that usually kills children at the age of two, and for which there is no known cure, can now be tested to tell if they are carriers. Many are Ashkenazi Jews of strong religious conviction. Their solution may be to break the engagement, if their fiancée is another carrier. People with milder views about abortion may decide not to carry a Tay Sachs or spina bifida child to term.

The list of techniques that are already here is frightening. A child can spring from a man's sperm, a woman's egg, another woman's womb, and socially recognized parents who are different again. This is extreme, because social parents will not usually pay for this rigma-role unless they contribute something, either genes or womb—the simpler solution is to adopt.[25]

Incidentally, "surrogacy" blatantly distances birth mothers. In the common case where the "surrogate" gives egg and womb, with sperm from the husband of a barren wife, the "surrogate" is the child's full biological mother. The barren wife is an adoptive mother. An adoptive mother has every right to be called the "mother," but a birth mother is not a surrogate. Of course, this arrangement long predates artificial insemination—it even predates Sarah, who, in the Bible, asked Abraham to go unto her handmaid Hagar that Hagar might bear them a child.

Today, two women could make a child with both their genes, if they could find a lab that would agree to do for humans what is easy enough in mice. They still need a father to contribute sperm. Because of genetic imprinting, some genes must come from a female, some from a male parent. However, if both women's eggs were inseminated, then cells of the two could be joined. A mosaic child, called a chimera, would grow up as a single person, with a mix of differing cells. Chimeras sometimes happen naturally, within one woman, when two of her fertilized eggs fuse. (It is the opposite of twinning, when a fertilized egg separates to form two people.)[26]

Cloning is coming. Agricultural animals will give drugs in their milk, through genetic engineering, and genetic engineering is easy on a cloned set of cells, difficult any other way. Agribusiness will make sure the techniques are developed.[27] Cloning produces an identical twin. Cloning from an adult gives a twin to someone whose adult form is already known. The clone is no closer than a twin. It will have its own life, its own experiences, and will be different in character as a result. If you feel it is somehow immoral to produce two people with the same DNA, are you dismissing all the identical twins who think that they have a right to their own lives? If you let parents have identical twins, why shouldn't they be allowed to have twins two years apart, or an identical younger sib of a child who has died?

Engineering "good genes" is perhaps further off. Would you consider adding a gene to your child? No? What if it were a gene for resistance to AIDS? It's the thin edge of the wedge. We are very far from knowing what genes predispose to intelligence or musical ability, but we know already much about genes for and against alcoholism, and more abstruse traits will follow. They are following fast: a marker that is more frequent in a group of mathematically gifted children has been announced.[28]

It is not so much cloning as the production of designer babies that frightens me—because of parents' desire to do the best thing for their children. That human urge is what may make it unstoppable.[29]

Greed, Surplus, Status, and Hubris

Four huge, new factors have entered human behavior in our own species' past. The first two are material greed and the storage, or transfer, of surplus production. To amass possessions beyond immediate needs for this day or this year sets us apart from most other animals. Only a few cache food, or build multiyear termite hills and beaver dams. A third factor is social status. All animals know inequalities between the sick and healthy, those with good or bad territories, those at the top or bottom of the pecking order. But we are almost the only ones besides ants, bees, wasps, and naked mole rats where some individuals actually command the labor of others and appropriate their surplus. Finally, there is the power of ideas—memes with their own reproduction inside our minds. So far as we know, there we are absolutely alone.

I tend to think that ideas and ideals came first. I can picture a slow build-up of language and the power of thought through australopithecine times and the first few species of *Homo*. The relationships of love for children, support for friends, and competition with rivals (which, as primates, we had been doing all along) surfaced into consciousness as ideas of right conduct. When they were put together with feelings of awe and incomprehension before the vagaries of nature, it would be natural to extend interpersonal rules toward the behavior of natural phenomena. Much later on, this quasi-animate natural world would crystallize into naiads and wood nymphs and thunder gods. Sometime in that development it would seem natural to propitiate those spirits, to offer them gifts. Then, of course, we are into the realm of true religion.

The amassing of possessions and the elaboration of social hierarchy seem only to date from the Great Leap Sideways of the Cro-Magnons. As I said in the last chapter, I favor a slow build-up among African *sapiens*. However, the earliest artifacts that show it for sure are the painted caves, the burial jewelry, the little carved statuettes. Cro-Magnons had ceremonial centers, and artists, probably also priests. This does not necessarily mean chiefs and viceroys. The man and children buried with all those beads may have been symbolic clan ancestors or sacrifices, not just personally rich. We do know that wealth and status can grow among hunter-gathers.

Agriculture then brought real power over the environment, and a surplus that could be appropriated by the rich to increase their status and found their cities. The rise of cities and kingdoms, with their multilayered hierarchies, came relatively late. By and large, it was the transition to settled agriculture that provoked larger-scale organization. People who might forfeit or consolidate their land were ripe for organization. If they happened to live where irrigation was essential—the valleys of the Yangtze, Mekong, Tigris, Euphrates, Nile—then they had to work together to control and share water. Heritable large-scale status apparently came later than either ideals or greed.

The rise of science and society has vastly increased our power over the environment, ballooned our bodies, and provided hi-tech tools to tinker with our own genes. Does this mean that we are actually in control of either our environment or ourselves? No, not yet—though the next chapter talks of what may be coming. For the moment, our species displays what the ancient Greeks called hubris—the overweening pride of the ignorant. We funnel the world's water into our mouths with minimal thought to the watersheds. We excrete carbon dioxide into the air, with talk but little action about the dangers of climate change. We prate about the importance of information and then demolish the world's biodiversity, burning the world's genetic libraries for firewood. Our search for control of nature has brought us to a point where our impact is out of hand.

One thing seems clear, though. Barring world catastrophe and collapse of civilization, we are reaching the same kind of transition point that gatherer-hunters reached at the Neolithic revolution, when they were forced to take responsibility for growing their own food. People's lives have changed before; they are about to change again.

18
The Global Organism

While I have written, bark has peeled from the trunk of the cauliflower-domed copper beech tree that stands between the sciences and the humanities. The wood beneath is streaked purplish-gray. Campus groundskeepers lop off branches as thick as three men and lay them in slices on the ground. The amputation bares a squirrel nest high among twigs, an offering raised on lifted hands to the baleful predators of the sky.

The beech is infected with purple canker. This is a fungus, injected by the mouthparts of an insect called the beech bark scale. Bracket mushrooms erupt from the flayed trunk, bouquets of contorted cones with delicate white gills beneath, the snow-white fruiting bodies of decay.[1]

I mourn for the tree, but not for its metaphor. Perhaps the frontier between art and science really is losing vital force. E. O. Wilson has written a book called *Consilience,* to claim in lambent prose that biology offers a basis for unity of understanding among us all, if only we consider humanity as a species. (So far, when Wilson has written a book with an odd title like *Sociobiology* or *Biodiversity,* that word enters our language.) He traces his stance to the clarity of the eighteenth-century Enlightenment, a period when intellectuals hoped that both science and society might be based on human reason, and all knowledge could interrelate as one vast encyclopedia. Wilson now foresees psychology, anthropology, even history and economics springing from evolutionary principles.[2] He allies his stance to clear, analytical thought, and the power to dissect complex wholes into simpler parts. It also implies a vision of progress. Seen in the context of evolutionary time, the frog prince and princess are transformed into humans and move toward a harmony of understanding, as they acknowledge their heritage. Nature did not "intend" to produce beings who recognize their own minds and others', who wield huge power over the biosphere, and who think in terms of purpose. But, having blindly achieved this transformation, the princess and prince now may travel still farther, endowed with the mentality to look forward on their path. Predictably, Wilson infuriates those colleagues in the arts who see it as academic imperialism. He accepts that "they will draw this indictment: *conflation, simplism, ontological reductionism, scientism* and other sins made official by the hissing suffix. To which I plead guilty, guilty, guilty."[3]

Cultural anthropologists, one might suppose, would be the people quickest to turn to evolutionary principles. After all, on many campuses they share a department with the physical anthropologists who piece together fossils, trail after monkey troops through the jungle, and trace the molecular family trees of African Eves and Neanderthal Nellies. However, influential schools of cultural anthropologists currently view culture as a whole separate realm from biology. The Social Science Straw Man almost exists.

Culture seen as apart from biology goes with a different outlook. In the first place, all generalizations about human nature are suspect, even conclusions about how humans diverged from chimps and still more from lemurs. What seems much more important is the differences between viewpoints: the views of various informants, and the viewpoint of the anthropologist herself, as an inevitably biased witness. The generalization most suspect of all is the idea of progress.

Progress is too often a code word for globalization, which is in turn a euphemism for American capitalism and American pop culture. (Watching the only television channel in Antananarivo, which was showing a B-run Hollywood action film badly dubbed in French, among children who speak only Malagasy, I see what they mean. Of course, I could have gone to a video stall down the street, to view a tape of the latest Jackie Chan movie, so Hong Kong is in the picture too.) The quasi-literary view of culture is a far cry from the assured generalities of E. O. Wilson, or even the cheerful certainties of Margaret Mead. It has something of the angst of the Romantics: not Culture as a single human capacity, but cultures, with all their hobgoblins and forest sprites, vulnerabilities and conquests, traditions, influences, emotions, pronouncements, interpenetrations, interpretations. Such cultures offer schemata that shape the mind of a growing child, but they can never be fully grasped, only approached as nodes of organization in all their emergent complexity.[4]

It is not that science is always reductionist, or art wholly and inherently emergent, but at the moment within anthropology they are at opposite poles. Small wonder that many anthropology departments have filed for divorce, while in others the cultural and physical anthropologists are so far estranged that they seem to have no more to say to each other, not even recriminations.

As usual, I wish to have it both ways. I never did see why emergence and reductionism, let alone culture and biology, need be set up in opposition. I want to both eat my cake and know the recipe. My own mental picture is of two somewhat octopus-shaped Klein bottles, each trying to swallow the other. The Klein Bottle is the three-dimensional analogue of the Möbius strip. It is just a bottle, but with a twist in the fourth dimension, which means that all the world is simultaneously inside and outside it. Scientists and humanists both think they hold the universe, the inside and outside of the human mind. To a scientist, the humanist's histories and cultural studies are a part of an older, larger story of evolution. To a humanist, the "larger story" biologists tell is just as idiosyncratic and socially constructed as any other cultural artifact. It would help if we do not claim that either view is a figment of the other's imagination, but rather that we approach each other beyond the dying copper beech.

We are moving into a new phase of history. I am about to argue that the changes are so great, those who look back on us may even claim our times ushered in a new phase of evolution. Academics are not going to direct societies' course, although ideologies of economists, discoveries of biologists, ingenuity of computer scientists are central to the change. But if our goal is to *understand* society, it is not a single discipline, or only the ivory tower, but a combination of us all who may fathom what humans are becoming next.

The Sponge Stage

We have traced the emergence of biological organization from primeval chemicals to bacterium to cell to body. At each stage a larger, coherent whole emerged from the linkage of independent parts. Each is a holon, simultaneously one and many, a single organism and yet a community of individuals.

People are now being joined into a global human society. Only a few of us so far live among the interlinked elite. All, however, are affected. It is not possible to opt out. A society that you cannot secede from is in some ways like an organism. The more the world integrates, the more our society becomes like a living thing.

The attributes of live organisms, as they dawned within the primeval slurry, are cooperation, communication, and differentiation of the parts, a boundary between self and not-self, and self-maintenance and reproduction. I shall treat these in turn, starting with the easy ones. We'll tick off communication, cooperation, and differentiation without much trouble—they are part of what people mean by "globalization." The boundary is much more ambiguous. Reproduction does not apply at all—though I will argue that in a meme-based society, it does not matter. Finally, self-maintenance seems to me the attribute of world society most like a living thing. Our species may or may not achieve real self-maintenance, but we are making a start.

Some people react with unease or even horror at this idea. Will we submerge our hard-won individuality into some worldwide human ant hill? Will we swarm anonymously together as (at this distance) we imagine medieval craftsmen joined to build Gothic cathedrals?[5] If so, that is a very long way off. It would take millions of years of biological evolution to restructure human genes enough to make us actually appreciate the life of an ant. I am talking about something much more immediate: the organization of memes, not genes, into a global unit over the next few centuries. Note the careful choice of time-scale. I do not wish to be proved wrong while anyone is still reading this book. There may well be pauses or reversals for a century or so. As of now, though, globalization is galloping toward us.

Biological evolution is speeding up, thanks to genetic engineering, reproductive technology, mass extinction. Social changes are also speeding up. Global change seems to be out of hand, beyond any one group to influence or control, despite our attempts to direct climate change, to bail out failing economies, and of course to sequester ever larger shares of world resources into the control of particular people and corporations.

Is global society more like an ecosystem or more like an organism? Right now, it is like an ecosystem. Each corporation, nation, billion-aire tries to survive and extend its sphere of influence. If a firm, or species of firm, goes extinct, the others are affected. However, plenty of others rush in to take its place if it dies and leaves a gap. There is little overall organization beyond the competition for resource flows.

However, we are not likely to leave world society in this chaotic state, now that we have the power to fix it. As I said back in Chapter

1, we do not want to recognize nature's indifference, its blind competition, because as social primates we ourselves do not work that way. People think. And feel. And organize.

Thomas Hobbes broached the same idea in 1651, with his "LEVIATHAN called a COMMON-WEALTH, or STATE . . . which is but an Artificiall Man; though of greater stature and strength than the Naturall."[6] Hobbes conceived human life in the state of nature as "nasty, brutish, and short." Therefore, he thought individuals must agree to the emergence of a sovereign state on purely rational grounds, as the only way to escape the worst evils of competitive anarchy and civil war. The modern primatologist sees instead that we evolved with a richly textured social life. Our brutish heritage includes cooperation as well as competition. Evolution gives us innate "appetites and aversions" (to use Hobbes's terms) that already include a tendency to identify with the larger group, and even to glorify it. People are not *less* likely to form alliances than chimpanzees.

Let me quickly say that I do not know how. Integration could be anything from a just and equal federation with respect for all, to a dictatorial cabal achieving Alexander's ambition to conquer the world. Much more likely, it will be some immensely complex mixture of the two.

As society grows more and more organized, more and more integrated, more and more interdependent, it is becoming more and more like an organism. Why do I blur the line between society and organism? Because it is blurry, not hard and fast—at least in the long view of the 3.8 billion years of life. If we are actually moving toward a global organism, we won't wake up some morning fused into a fait accompli. There must be a period of transitional limbo that could unite us further or, perhaps, for a while, disintegrate.

A better analogy than an anthill is the humble sponge. Sieve a live sponge through a cheese-cloth mesh. This separates its individual cells, which then crawl back together over the bottom of the petri dish to reconstruct their communal form. Each cell retains the power to reproduce new buds of sponge, but they would rather be together. If we are moving toward a species-wide unit, we are merely in the process of transition, like cells only just coalescing into something worth granting the selfhood of a sponge.

Global Communication: Sex, Language, and IT

Information technology, IT, is the communicative core of globalization. Its nerve pulses underlie all the other possibilities. There has been a long slow run-up to IT: the camel caravan of Marco Polo, the fleet of junks that brought Cheng Ho from China to Africa before Columbus reached America, cuneiform tablets, the Gutenberg Bible. But pulses of electrons now carry information between people in "real time." Real time is the timescale of the human attention span, as it evolved to deal with face-to-face personal interactions. Until the advent of electronic communication, we could have empires but nothing approaching planetwide social organization of individuals. IT works at scales as large as the multinational corporation, as small as the family. While I wrote this book, my immediate family lived in New York, Edinburgh, Brighton, Beijing. Crises and triumphs shared in real time keep the emotional links; the family member out of touch was the one whose lifestyle precluded e-mail.

The computer historian and visionary George Dyson sums up the change: "In the 1950's computers demonstrated their dexterity at manipulating very large numbers over minute increments of time . . . In the 1970's computers began to reproduce themselves in automated factories . . . but we decided that mere spreadsheets and word processors did not merit raising the alarm. Another twenty years passed. Computers, now teeming like herrings in the early spring, began pooling their intelligence, exchanging states of mind in the blink of an eye, half a dozen languages removed from those we can comprehend. Only an esoteric fraternity, uttering one line of code at a time, still holds congress with the machines . . . We stand transfixed, like monkeys given a mirror, by the novelty of our own image reflected in the surface of the web. When the smoke clears . . . the computer as disembodied head will have disappeared, replaced by a diffuse tissue enveloping us in nebulous bits of meaning, as neurons are enveloped in electrolyte by the brain."[7]

As of February 1998, about one of every six persons in North America and Europe was on the Internet—up to 70 million internet users in the United States and Canada alone. There were far fewer in the developing world. All Asia had about 14 million, 9 million of them in Japan. As with all else economic, Africa lagged, with less than 1 million Internet users in the whole continent, and in fact only one telephone per 200 African inhabitants. If this is something like a world nervous system, it is very unequally distributed. However, in almost all countries, the rate of growth increases exponentially. Usage in Latin America went up 788 percent between 1995 and 1997. In Africa, half the countries had no Internet in May 1996; essentially all were connected by October 1998.[8]

IT grows by an evolutionary process that resembles natural selection, not by planned design. Although subassemblies of hardware and software are the product of intense human planning, the technology jostles, competes, survives, reproduces, or dies young, in the environment of other ideas. (Like organic evolution, the outcome is not perfect optimization of the best techniques. We are still stuck with the QWERTY keyboard, which dates from the days when mechanical typewriter keys got in one another's way, with no relation to how human fingers optimally type English.) Chunks of the nexus, including ads, mail-order houses, and the ambitions of Mr. Gates, proliferate in their own environments or die. Our own brain seems to grow in the same Darwinian way, by the proliferation and pruning back of neurons: neurons that find no functions wither away.[9]

IT is in the tradition of accelerated information transfer in evolution. First came life reproducing a genetic code. That was the dawn of biological information transmission, change brought by random mutation, pruned by natural selection. Then came sex. Change speeded up when organisms could reshuffle and transfer genes to one another, guaranteeing that each offspring would differ from its parents. Then came human intelligence and language, and the Great Leap, when humans began consciously to seek understanding and control of the world. Life—our own and the biosphere's—began to change at the accelerated pace of human culture.

IT is a transition on this grand scale. If human brain power and language were as big an innovation, biologically speaking, as the evolution of sex, then worldwide linking of human and electronic brain power is potentially the same kind of step into the unknown. Of course, war and social collapse could obliterate much of IT's infrastructure, as the "dark age" eclipsed the technology of Rome. Even so, for the next few centuries, in a few monasteries, tonsured computer nerds would copy the wisdom of ancient disks onto new ones with illuminated capitals, until an eventual Renaissance reinvents the Internet.

Global Cooperation and Differentiation

Two criteria of any organism or integrated society are internal cooperation between the parts and internal differentiation of parts from one another. In a globalizing society, the first is economic integration, the second international division of labor. As in all the previous chapters, I say "cooperation" to mean growing interdependence. Individuals join for their self-interest, which does not mean the process is kind or cuddly.

Multinational corporations are becoming a global circulatory system, shipping metabolic products from one place to another. Between 1820 and 1992, the world's human population increased five-fold, per capita production increased eight-fold, but world trade multiplied 540-fold. Much of the growth has been very recent.[10] The precursor was local trade. Nothing wrong with local trade, but it is slow and by definition local, like osmosis. If you are going to be nourished by osmosis, you cannot grow much bigger than a flatworm. By the nineteenth century, trade was global but still slow. To respond to the requirements in different parts of a far-flung body, you need both a means of moving things from place to place and the ability to sense distant demands. In other words, you need transportation and communication systems. It is not even the size but the speed of response that matters most for this story. With IT, a factory in the Philippines produces on Monday the dresses that sold out in Wisconsin on Saturday.

An organism needs an energy source. At the moment this is overwhelmingly oil and natural gas—discovered, shipped, and sold globally. The oil companies are also exploring and staking their claim to renewable energy sources. Companies now rapidly buy up future options to other resources likely to be managed across national boundaries—in particular, regional supplies of fresh water. The resources that each part of society needs to survive are produced in one area and shipped to others, like the cooperating, but different, organs of the body.

Is there any analogy to a central brain? Or, to be less ambitious, even the resonating nerve-net of a jellyfish? Or does each trading company, nation, and power bloc operate on its own in competitive chaos? There are central systems, though not as coherent as a jellyfish's. Nations make trading agreements. Companies do too, though antitrust laws try to discourage them.

Some 35 world organizations are loosely grouped under the United Nations. The International Monetary Fund, World Bank, and World Trade Organization are three of them. They set ground rules, lend and also apply pressure to debtor nations, and support an open world economy. In the fall-out of the Asian financial meltdown of 1998, it is not hard to predict that the world will try to tighten global financial safeguards, as national governments did after the Great Depression of the 1930s. Other less visible UN bodies operate because all parties see the need for consensus. The International Telecommunications Union, first started in 1865, allocates radio frequencies. The International Civil Aviation Authority coordinates air traffic. The International Meteorological Authority tracks weather systems across the globe.

National governments are still major players on the global scene, particularly governments of large, rich countries, but their power is waning and increasingly porous. Every day makes it clear that the nuclei of local power represented by national governments more and more respond to international forces. They are also prodded by nongovernmental organizations among their own people—who in turn communicate with like-minded people across frontiers. Governments are local entities, where local means having a geographical existence, rather than a functional existence. Humans will continue to belong to local groups in the same sense that a liver cell belongs to the liver. But the liver, like the nation, does not live or die alone.[11]

Finally, we have global parasites—traffic in guns and drugs, in wildlife, in sexual slavery. These enterprises follow the same pattern as legal multinational trade, except that they are even riskier and more lucrative. Of course some businesses now thought legal and backed by national governments would be equally objectionable if considered as internal to a world society, especially the international sale of armaments.

In short, the regions of the world are ever more closely interdependent through trade, formal agreements, the media, and nongovernmental networking. At the same time they differentiate through global division of labor. Two of the central attributes of an organism, internal cooperation and differentiation, are increasingly being met by world society.

Boundaries

A boundary is defined by exclusion. Would a global organism include our whole species but exclude AIDS and ebola viruses, cows, corn, and computers, and any passing extraterrestrials? Would it exclude other parts of humanity? The global society might be merely the connected elite, not the lower strata, that is, most of the world's people who don't own computers.

Incomes are growing rapidly more unequal. In 1820 the gap in per capita income between the richest and poorest nations was 3:1; in 1960 it was 30:1; in 1995, about 85:1. Income gaps between rich and poor also widen within many nations, notably the United States. The world is divided today like the French aristocrats and peasants before their Revolution, like slaveowners and slaves in the United States before the Civil War. The global underclass could become ever more marginalized, mere serfs to the Old Boy Internet.[12]

The terms of the linkage may be as unequal as slave and slaveowner, but both are enmeshed in each other. Social classes are not distinct like species in an ecosystem, not predators or prey or even livestock. So long as we are all capable of interbreeding and addressing one another with moving words, the coordination and division of labor (including labor's exploitation) within global society makes it one large whole.

Local boundaries arise when people deliberately shield their own cultures from the world around. Peaceable Amish and the surly survivalists and nationalists of all ilks try to do so. Perhaps the world will break apart into feuding religious power blocs. However, to isolate themselves, they must keep control of IT, as the Taliban of Afghanistan are trying to do in banning television. For the rest of us, televi-

Figure 20.   Dr. Ian Wilmott with Dolly the clone. Dolly, like other mammals, sniffs her acquaintances—in this case her creator. (After Stephen Ferry, *Life Magazine,* January 1998)

sion homogenizes the world. Backlashes of fragmentation, or outreach of understanding, will also happen on the box. All the boxes, everywhere. It is all too easy for propaganda to whip up tribal hatreds, but if people gain access to information from beyond their own tribe, there is a powerful force for globalization to go ahead.[13]

A wilder thought: could we fission into new biological species? The desire of every parent to provide the best for their children points toward genetic engineering, first to defeat Tay Sachs disease and cerebral palsy, then to avoid schizophrenia and alcoholism, then to add in math skills or a musical ear or athletic prowess. Suppose the rich could stuff an extra chromosome full of desirable genes into their kids. That would definitely stop them from interbreeding with old-style unenhanced humans. (Most of the technology to do so is already around today.) As the world divides into the gene-enriched and the unimproved, the distrust that each ethnicity has always felt for others could finally find some objective basis in fact. H. G. Wells had it wrong: the time machine moves faster than he thought. It might take us mere centuries, not millions of years, to diverge, though not, I hope, into Morlocks and Eloi, one eating the other.[14]

We may never manage to act as one species in the face of evolved tribalism. Species before us have never acted as a unit. Natural selection chooses between individuals. If animals evolve to cooperate in social groups, it is because this serves individuals' genes. The individual calibrates the sacrifices it will make for a particular group by the likelihood of migrating out—finding a way to do better alone, or in a different group. The more tightly it is bound, the more tightly its interests coincide with the group as a whole. We are all too ready to devote allegiance to our tribe, but if society does not fragment (and indeed present regional fragments are joining up—witness the European Union), we begin to share as a whole species the most fundamental reason that individuals support their groups: we can't leave.

Reproduction? Or Conscious Purpose?

We doubted the idea of Gaia, Earth as a live organism, back in the first chapter, on the grounds that there is only one Earth. The only unplanned mechanism we know to produce a complex organism out of simpler components is natural selection. Natural selection happens between a multitude of varied organisms. How could an integrated single Earth organism arise without natural selection? Unless you believe that complexity and unity derive from a purposeful Creator, the problem of lack of reproduction leaves insentient Earth no way to become Gaia.

Could a global human society reproduce? It might colonize other planets, budding off new planetary creatures. Or it might fission into power blocs, even into two biological species. If two human species each felt themselves—itself—a unit, the two might then continue in sibling rivalry. There is not a lot of scope, since all this takes place on a full planet, not as trillions of specks in a primeval sea. Natural selection between just two hardly counts.

Reproduction of a global human organism, in short, is either not in the cards or works under very different rules in an environment that is already full. Without reproduction there can be no natural selection at the organism level, and no selection for tighter integration in the face of competition between rival organisms. On this criterion, global society could not become a single organism after all, by the only way that insentient matter has ever done so.

Selection on memes is different. Memes are indeed the creation of sentient beings. Memes do not merely *mimic* design, they *are* design and purpose. Conscious purpose can include the idea that personal survival and welfare depends on getting things right for the species. It is the old argument about group selection: group selection only overcomes individual interest if there is vanishingly small possibility for the individual to migrate to another group. We are well and truly stuck within our species and on our planet, with no chance to opt out. Many people now see that our species has a new environment, where some costs and benefits reach most of us at once. This means that the species can evolve by extension of memes that concern the species' own progress and survival. In an organism with conscious purpose, reproduction is not necessary for survival or progressive change.

Of course, prevailing ideas of how to achieve survival may be wrong. If they go massively wrong, they may be massively dangerous, even lethal to people on a global scale. But ideas of economics, health, and global responsibility sweep through the earth as a whole, borne on ever-more powerful media and organizations. (World Cup

Soccer may be showing the way.) If enough people share the same idea, or act under the influence of the same idea, is there a sense in which the species itself has the idea? Maybe then it could be considered alive?

Species Self-Maintenance

Self-maintenance as a species is what I see as the attribute of global society most like a living thing. We begin to achieve this through memes in action.

There is no need to dwell on what happens if globalization goes wrong. Global misery, the elite in their bastions, anarchic parts of the world written off as no-go zones, poor regions enslaved, environmentally devastated regions hungry and thirsty—all this is happening now. We have only to imagine it in large-screen technicolor. Of course we could choose death, not mere devastation. Mutually assured destruction by hydrogen bombs and arsenals of anthrax could go far toward the biological insanity of species suicide. The story less often told is our first steps toward global housekeeping: conscious responsibility for the health of our planet.[15]

The idea of a species-organism is not new. Neither is helping other people. However, setting concrete goals for global action on human welfare is new. Two days after John Kennedy announced the race to the moon, he visited the United Nations. Speaking to the General Assembly, he declared that if humanity could send a man to the moon, it could act on a world scale to see that no one goes hungry.[16] The United Nations, which was founded to be a peacekeeping organization, had never before contemplated a universal *economic* goal, in spite of the words in its charter about promoting social progress. There were regional precedents, of course, including UNICEF's aid to the children of Europe, and the Marshall Plan to underwrite European recovery and development. But there were no precedents for formulating an economic goal for the entire world.

The UN established a working group with economist Hans Singer as its secretary to coordinate the economic actions of its agencies. The group sketched the elements of the first "development decade," 1961–1970, to accelerate economic and social development. During the development decade the world more than reached its stated economic goal: 5 percent growth of GNP in developing countries. This set the stage for new goals for development in the 1970s, 80s, 90s.[17]

In 1972 Maurice Strong and Barbara Ward chaired the first World Conference on the Environment, in Stockholm. This linked our species to its ecology, again in terms of action, and officially launched the environmental consciousness that has been growing ever since. To leap to September 1990, the present decade opened with the World Summit for Children, brainchild of James Grant of UNICEF, which set goals for global child health improvement, most now achieved.[18]

This is a beginning of active self-maintenance *as a species*. Being responsible for our planet is not just good sense but a step toward becoming a species that might consciously act for its own welfare and survival.

Health. Specieswide interventions in health are already famous. Smallpox was the first disease eradicated from the wild. The World Health Organization coordinated the effort. The last man to catch natural smallpox was 23-year-old Ali Maow Maalin, of Merka, Somalia. On October 13, 1997, he was kind enough to escort a family of out-of-town nomads taking their two sick children to Merka's smallpox isolation center—about 10 or 15 minutes' contact time. The family's 4-year-old daughter died two days later. Maow fell ill after nine days but was first diagnosed with malaria, then chicken-pox; meanwhile, he met other sick in hospital, and a series of friends stopped in to commiserate and to help him spend his pay as a hospital restaurant worker. By the time a friend, a male nurse, reported Maow's case, the World Health Organization felt obliged to vaccinate 54,777 people who might be infected by his contacts' contacts. Maow survived; in six months his scars were fading and he went back to work.[19] After that, wild smallpox was gone forever in human populations, though it still exists in a few carefully controlled laboratories. There is still the fearful prospect of its deliberate use in warfare if humans decide to release the scourge once more.[20]

The next disease on the hit list is polio. In my childhood, we were barred from our swimming holes in the heat of August, from terror of life in an iron lung. I can just remember Franklin D. Roosevelt as president, crippled while in office. Polio is gone from North and South America, but still in Madagascar children with wasted legs maneuver homemade wooden trolleys among the wheels of city traffic and reach up to beg at car windows. Yet even in Madagascar there were only eight confirmed new polio cases in 1997. Hand-lettered banners over the dirt streets urge village mothers to bring their children for vaccination, and health teams trek to remote villages carrying vaccine in insulated picnic coolers. The most massive success story is India, where for a few days of December 1996 three million health workers were deployed throughout the country to immunize all vulnerable children. It is likely that polio will disappear from Earth on the target date of 2000 or soon after.[21]

The World Health Organization and UNICEF have adopted a target of immunizing 80 percent of the world's children against other common diseases. That is enough to slow the spread of disease in most places to a standstill by the so-called herd effect—if enough members of a group are inoculated, the infectious disease will die out for lack of a pool of vulnerable people to transmit new cases. In 1980, coverage was about 20 percent; by 1990 it exceeded 80 percent. (Coverage is uneven, of course, with neglected spots like the

Sudan and Somalia and Washington, D.C.) Presidents and priests and imams and journalists and pop stars and school teachers promote the campaigns. Death rates have plummeted. By 1992, UNICEF estimated that abatement of measles, whooping cough, and tetanus was preventing 3.2 million child deaths a year, 9,000 a day. Of course immunization must be maintained. Each new cohort of babies needs vaccination, at least until the disease is eradicated from all populations. But it is happening.[22]

One more example of intervention on a global scale is iodized salt. Lack of iodine produces the world's commonest form of preventable mental deficiency. At an extreme, women miscarry or give birth to cretins—the medical term for children who suffer neurological damage because their mother's diet during her pregnancy was severely deficient in iodine. In 1995 some 5.7 million people were born with this condition; 655 million more had goiter, the outward sign of a thyroid overworking to try to gain iodine. Minor iodine deficiency leads to minor mental deficit—a loss of a few IQ points, a dullness in proportion to the mineral lack. The swathe of New York State where I grew up was once called the Goiter Belt. We shared that risk with about a third of the world's population. Some 1.6 billion people live on iodine-poor continental rock, from the Adirondacks, Alps, and Andes to the Yangtze and Zaire Rivers. The year that my daughter lived on student dormitory food beside the Yangtze, she and her Chinese roommate both grew "thick necks"—until the doctor advised them to start eating seaweed.

The prevention for cretinism, goiter, and that needless mental dullness is simply to iodize salt. Switzerland began it in 1892, the United States in 1924. The Incas knew the principle before Cortés. They sent for their salt to iodine-rich springs. A Swiss doctor copied them; then the technique returned home to the Andes from Switzerland. The global campaign started with the Summit for Children in 1990. By 1997 some 60 percent of the world's salt was iodized. It is a simple idea: so simple that at one meeting of African ministers of health, UNICEF provided each of them with a saucer, packages of salt bought in their own countries, and a few drops of starch to see if the salt's iodine turned the mixture blue. Any grade-school child can do the test; so can a government minister. Several who found their own countries' salt was lacking went home to change things.[23]

This is not to pretend that all the world's health problems are on the way to being solved. Malaria is on the rise, the target of a newly announced world campaign by WHO. AIDS still outstrips any medicinal cure; changing behavior is the only current defense. At least twenty new diseases have appeared in the last two decades, in the rich substrate of an interlinked population of 5–6 billion humans. World travel has increased exposure; AIDS jumped from chimpanzees to the human population in the 1920s but just in the last couple of decades became worldwide. Other diseases like ebola are lurking in the tropics, waiting to spread like wildfire. By 2020 increased tobacco use is expected to cause more than 10 million premature deaths per year and more disability than any other single disease, including AIDS, tuberculosis, or the greatest killer, simple dehydration from diarrhea.[24]

Again, efforts toward response are international, in medical research, in young people's education, in subsidizing the price of condoms, in the politics of smoking. Our answer is no longer confined to evolving a better immune system, or sexual reproduction with its promise of a newly defended generation of children. It is not even confined to what we can do to clean or quarantine our local neighborhood. We can plan and mount global responses—the beginnings of an immune system for the species.

Wealth. Global public health has had some notable successes. In contrast, the management of global wealth is in its infancy. We are torn between interventionist and hands-off philosophies, and not at all clear about the results of either approach. The underfunded IMF is no match in power for the 1.5 trillion dollars of hot money electronically dispatched across national boundaries every day, whose flight can potentially bring any country's economy to its knees.

The world's 225 billionaires had a combined wealth of over a trillion dollars in 1997, equal to the yearly income of the poorest 47 percent of the world—2.5 billion people. The billionaires invest most of those trillions in factories, banks, oil rigs, and so forth, not so much in yachts and houses and artwork. Still, for less than 4 percent of the combined total wealth of those 225 billionaires, or well under 1 percent of global yearly income, the world could achieve *and maintain* universal basic education, plus universal basic health care, plus reproductive health care for all women, plus adequate basic food, safe water, and sanitation for all. It sounds as though humans could actually afford some progress, if we had the political will—and the knowledge to bring it about.[25]

Woman power and population. An organism has some control over the growth of the parts of its own body. This is not done by central planning. Each cell or organ responds to what it senses in its own environment. Is human population growth adjusting to the pressures and possibilities of an interlinked world?

The largest annual cohort of babies ever born appeared sometime in the late 1980s, around 130 million of them. Since then the accelerating fall in birthrate has more than counterbalanced the still-rising number of adolescents reaching reproductive age. World population still grows, but year by year it grows more slowly.[26]

The change came sooner than people expected. Even ten years ago the UN Population Fund estimated that in 2050 world population would reach 10–14 billion, still sharply rising. By 1995 the estimate dropped to 7.8–12.5 billion, perhaps stabilized or falling by midcentury. As I started to write this book, the largest cohort of babies was predicted for the year 2001; now as I finish, statistics show the watershed passed before anyone knew.

Why has population growth slowed? Not from deaths. Killing people is a very inefficient population check. The 250 wars worldwide in this century killed off about 110 million people, averaging a mere 1.1 million a year. World War II alone accounted for 40 million of those, and there has been ample capacity to rebound in numbers since then. In 1980, 3–4 million children under five died of diarrhea, childhood's biggest killer. The spread of oral rehydration therapy has brought that down to 2.2 million per year in the late 1990s. Deaths from disease and war in this century, while horrible, have been a relatively trivial brake on soaring population growth, as compared to the 130 million births of the largest cohort.[27]

To check population, one must bring down the birth rate. Birth rate does not fall in conditions of misery. It falls when children are more likely to survive. In country after country, it has taken ten to twenty years after the first steep fall in child death rate for birth rate to follow downward. The trend is accelerating. In 1940, mortality of under-5-year-olds in the United Kingdom and United States was 70 deaths per thousand births. Today, only 70 countries out of 189 with data do as badly as that. The United States, ranked twenty-ninth from the top, has eight deaths per thousand, the United Kingdom has 7, Sweden, Singapore, and Finland just four.[28]

Limiting population growth does not come by fiat, or even just by contraceptive technology. The measures that correlates most strongly with both infant survival and lowered fertility are women's education, gender equity, and reproductive health. Then people (like dandelions) have fewer and stronger children. People, unlike dandelions, consciously choose to do so. And among people, very unlike dandelions, different parts of the species can help other parts to make that choice.

One of the most extraordinary changes of this half-century has been the emergence of the women's movement. The conferences of Nairobi and Beijing, and the Population Conference of Cairo, put forward an agenda of women's control over their own bodies that would have been unthinkable a hundred years ago. Cairo, in 1994, brought a dramatic three-way conflict. On one hand stood traditional opponents of birth control, a mix of different fundamentalisms not usually found as bedfellows. A second group of technocratic birth-control advocates thought largely in terms of handing out condoms or implanting time-release contraceptives under the skin. The third side was the vocal women's group who staked out the importance of women making their own decisions. The third group won.[29]

By 1998, most national governments have adopted the Beijing Women's' Conference Platform for Action. More than 70 percent of them have drawn up national action plans to address the needs and priorities of women in their countries. Of course some people and some societies have always recognized women's equality. It is one option of our species' behavior. It is now a specieswide option, promoted by feminists and the power of global television—a conscious international movement. To stop a woman knowing what other women think, you'd need thought police. In the long run, the thoughts leak through.[30]

> She is holding up the whole world.
> What you gonna do?
> You can't stop her
> You gonna just stand there and watch her with your
>   mouth open?
> Or are you gonna try to get down?
> You can't stop her
> She is holding up the whole world.[31]

Shaping the environment.   Weather does not know boundaries. Neither does pollution, water, even, to an extent, topsoil. Odd that the land itself can move downstream to another country, but it does. We are struggling with the Law of the Sea, the Ozone treaty, vague promises made in Kyoto and arguments in Argentina to deal with global warming. However vague, it would have been unthinkable even half a century ago that nations would see industrial efficiency and land devoted to forests as a global matter, not a local one, or recognize that we have imposed more demands on the world's environment between 1950 and 1990 than in all previous human history and prehistory. Many people have not yet got that message, but they will.[32]

Of course we prefer to ignore long-term environmental changes that are beyond individual memory, and far beyond the cycle of political elections. If sea level rises in New York and London over a century, those cities will simply find ways to tuck up their skirts and attempt to ignore the effect on Bangladesh. What impresses people is the smoke of forest fires that kills babies and helps topple governments in the streets of Jakarta, or drops airplanes out of the air in Singapore. Fires and floods in Australia, mudslides in Peru, storms that batter California and obliterate Nicaragua, do make us notice. If we react sooner rather than later, we might avoid some future misery, but sooner or later we will react. We do not simply have to lie down and take it. We can clean up our own mess.

We have the wit and power to save water, and to let forests and agro-forestry return to the headwaters of our water supply. (After the floods of 1998, China announced a forestry plan to make the Yangtze run clearer and the Yellow River less yellow. It is one of the few countries whose leaders admit at the start that cleanup must take generations.) After the green revolution, world food supply is now actually 20 percent higher per capita than it was in 1970, in spite of population growth. If we grow hungry, we can decide to control the worst of international over-fishing, or eat the 70 percent of world fish catch and majority of world grain production that now feeds livestock. We can in large part avoid local famines, as India and southern

Africa have done, through information, inter-regional transport, and preplanning.[33]

In ecology, as in economics, we can make no simple decision to steer a set course. Politics and economic interests won't let us. Even if they did, we are far too ignorant of environmental complexity. We must steer the ship gingerly, looking out for tides and icebergs. Still, for problem after problem of the environment, political will is lacking, not means or knowledge. An estimated $1.4 trillion worldwide of perverse governmental subsidies promote environmental pollution and other environmental damage—twice global military spending. Removing those misdirected subsidies would free up government funds and rechannel activity toward more useful ends.[34]

If we do not do any of this, but abandon our food, air, and water to the tragedy of the commons, that is also a decision. Trends now are not toward abandonment. They are toward control, either by powerful exploiters or by coalitions of environmentalists working through their governments. Or, occasionally, coalitions of both sides, when companies see their own benefit in environmental protection and improvement.

Other animals. The first true cells were symbioses of radically different bacterial species. The human species-organism also has its symbionts. We have lived in association with cows and wheat and rice and millet and house cats since the Neolithic period. Their genes have evolved along with our own. Now we insert genes one by one, so that the potato beetle falls poisoned from the potato, enriching the owners of the genes, and with unknown biological and economic effects on the rest of us. It might be too much to say of our symbionts that they are a part of us, but they are wholly involved in our species' system of survival, as we are in theirs.[35]

Wild species are not outside of us, either. Some take without giving, like the resourceful kitchen cockroach. Some are hitchhikers: zebra mussels, American starlings, Australian rabbits. Some are used:

ocean fishes, timber trees. We do not manage such resources efficiently, but the days are about gone when they were global commons with unlimited access for anyone with an ax or a boat. Some are creatures of beauty and wonder: the majestic tiger, its bones ground up for medicine, or the chimpanzee and bonobo, whose individuality rivals our own but whose end, too often, is bushmeat. Some are simply obliterated, without even a name. Our management decisions affect them all.

Again, the decisions are international, by nature for ocean fish, by economics for the beetle-eating potato, or—for the chimp, bonobo, tiger—by an intercontinental conflict of aesthetics versus a mistaken belief in aphrodisiacs. (If Viagra saves the tiger, it really is a great invention!) Decisions in the end reflect our love for wild animals and natural landscapes—open vistas and tantalizing trails to explore. That fascination for nature may be as deep in us as any other instinct—E. O. Wilson called it "biophilia"—dating from our long past on the African savanna, when nature was where we lived.[36]

Naturalists like me once went to the wild hoping to observe, admire, and leave no footprint. It is no longer possible to pretend we can, or only for a moment, like the poet who lets a butterfly alight upon her palm in late autumn sun:

> . . . Top-heavy, ticklish, nourished on a weed
> Dotted and dashed with signals I can't read,
>
> It comes in black, white, orange, blue and brown,
> Topples a moment and settles blandly down.
>
> Calm in the sun that made today its day.
>
> Be off, you.
> Do whatever it is you have to do.
>
> I do not kill, nor spare, nor pardon.
> There is no god walking in this garden.[37]

We are not gods, but the world has become our garden, where our species willy-nilly will kill or spare or pardon.

Specieswide health initiatives start to show us the path. Specieswide politics, from economic planning to women's movements to environmentalism, tentatively follow after. The point here is not to predict whether we create utopia or dystopia. The point is that to a very large extent, and for the first time in history, we know on a global scale that we are creating *something*.

We know we share one circumscribed land, giver of air, water, food, and outer beauty and knowledge. Our *oikos*, our house, our home, the planet where our species lives, our only Earth.

Purpose Enters Evolution

This book opened by saying that with the evolution of human consciousness, purpose came into the world. Natural selection mimics purpose. It blindly creates the exquisite complexity of living things. When Darwin showed how a purposeless mechanism could have done so, that brought existential despair. Darwinian nature seemed even worse than cruel. Darwinian nature was indifferent. Human hopes and loves weighed nothing in the scales of natural selection. The less fit died, or died childless, culled by their immediate environment. The laws acting around us include no law of morality or altruism.

But particulate individualism is only a part of the story. Cooperation between entities evolved even before life itself. The major transitions in evolution, the major increases in complexity, all arose from cooperation. Chemicals bonded in the prebiotic soup and elaborated as bacteria, then bacteria joined in the cellular community of one, then cells cloned multicellular bodies. Social groups of such bodies coalesced among insects and vertebrates. One social species communicates through speech, writing, and now electronic impulses. This is leading us to a fifth level of cooperation: specieswide, planetwide.

We examined the interplay of individualism and cooperation in the evolution of sex—an elaborate sharing of genes with another individual, to produce offspring unlike either parent. We looked at the minuet of courtship, the displays and contests of one sex to attract and impress the other. We looked at our own displays, and how they bring our own sexes together. This led to a scenario of humans as evolving primates, sharing traits with our nearest cousins, the bonobos and chimpanzees. Our species diverged from our primate kin on the threshold of language and culture.

We narrowed the focus down to the development of one human being, in the womb and as baby and toddler. That person has all the biological inheritance of human intelligence, language instinct, potential gender identity. She/he forges a personality out of the interplay of gene and meme. In the growing child, as in evolution, intelligence and culture gradually become part of our species' essential traits. Like sexual reproduction, intelligence trades certainty for variability and the chance to change.

Now humans join in a whole new form of communication. Language allowed the ape to be human, a transitional creature with layers of social relations extending outward to larger and larger groups. The information revolution makes those links global. It is not simply language but electronic communication of language that gives humanity the potential to become a new biological entity.

Humanity's power over the rest of nature appeared in an evolutionary eye blink—at most 500 of our generations since the invention of agriculture. The agricultural revolution of the Neolithic almost implied the industrial revolution. Control of plants and animals led straight on to control of energy and machines. We now extend our interference into massive effect upon the oceans and the atmosphere and the remaining "wild" places. We are beginning to tinker with

Figure 21. Opening the Women's Global Forum, Beijing, 1995, with a torch and flame brought from newly democratic South Africa. Left to right: Dr. Achola pala Okeyo, Kenya; Her Excellency Tadelech Haile Mikael, Ethiopia; Dr. Inonge M. Lewanika, Zambia; Dr. Noeleen Heyser, Singapore; Dr. Laketch Dirasse, Ethiopia. (After *Women of China,* Special Issue. Courtesy of Dr. Noeleen Heyser, Executive Director of UNIFEM, The United Nations Development Fund for Women)

ourselves, vaccine by vaccine, gene by gene, consciously globalizing the kind of choice we have always made as individuals, in mating, in survival, in allowing others to survive.

It remains to be seen what role our evolved conscious purpose will play in using our new global power. We may choose despotism, ecological blight, death for other species, destruction for our own. Or else we may successfully improve our lot, stabilize our demands, preserve and enrich the biosphere. Biology has nothing to predict about which course we take. It only says that we are something new under the sun. We are bringing changes to the biosphere as important as the first true cell that survived to leave progeny to inhabit the earth, almost as drastic as the first green mote that made energy from sunlight. Just perhaps, we may become even more important, not as individuals but as a global organism.

Human individuals are the first of all earthly life forms to evolve a sense of purpose. The primates that would become humans lived in social neighborhoods for tens of millions of years; our neighborhood has now become the globe. It is not just humans but humanity that has the power to choose.

Gaia is not our mother. She could be our daughter.

NOTES / BIBLIOGRAPHY / INDEX

NOTES

PROLOGUE: BEYOND THE COPPER BEECH TREE

1. Maynard Smith, J., and Szathmáry, E. (1995).
2. *Sociobiology:* Wilson, E. O. (1975); *The Selfish Gene:* Dawkins, R. (1976).

1. THE EVOLUTION OF PURPOSE

1. Chance, M. R. A., and Mead, A. P. (1953); Jolly, A. (1966); Humphrey, N. (1976); Byrne, R. W., and Whiten, A., eds. (1988); Whiten, A., and Byrne, R. W., eds. (1997).
2. Savage-Rumbaugh, Sue (personal communication).
3. Dawkins, R. (1986), p. ix.
4. Tennyson, A. L., *In Memoriam* (1850), verse 56.
5. Gould, S. J. (1977).
6. Darwin, C. (1859).
7. Desmond, A., and Moore, J. (1991).
8. Dennett, D. C. (1995).
9. Desmond, A., and Moore, J. (1991), p. 236.
10. Paley, W. (1802).
11. Malthus, T. R. (1826); Martineau, H. (1832); Desmond, A., and Moore, J. (1991).
12. Desmond, A., and Moore, J. (1991).
13. Desmond, A., and Moore, J. (1991).
14. Aristotle (350 BC/1961); Dawkins, R. (1986).
15. Darwin, C. (1859), pp. 489–490.
16. Maynard Smith, J., and Szathmáry, E. (1995).

17. Lovelock, J. E. (1979); Hamilton, W. D., and Lenton, T. M. (1998); Hunt, L. (1998).
18. Dawkins, R. (1982).
19. Mayr, E., and Provine, W. (1980).
20. Everdell, W. R. (1997).
21. Mayr, Ernst (1991).
22. Weismann, A., in Maynard Smith (1989).
23. Adams, M. B. (1980).
24. Malthus, T. R. (1826).
25. Maynard Smith, J., and Szathmáry, E. (1995).
26. Bonner, J. T. (1988).
27. Knight, R. D., Freeland, S. J., and Landweber, L. F. (1999).
28. Darwin, C. (1859), pp. 489–490.
29. Dawkins, R. (1986); Dennett, D. C. (1995).
30. Diamond, J. M. (1996).
31. Beer, G. (1983).

2. LIFE, SEX, AND COOPERATION

1. Maynard Smith, J., and Szathmáry, E. (1995).
2. Simon, H. A. (1962).
3. Koestler, A. (1968).
4. Maynard Smith, J., and Szathmáry, E. (1995).
5. Maynard Smith, J., and Szathmáry, E. (1995).
6. Hayes, J. M. (1996); Mojzsis, S. J., Arrhenius, G., McKeegan, K. D., Harrison, T. M., Nutman, A. P., and Friend, C. R. L. (1996).
7. Maynard Smith, J., and Szathmáry, E. (1995).
8. Maynard Smith, J., and Szathmáry, E. (1995).
9. Maynard Smith, J., and Szathmáry, E. (1995).
10. Stearns, S. C. (1987).
11. Margulis, L., and Sagan, D. (1986).
12. Gould, J. L., and Gould, C. G. (1997).
13. Gould, J. L., and Gould, C. G. (1997).
14. Margulis, L., and Sagan, D. (1986).
15. Margulis, L., and Sagan, D. (1986).
16. Rozema, J., van de Staaij, J., Björn, L. O., and Caldwell, M. (1997).
17. Gould, J. L., and Gould, C. G. (1997).
18. Margulis, L., and Sagan, D. (1986).

19. Margulis, L. (1970); Wilson, E. B. (1925), quoted in Margulis, L. (1981).
20. Fenchel, T., and Finlay, B. T. (1994).
21. De Duve, C. (1996).
22. Kabnick, K. S., and Peattie, D. A. (1991); Fenchel, T., and Finlay, B. T. (1994).
23. Cloud, P. (1978).
24. Margulis, L., and Sagan, D. (1986).
25. Margulis, L., and Sagan, D. (1986).
26. Margulis, L., and Sagan, D. (1986).
27. Stearns, S. C. (1987); Gould, J. L., and Gould, C. G. (1997).
28. Hurst, L., and Hamilton, W. D. (1992).
29. Ankel-Simons, F., and Cummins, J. M. (1996); Eyre-Walker, A., Smith, N. H., and Maynard Smith, J. (1999).
30. Hurst, L. D., and McVean, G. T. (1996).
31. Biology and Gender Study Group (1989).
32. Martin, E. (1991).
33. Huxley, Aldous, "Fifth Philosopher's Song," in Huxley, A. (1967). Reprinted by permission of Random House UK Ltd on behalf of Chatto & Windus and of Harper Collins USA.
34. Hubbard, R. (1979); Martin, E. (1991).
35. Martin, E. (1991).

3. SEX—WHY BOTHER?

1. Stearns, S. C. (1987).
2. Gould, J. L., and Gould, C. G. (1997).
3. Stearns, S. C. (1987).
4. Stearns, S. C. (1987).
5. Williams, G. C. (1975); Maynard Smith, J. (1978).
6. Gould, J. L. (personal communication).
7. Muller, H. J. (1964); Kondrashov, A. S. (1988); Ridley, M. (1994a).
8. Crow (1999); Eyre-Walker and Keightley (1999).
9. Williams, G. C. (1975).
10. Bell, G. (1982); Ridley, M. (1994a).
11. Andrew, R. J. (1962b); Van Valen, L. (1973).
12. Hamilton, W. D., and Zuk, M. (1982).
13. Bell, G. (1988); Ridley, M. (1994a).
14. Jolly, A., "Homage to Hamilton and Zuk."

15. Lively, C. M. (1987); Ridley, M. (1994a); Gould, J. L., and Gould, C. G. (1997).
16. Crews, D. (1994).
17. Browne, R. A. (1993).
18. Concar, D. (1996); Anderson, J. B., and Kohn, L. M. (1998).
19. Gould, James L. (personal communication).
20. Stearns, S. C. (1987).
21. Stearns, S. C. (1987).
22. Forbes, L. S. (1997).
23. Forbes, L. S. (1997).
24. Bruner, J. S., Jolly, A., and Sylva, K., eds. (1976).
25. Jolly, A. (1997).
26. Dawkins, R. (1976).

4. COURTSHIP AND CHOICE

1. Trivers, R. L. (1972); Trivers, R. L. (1985).
2. Scheibinger, L. (1999).
3. Bishop, Morris, "Limericks Long after Lear," in Bishop, M. (1942), p. 141.
4. Bateman, A. J. (1948).
5. Trivers, R. L. (1985); Jones, I. L., and Hunter, F. M. (1993).
6. Lessing, D. (1994).
7. Gibson, R. M., and Langen, T. A. (1996).
8. Jolly, A. (1980); Wasserthal, L. T. (1998); Nilsson, L. A., Rabakonandrian-ina, E., and Pettersson, B. (1992).
9. Birkhead, T., and Møller, A. (1993).
10. Gibson, R. M., and Langen, T. A. (1996).
11. Erikson, C. J., and Zenone, P. G. (1976).
12. Gould, J. L., and Gould, C. G. (1997).
13. Lorenz, K. Z. (1952); Lorenz, K. (1966); Forsyth, A. (1986).
14. Goodall, J. (1986); de Waal, F. B. M. (1989); Nishida, T., Hosaka, K., Naka-mura, M., and Hamai, M. (1995); Nishida, T., and Hosaka, K. (1996).
15. Nishida, T., Hiraiwa-Hasegawa, M., Hasegawa, T., and Takanata, Y. (1985); Goodall, J. (1986).
16. de Waal, F. B. M. (1989), p. 69.
17. Reynolds, J. D., and Harvey, P. H. (1994).
18. Trivers, R. L. (1985); Gould, J. L., and Gould, C. G. (1997).
19. Smuts, B. (1992); Smuts, B. B., and Smuts, R. W. (1992).

20. Tutin, C. E. G. (1979).
21. Watts, D. P. (1998).
22. Smuts, B. (1995); United Nations (1995).
23. Rubenstein, D. (1994).
24. Barash, D. P. (1978).
25. MacKinnon, J. (1979); Galdikas, B. M. F. (1981).
26. Gould, J. L., and Gould, C. G. (1997); Wang, H., Paesen, G. C., Nuttall, P. A., and Barbour, A. G. (1998).
27. Andrade, M. C. B. (1996).
28. Stanford, C. B. (1995a); de Waal, F. (1996); de Waal, F. (1997).
29. Parker, Dorothy, "One Perfect Rose," copyright 1926, renewed ©1954 by Dorothy Parker, from *The Portable Dorothy Parker*. Used by permission of Penguin Putnam Inc and Gerald Duckworth & Co. Ltd.
30. Darwin, C. (1871).
31. Hamilton, W. D., and Zuk, M. (1982); Houde (1997).
32. Zahavi, A. (1975).
33. Diamond, J. (1992).
34. Fisher, R. A. (1930).
35. Ryan, M. J., Fox, J. H., Wilczynski, W., and Rand, A. S. (1990); Ryan, M. J. (1991).
36. Jolly, A., "The Tungára Frog."
37. Andersson, M. (1982).
38. Møller, A. P. (1992).
39. Maynard Smith, J. (1982).
40. Sinervo, B., and Lively, C. M. (1996).
41. Trivers, R. L. (1985).
42. Hatziolos, M., and Caldwell, R. L. (1983).
43. Strier, K. (1994).
44. Harcourt, A. H., Harvey, P. H., Larson, S. G., and Short, R. (1981); Harcourt, A. H., Purvis, A., and Liles, L. (1995); Forsyth, A. (1986).
45. Baker, R. R., and Bellis, M. A. (1993); Baker, R. R., and Bellis, M. A.(1995).
46. Baker, R. R., and Bellis, M. A. (1993).
47. Low, B. S., Alexander, R. D., and Noonan, K. M. (1987); Pond, C. (1997).
48. Frisch, R. E., and MacArthur, J. W. (1974); Frisch, R. E. (1976).
49. de Waal, F. B. M. (1995).
50. Diamond, J. (1992).
51. Diamond, J. (1992).
52. Bateson, P. (1980).

53. Wedekind, C., Seebeck, T., Bettens, F., and Paepke, A. J. (1995); Wedekind, C., and Furi, S. (1997).

54. Buss, D. M. (1994a); Buss, D. M. (1994b).

55. Berman, J. C. (in press).

56. Payne, K., and Payne, R. (1983); Payne, K., Tyack, P., and Payne, R. (1983); Guinee, L., and Payne, K. (1988).

57. Chapman, T., and Partridge, L. (1996); Rice, W. (1996).

5. CALCULATING LOVE

1. Darwin, C. (1859), p. 236.

2. Williams, G. C. (1966); Maynard Smith, J. (1978).

3. Hamilton, W. D. (1964); Hamilton, W. D. (1996).

4. Wilson, E. O. (1994), p. 319–320.

5. Wilson, E. O. (1994), p. 325.

6. Randall (1958), in Trivers, R. L. (1971), p. 35.

7. Trivers, R. L. (1971), p. 35.

8. Archílochus of Paros, p. 5, in Lattimore (1960). Reprinted by permission of the University of Chicago Press.

9. Hamilton, W. D. (1996).

10. Sober, E., and Wilson, D. S. (1998).

11. Wynne-Edwards, V. C. (1962); Williams, G. (1966); Maynard Smith, J. (1976).

12. Wilson, E. O. (1975); Dawkins, R. (1976).

13. Kilner, R., and Johnstone, R. A. (1997).

14. Trivers, R. L. (1974).

15. Haig, D. (1993).

16. Haig, D., and Trivers, R. (1995); Trivers, R. L. (1997).

17. Vrana, P. B., Guan, X.-J., Ingram, R. S., and Tiglman, S. M. (1998).

18. Sugiyama, Y. (1965); Sugiyama, Y. (1967); Mohnot, S. M. (1971).

19. Hrdy, S. B. (1977), p. 3.

20. Hrdy, S. B. (1977).

21. Hrdy, S. B. (1977).

22. Schaller, G. (1972); Sussman, R. W., Cheverud, J. M., and Bartlett, T. Q. (1995).

23. Daly, M., and Wilson, M. (1988).

24. Harcourt, A. H. (1988).

25. Scott, M. P. (1990); van Schaik, C. P., and Dunbar, R. I. M. (1990); van Schaik, C. P., and Kappeler, P. M. (1993); Dunbar, R. I. M. (1995a); Suss-

man, R. W., Cheverud, J. M., and Bartlett, T. Q. (1995); van Schaik, C. (1995); van Schaik, C. P. (1996).

26. Jakubowski, M., and Terkel, J. (1982); Perrigo, G., and vom Saal, F. (1994); Pusey, A. E., and Packer, C. (1994).

27. Goodall, J. (1986); Jolly, A. (1985); de Waal, F. (1996).

28. Eisenberg, J. F. (1981); Fossey, D. (1983); Hausfater, G., and Hrdy, S. B., eds. (1984); Parmigiani, S., and vom Saal, F., eds. (1994).

29. Shostack, M. (1981).

30. United Nations (1995).

31. Gadgil, M., and Bossert, W. (1970)

32. MacArthur, R. H., and Wilson, E. O. (1967).

33. MacArthur, R. H., and Wilson, E. O. (1967); Stearns, S. C. (1992).

34. Fisher, R. A. (1930).

35. Clutton-Brock, T. H., Guinness, F. E., and Albon, S. D. (1982); Clutton-Brock, T. H., Albon, S. D., and Guinness, F. . (1984).

36. UNICEF (1997).

37. Komdeur, J., Daan, S., Tinbergen, J., and Mateman, C. (1997).

38. Altmann, J. (1980); Altmann, J., Hausfater, G., and Altmann, S. A. (1988).

39. van Schaik, C. P., and Hrdy, S. B. (1991); Dittus, W. P. J. (1998).

40. Trivers, R. L., and Willard, D. E. (1973); Clutton-Brock, T. H., Guinness, F. E., and Albon, S. D. (1982); Clutton-Brock, T. H., Albon, S. D., and Guinness, F. E. (1984).

41. Wasser, S. K., and Norton, G. (1993); Komdeur, J., Daan, S., Tinbergen, J., and Mateman, C. (1997).

42. Snyder, R. G. (1961); Trivers, R. L. (1985); Mealey, L., and Mackey, W. (1990); Ridley, M. (1994b).

43. Diamond, J. (1992).

44. Shakespeare, W. (1599), *Henry V,* act IV, scene 3.

45. Huxley, T. H. (1863), p. 103.

46. *New York Review of Books,* November 13, 1975, in Wilson, E. O. (1994), pp. 337–338.

47. Dawkins, R. (1976).

48. Wilson, E. O., ed. (1988); Wilson, E. O. (1998).

49. Gould, S. J. (1997a); Gould, S. J. (1997b).

50. Gould, S. J. (1989).

51. Gould, S. J., and Lewontin, R. C. (1979).

52. Eldredge, N., and Gould, S. J. (1982); Gould, S. J., and Eldredge, N. (1993).

53. Jones, J. S. (1981).

54. Grant, P. R. (1986); Grant, B. R., and Gant, P. (1989).
55. Gould, S. J. (1989).
56. Bonner, J. T. (1998).
57. Kitcher, P. (1985); Gould, S. J. (1987).
58. George Williams, in Hausfater, G., and Hrdy, S. B., eds. (1984), foreword.

6. WOMEN IN THE WILD

1. Haraway, D. (1989).
2. Lawick-Goodall, J. v. (1971); Fossey, D. (1983); Fedigan, L. M. (1997); Scheibinger, L. (1999).
3. Burton, F. (1972); Pereira, M. E., and Fairbanks, L., eds. (1993).
4. Haraway, D. (1989).
5. Montgomery, S. (1991).
6. Wilson, E. O. (1984); Ramanunjan, A. K. (1985); Haraway, D. (1989); Said, E. (1978).
7. Said, E. (1978).
8. Fedigan, L. M. (1997).
9. Cartmill, M. (1991), p. 67–69.
10. Haraway, D. (personal communication).
11. Haraway, D. (1989), p. 1.
12. Goodall, J. (1964).
13. Haraway, D. (1989), p. 169.
14. Jane Goodall Institute (1998).
15. Jolly, A., and Jolly, M. (1990).
16. Haraway, D. (1991); Haraway, D. (1997).
17. White, T. H. (1938/1971), p. 52.
18. Fedigan, L., and Strum, S., eds. (in press).
19. Cartmill, M. (1991), pp. 74–75.
20. Goodall, J. (1968); Goodall, J. (1986).
21. Imanishi, K. (1965); Haraway, D. (1989); Asquith, P. (1991); Asquith, P. (1996); Takasaki, H. (1996); Hiraiwa-Hasegawa, M. (personal communication).
22. Itani, J., Tokuda, K., Furuya, Y., Kano, K., and Shin, Y. (1963), pp. 7–8.
23. Haraway, D. (1989); Fedigan, L. M. (1997).
24. Wrangham, R. W. (1980).
25. Crook, J. H. (1965); Crook, J. H., and Gartlan, J. S. (1966); Eisenberg, J. F. (1966); Eisenberg, J. F. (1981).
26. Wrangham, R. W. (1980); Altmann, J. (1990).

27. Altmann, J. (1974); Hrdy, S. B. (1977); Altmann, J. (1980); Wrangham, R. W. (1980); Hrdy, S. B. (1981).
28. Small, M., ed. (1984); Goodall, J. (1986); Haraway, D. (1989); Scheibinger, L. (in press).
29. Haraway, D. (1989); Tuttle, R. H. (1998).
30. Nelson, Howard, "For Dian Fossey," Nelson, H. (1997), p. 8. Reprinted by permission of the author.

7. LEMURS, MONKEYS, APES

1. Fleagle, J. G. (1988); Simons, E. (1993).
2. Napier, J. R. (1961); Cartmill, M. (1974); Fleagle, J. G. (1988).
3. Huxley, T. H. (1861).
4. Huxley, T. H. (1863), p. 98.
5. Jolly, A. (1980); Schmid, J., and Kappeler, P. M. (1994).
6. Alterman, L. (1995).
7. Richard, A. F., Goldstein, S. J., and Dewar, R. E. (1988); Kummer, H. (1995); Zhao, Q.-K. (1997).
8. Ennius, *Satires*, 2nd C BC: *Simia quam similis turpissima bestia nobis*, tr. Harold Coolidge. Quoted by Cicero, *De Natura Deorum* I.97. Cicero made the cladists' point that just because things resemble each other, they are not necessarily much alike. He suggested that elephants are actually more intelligent than monkeys, and thus more like humans, in spite of superficial appearances. Ennius' phrase "a nasty beast" may have become a long-lived cliché: Quintus Serenus Samnonicus used "turpissima bestia" in his 3rd century AD prescription for treating infected monkey or human bites. He suggested dressing them with tincture of spurge, still a medicinal plant—or just drinking wine of the proper vintage. With thanks to D. Hurley for tracking this down.
9. Diamond, J. (1992); Waddell, P. J., and Penny, D. (1996).
10. Nelson, Howard, "Gorilla Blessing," Nelson, H. (1993), p. 1. Reprinted by permission of the author.
11. Jolly, A. (1985); Foley, R. A., and Lee, P. C. (1989).
12. Charles-Dominique, P. (1977).
13. Rodman, P. (1979); Galdikas, B. M. F. (1988); Charles-Dominique (1995); Normile, D. (1998).
14. MacKinnon, J. (1974).
15. Galdikas, B. M. F. (1995b).
16. Bearder, S. K., and Martin, R. D. (1979); Galdikas, B. M. F. (1981); Sterling, E. J. (1994).

17. Reichard, U. (1995).
18. Garber, P. A., Encarnacion, F., Moya, L., and Pruetz, J. D. (1993).
19. van Schaik, C. P., and Dunbar, R. I. M. (1990); Dunbar, R. I. M. (1995a); van Scaik, C. (1996); van Schaik, C. P., and Kappeler, P. M. (1996).
20. Sussman, R. W., Cheverud, J. M., and Bartlett, T. Q. (1995); Dunbar, R. I. M. (1995a); Stanford, C. B. (1995a).
21. van Scaik, C. (1996); Pollock, J. I. (1979); Wright, P. C. (1990).
22. Goldizen, A. W. (1990); Garber, P. A. (1997).
23. Gould, J. L., and Gould, C. G. (1997).
24. Kummer, H. (1995).
25. Smuts, B. (1992); Kummer, H. (1995); Smuts, B. (1995).
26. Doran, D., and McNeilage, A. (1998).
27. Watts, D. P. (1989).
28. Fossey, D. (1983); Watts, D. P. (1985); Watts, D. P. (1990).
29. Marsh, C. W. (1979); Stanford, C. B. (1995b).
30. Engels, in Hubbard, R. (1979), p. 56.
31. Darwin, in Hubbard, R. (1979), p. 56.
32. Jolly, A. (1985); Hood, L. C., and Jolly, A. (1995).
33. Thanks for the phrase to Kathryn Clark.
34. Kawai, M. (1958); Altmann, J. (1980); Altmann, J., Hausfater, G., and Altmann, S. A. (1988).
35. Mori, A. (1979); Altmann, J., Hausfater, G., and Altmann, S. A. (1988).
36. Sugiyama, Y., and Ohsawa, H. (1982).
37. Sapolsky, R. M. (1980); Sapolsky, R. M. (1990).
38. Berard, J. D., Nürnberg, P., Epplen, J. T., and Schmidtke, J. (1993); Bercovitch, F. B. (1995).
39. Pereira, M. E., and Weiss, M. L. (1991); Fedigan, L. M., and Rose, L. M. (1995); Jolly, A. (1998).
40. Hrdy, S. B. (1981).
41. Itani, J., Tokuda, K., Furuya, Y., Kano, K., and Shin, Y. (1963); Wrangham, R. W. (1980).
42. de Waal, F. B. M. (1982); de Waal, F. B. M. (1989); Nishida, T., and Hosaka, K. (1996); Stanford, C. B. (1999).
43. Nishida, T., Hiraiwa-Hasegawa, M., Hasegawa, T., and Takanata, Y. (1985); Goodall, J. (1986); Wrangham, R., and Peterson, D. (1996).
44. Drury, R. (1729) Chagnon, N. A. (1988).
45. Milton, K. (1985); Strier, K. (1994).
46. Tutin, C. E. G. (1979); Goodall, J. (1986).
47. Gagneux, P., Boesch, C., and Woodruff, D. S. (1999).

48. Coolidge, H. J. (1933); Zihlman, A. J., Cronin, J. E., Cramer, D. L., and Sarich, V. M. (1978); Kano, T. (1992); de Waal, F., and Lanting, F. (1997).

49. Wrangham, R., and Peterson, D. (1996); de Waal, F., and Lanting, F. (1997).

50. Furuichi, T. (1987); Furuichi, T. (1992); Takahata, Y., Ihobe, H., and Idani, G. (1996).

51. Kano, T. (1992).

52. Wrangham, R. W., McGrew, W. C., de Waal, F. B. M., and Heltne, P. G. (1994); Hrdy, S. (personal communication).

53. Fossey, D. (1983), Warren Thomas (personal communication).

54. Normile, D. (1998); Takahata, H., and Takahata, Y. (1989).

55. Goodall, J. (1986); Takahata, H., and Takahata, Y. (1989); Normile, D. (1998).

8. HUMAN APES

1. Harcourt, A. H., Harvey, P. H., Larson, S. G., and Short, R. (1981); Fossey, D. (1983); Watts, D. P. (1990); Harcourt, A. H., Purvis, A., and Liles, L. (1995).

2. Baker, R. R., and Bellis, M. A. (1993).

3. Short, R. V. (1976).

4. Frank Beach, in Wolfe, L. D., Gray, J. P., Robinson, J. G., Lieberman, L. S., and Peters, E. H. (1982), p. 302.

5. Hrdy, S. B. (1977); Hrdy, S. B. (1993); Wallis, J. (1995).

6. Sillén-Tullberg, B., and Møller, A. P. (1993).

7. Stanford, C. B. (1995b); Stanford, C. B. (in press).

8. Diamond, J. (1997).

9. Morris, D. (1967).

10. Hrdy, S. B. (1981); Furuichi, T. (1987); Hrdy, S. B. (1993); Takahata, Y., Ihobe, H., and Idani, G. (1996).

11. Tutin, C. E. G. (1979).

12. Diamond, J. (1992); Diamond, J. (1997).

13. Kummer, H. (1995).

14. Smuts, B. (1992); Smuts, B. (1995).

15. Daly, M., and Wilson, M. (1983).

16. Nichols, Grace "Like a Flame," in Nichols, G. (1984). Reprinted by permission of Little, Brown for Virago Press.

17. Galdikas, B. M. F. (1981); Galdikas, B. (1995a).

18. Goodall, J. (1986); Watts, D. P. (1990).
19. Goodall, J. (1986); Bass, K. (1997).
20. Short, R. V. (1976).
21. Short, R. V. (1976); Blurton-Jones, N. G., Hawkes, K., and O'Connell, J. (1989).
22. White, T. H., in Warner, S. T. (1967/1989), p. 169.
23. Cheney, D. L. (1981); Wrangham, R., and Peterson, D. (1996).
24. Packer, C., Tatar, M., and Collins, A. (1998).
25. Diamond, J. (1992); Caro, T. M., Sellen, D. W., Parish, A., Frank, R., Brown, D. M., Voland, E., and Borgerhoff Mulder, M. (1995); Diamond, J. (1997).
26. Allman, J. M. (1999).
27. Diamond, J. (1997), pp. 131–132.
28. Waser, P. M. (1978); Foley, Charles (personal communication).
29. Swirszczynska, Anna, "Two Old Women," trans. Marshment, M., and Baran, G., in Rumens, C., ed. (1985), p. 52.

9. APISH INTELLIGENCE

1. Hubbard, R. (1979).
2. Darwin, C. (1871), in Hubbard, R. (1979), pp. 873–874.
3. Oakley, K. P. (1961); Goodall, J. (1963).
4. McGrew, W. C. (1981).
5. Byrne, R. W. (1997).
6. Byrne, R. W. (1993a); Byrne, R. W. (1996).
7. Parker, S. T., and Gibson, K. R. (1979).
8. Parker, S. T., and Gibson, K. R. (1979); Byrne, R. W. (1997).
9. Sugiyama, Y., and Koman, J. (1979); McGrew, W. C. (1992); Sugiyama, Y. (1995b); Sugiyama, Y. (1995a).
10. Robinson, J. G. (1986); Milton, K. (1988).
11. Galdikas, B. M. F. (1988).
12. Vander Wall, S. B. (1982); Smulders, T. V. (1997).
13. Menzel, C. R. (1997).
14. Washburn, S. L., and Lancaster, C. S. (1968); Zihlman, A. (1997); Stanford, C. B. (1999).
15. Stanford, C. B. (1995a); Diamond, J. (1997).
16. Boesch, C., and Boesch, H. (1989); Boesch, C. (1994); Stanford, C. B. (1999).
17. Stanford, C. B. (1999), p. 118.

18. Nishida, T. (1983); Griffin, D. R. (1984); Boesch, C. (1994); Nishida, T., and Hosaka, K. (1996); Stanford, C. B. (1999).
19. Gigerenzer, G. (1997).
20. Menzel, C. R. (1997).
21. Gould, Lisa (personal communication).
22. de Waal, F. B. M. (1982), p. 212.
23. Byrne, R., and Whiten, A., eds. (1988); Whiten, A., and Byrne, R. W., eds. (1997).
24. Byrne, R. W., and Whiten, A., eds. (1985).
25. Lawick-Goodall, J. v. (1971); de Waal, F. B. M. (1982); Goodall, J. (1986); Byrne, R. W., and Whiten, A. (1990).
26. Goodall, J. (1986); de Waal, F. B. M. (1982); Premack, D., and Woodruff, G. (1978).
27. Kummer, H. (1967); Kummer, H. (1995).
28. Jolly, A. (1996); Byrne, R. W., and Whiten, A. (1997).
29. Machiavelli, N. (1513/1961), in Byrne, R. W., and Whiten, A. (1997), p. 13.
30. de Waal, F. B. M. (1989); de Waal, F. (1996); Cords, M. (1997).
31. de Waal, F. B. M. (1989).
32. Kummer, H., Daston, L., Gegerenzer, G., and Silk, J. (1997).
33. Jolly, A. (1966); Kummer, H. (1967).
34. Kappeler, P. M. (1993).
35. Gouzoules, S., Gouzoules, H., and Marler, P. (1984); Pereira, M. E. (1993); Pereira, M. E. (1995); Chapais, B., Gauthier, C., Prud'homme, J., and Vasey, P. (1997); Nakamichi, M., and Koyama, N. (1997).
36. Vick, L. G., and Pereira, M. E. (1989); , L. (1996).
37. Pereira, Michael E. (personal communication), in Jolly, A. (1998).
38. Dunbar, R. I. M. (1992); Bass, K. (1995); Dunbar, R. I. M. (1995b); Dunbar, R. I. M. (1996); Rogers, E. M., Abernethy, K. A., Fontaine, B., Wickings, J. E., White, L. J. T., and Tutin, C. E. G. (1996).
39. Clutton-Brock, T. H., and Harvey, P. H. (1980); Byrne, R. W. (1993b).
40. Barton, R. A. (1996); Barton, R. A., and Dunbar, R. I. M. (1997).
41. Barton, R. A., and Dunbar, R. I. M. (1997); Dunbar, R. (1998b).
42. Simon, H. A. (1982); Gigerenzer, G. (1997).
43. Jolly, A. (1998).
44. Su Tung-p'o, "On the Birth of His Son," in Waley (1919). Copyright 1919 and renewed 1947 by Arthur Waley. Reprinted by permission of Alfred A. Knopf, Inc., and the Arthur Waley Estate.
45. Sugiyama, Y., and Koman, J. (1979); Jolly, A. (1985).
46. Fogel, R. W. (1997).

47. Hutchinson, G. E. (1981).
48. Gigerenzer, G. (1997).
49. Crick, F. (1994), in Barton, R. A., and Dunbar, R. I. M. (1997).
50. Simon, H. A. (1962); Dennett, D. C. (1991); Pinker, S. (1997).
51. Jolly, A. (1964); Smith, P. K. (1988); Cheney, D. L., and Seyfarth, R. M. (1990); Raps, S., and White, F. J. (1996).
52. Cosmides, L., and Tooby, J. (1992).
53. Cosmides, L., and Tooby, J. (1992); Gigerenzer, G. (1997).
54. Cosmides, L., and Tooby, J. (1992).
55. Gigerenzer, G. (1997); Pinker, S. (1997).

10. ORGANIC WHOLES

1. Leigh, E. G. (1971).
2. Webster, G., and Goodwin, B. (1996); Goodwin, B. (1997).
3. Kauffman, S. (1992); Lewin, R. (1992).
4. Reidl, R. (1978); Goldschmidt, K. (1995).
5. Fox Keller, E. (1994).
6. Koestler, A. (1968).
7. Shore, B. (1996); Everdell, W. R. (1997).
8. Sewell, E. (1951); Everdell, W. R. (1997).
9. Nyssen, Anne (personal communication); Everdell, W. R. (1997).
10. Everdell, W. R. (1997).
11. Sewell, E. (1951); Bishop, M. G. (1965).
12. Mallarmé, Stéphane, "Un Coup de Des," in Sewell, E. (1951), p. 10.
13. Rimbaud, Artur, "Le Bateau Ivre," in Bishop, M. G. (1965). Reprinted by permission of Harcourt, Brace & Co.
14. Swift, J. (1782).

11. A GENDERED BODY

1. Bush, S. (1978).
2. Bush, S. (1978).
3. Berta, P., Hawkins, J. R., Sinclair, A. H., Taylor, A., Griffiths, B. L., Goodfellow, R. N., and Fellous, M. (1990); Koopman, P., Gubbay, J., Vivian, N., Goodfellow, P., and Lovell-Badge, R. (1991).
4. Marx, J. (1995).
5. Weismann, A. (1891); Hunter, R. H. F. (1995).
6. Daly, M., and Wilson, M. (1983).

7. Behringer, R. R., Cafe, R. L., Froelick, G. J., Palmiter, R. D., and Brinster, R. L. (1990); Haqq, C. M., King, C.-Y., Ukiyama, E., Falsafi, S., Haqq, T. N., Donahoe, R. K., and Weiss, M. A. (1994).

8. Gahr, M. (1994).

9. Pinker, S. (1997).

10. Swain, A., Narvaez, V., Burgoyne, P., Camerino, G., and Lovell-Badge, R. (1988); Korach, K. S. (1994).

11. Hunter, R. H. F. (1995).

12. Swain, A., Narvaez, V., Burgoyne, P., Camerino, G., and Lovell-Badge, R. (1998).

13. O'Connell, H. E., Hutson, J. M., Anderson, C. R., and Plenter, R. J. (1998); Williamson (1998).

14. Money, J., and Ehrhardt, A. (1972); Fausto-Sterling, A. (1992).

15. Wilson, J. D. (1994); Hunter, R. H. F. (1995).

16. Daly, M., and Wilson, M. (1983); Hunter, R. H. F. (1995).

17. Money, J., and Ehrhardt, A. (1972); Daly, M., and Wilson, M. (1983).

18. Diamond, M. (1997); Diamond, M., and Sigmundson, M. D. (1997).

19. Diamond, M., and Sigmundson, M. D. (1997).

20. Diamond, M., and Sigmundson, M. D. (1997).

21. Angier, N. (1997a); Angier, N. (1997b).

22. Daly, M., and Wilson, M. (1983).

23. Morris, J. (1974), pp. 1–2, 20.

24. Hall, C. Stuart (personal communication).

25. Morris, J. (1974), p. 25.

26. Vines, G. (1992), p. 39.

27. Carlson, A. (1988).

28. Carlson, A. (1991).

29. Carlson, A. (1991).

30. Simpson, J. L., Ljungqvist, A., de la Chappelle, A., Ferguson-Smith, M. A., Genel, M., Carlson, A., Ehrhardt, A. A., and Ferris, E. (1993); Carlson, A. (1994); Carlson, A., Ehrhardt, A. A., and Ferris, E. (1993).

31. Vines, G. (1992); Helmstaedt, K., and Freeman, A. (1998).

32. Ward, Barbara (personal communication).

33. Profet, M. (1993).

34. Strassman, B. I. (1996a); Strassman, B. L. (1996b).

35. Dalton, K. (1964).

36. Graham, C. E., ed. (1981).

37. Sappho of Mytiléne, in Lattimore, R. (1960), pp. 39–40.

38. Greenwood, P. J., and Adams, J. (1987).
39. Forsyth, A. (1986).
40. Greenwood, P. J., and Adams, J. (1987).
41. Leigh, E. G. (1971).

12. INSTINCT, LEARNING, AND FATE

1. Tooby, J., and Cosmides, L. (1992), p. 33.
2. Tooby, J., and Cosmides, L. (1992), p. 27.
3. Tooby, J., and Cosmides, L. (1992), pp. 28–29.
4. Cornell University (1997–1998), p. 559.
5. Whitman, Walt, "I think that I could turn, and live with animals," in Whitman, W. (1855/1986), p. 32.
6. Pinker, S. (1997).
7. Pear, R. (1997).
8. Pinker, S. (1997).
9. James, W. (1892/1920), pp. 393–394, in Pinker, S. (1997), p. 185.
10. Lorenz (1935/1970).
11. Lorenz (1935/1970); Tinbergen (1951).
12. Burghardt, G. M. (1991).
13. Griffin, D. R. (1984).
14. Huxley, J. S. (1914), p. 500.
15. Huxley, J. S. (1914).
16. Eibl-Eibesfeldt, I. (1972); Eibl-Eibesfeldt, I. (1973); Meltzoff, A. N., and Moore, M. K. (1977); Jolly, A. (1985).
17. Darwin, C. (1872).
18. Tinbergen, N. (1951).
19. Morris, D. (1957/1970).
20. Moynihan, M. (1970).
21. Plomin, R. (1990); Plomin, R., Owen, M. J., and McGuffin, P. (1994).
22. Plomin, R. (1990).
23. Miller, S. K. (1993), p. 37.
24. Miller, S. K. (1993), p. 37.
25. Hamer, D. H., Hu, S., Magnuson, V. L., Hu, N., and Pattatucci, A. M. (1993); Pool, R. (1993).
26. King, M.-C. (1993).
27. King, M.-C. (1993), p. 289; Fausto-Sterling, A., and Balaban, E. (1993), p. 1257.

28. Holden, C. (1992); Holden, C. (1995).
29. Fay, R. E., Turner, C. F., Klassen, A. D., and Gagnon, J. H. (1989); Ferveur, J.-F., Störtkuhl, K. F., Stocker, R. F., and Greenspan, R. J. (1995).
30. Kinsey, C., Pomeroy, W. B., and Martin, C. E. (1949); Hamer, D. H., Hu, S., Magnuson, V. L., Hu, N., and Pattatucci, A. M. (1993).
31. Vasey, P. L. (1995); Vasey P. L. (1998).
32. Fildes, V. (1986); Fildes, V. (1988).
33. Miller, G. F. (1997).
34. Lorenz, K. Z. (1952); Lorenz, K. (1975).
35. Shore, B. (1996).
36. Feldman, M., and Laland, K. (1996).
37. Boyd, R., and Richerson, P. J. (1985); Boyd, R., and Richerson, P. J. (1996); Feldman, M., and Laland, K. (1996).
38. Whitman, Walt, "There Was a Child Went Forth," Whitman, W. (1855/1986), pp. 138–139.

13. THINKING

1. Hofstadter, D. R. (1979); Pinker, S. (1997).
2. Dennett, D. (1978), p. 124; Dennett, D. C. (1991).
3. Dennett, D. C. (1991), p. 177. Dennett credits the tenure analogy (a meme now widespread in academic circles) to neuroscientist Rodolfo Llinás.
4. Dennett, D. C. (1991); Pinker, S. (1997).
5. Descartes, R. (1637/1993); Dennett, D. C. (1991).
6. Hoffmann, E. T. A. (1816/1971); Hofstadter, D. R. (1979); Gibson, W. (1996); *The Stepford Wives*, directed by Bryan Forbes, screenplay by William Goldman.
7. Griffin, D. R. (1976); Griffin, D. R. (1984).
8. "The Centipede," attr. Mrs. Edmund Craster, d. 1874.
9. Hyman, S. E. (1998); Morris, J. S., Öhman, A., and Dolan, R. J. (1998).
10. Pinker, S. (1997).
11. Nagel, T. (1974); Crook, J. H. (1988); Jolly, A. (1991).
12. Bateson, G. (1955); Bruner, J. S., Jolly, A., and Sylva, K., eds. (1976).
13. Hayes, K., and Hayes, C. (1952).
14. Temerlin, M. K. (1975); Herrnstein, R. J., Loveland, D. H., and Cable, C. (1976).
15. Dasser, V. (1988).

16. Cheney, D. L., and Seyfarth, R. M. (1990).
17. Burton, F. (1972); Patterson, F., and Linden, E. (1981).
18. Hayaki, H. (1985).
19. Wrangham, R., and Peterson, D. (1996), pp. 254–255.
20. Gardner, R. A., and Gardner, B. T. (1969).
21. Hayes, C. (1951), in Jolly, A. (1985), p. 386.
22. Hayes, C. (1951), in Jolly, A. (1985), p. 386.
23. Kohler, W., in Bruner, J. S., Jolly, A., and Sylva, K., eds. (1976), p. 35.
24. Westergaard, G. C., and Suomi, S. J. (1997).
25. Morris, J. (1962).
26. Schiller, P. H. (1951); Morris, J. (1962).
27. Gardner, R. A., and Gardner, B. T. (1978).
28. Wrangham, R., and Peterson, D. (1996), p. 255.
29. Gallup, G. G. J. (1982).
30. Jolly, A. (1985).
31. Gallup, G. G. J. (1970); Gallup, G. G. J. (1982); Savage-Rumbaugh, E. S. (1986); Povinelli, D. J., Rulf, A. R., Landau, K. R., and Beirschwale, D. T. (1993); Patterson, F. G. P., and Cohn, R. H. (1994); Bass, K. (1995); Gallup, G. G. J., Povinelli, D. J., Suarez, S. D., Anderson, J. R., Lethmate, J., and Menzel, E. W., Jr. (1995).
32. Pinker, S. (1997).
33. Cheney, D. L., Seyfarth, R. M., and Palombit, R. (1996).
34. Wimmer, H., and Perner, J. (1983).
35. Whiten, A. (1997).
36. Premack, D., and Woodruff, G. (1978); Premack, D. (1988).
37. Whiten, A. (1997).
38. Povinelli, D. J., Nelson, K. E., and Boysen, S. T. (1990); Povinelli, D. J., Nelson, K. E., and Boyson, S. T. (1992); Povinelli, D. J., Parks, K. A., and Novack, M. A. (1992); Povinelli, D. J., and Eddy, T. J. (1996).
39. Byrne, R. (1995).
40. Tomasello, M., Savage-Rumbaugh, E. S., and Kruger, A. C. (1993).
41. Inoue, N., and Matsuzawa, T. (in press).
42. Hayes, K., and Hayes, C. (1952); Miles, H. L., Mitchell, R. W., and Harper, S. (1996).
43. Russon, A. E., and Galdikas, B. M. F. (1993); Bass, K. (1995); Russon, A. E., and Galdikas, B. M. F. (1995).
44. Caro, T. M., and Hauser, M. D. (1992).
45. Caro, T. M., and Hauser, M. D. (1992).

46. Russon, A. E. (1997), p. 196, after Boesch, D. (1991).
47. Lorenz, K. Z. (1952); Heinrich, B. (1989); Galef, B. G. (1992).
48. Jolly, A. (1985); McGrew, W. C. (1992).
49. McGrew, W. C. (1992); McGrew, W. C., Ham, R. M., White, L. J. T., Tutin, C. E. G., and Fernandez, M. (1997).
50. Boyd, R., and Richerson, P. J. (1996).
51. Hartung, J. (1976).
52. Heurtebize, G. (1986).
53. Hrdy, S. B. (1981); Gaulin, S. J. C., and Boster, J. S. (1990).
54. Dickmann, M. (1981).
55. Dickmann, M. (1981); Hrdy, S. B. (1981).
56. United Nations (1995); UNICEF (1997).
57. Fraser, G. M. (1994).

14. SPEAKING ADAPTIVELY

1. Pollack, R. (1994).
2. Liberman, A. M., and Mattingly, I. (1989); Pinker, S. (1994).
3. Kuhl, P. K., Williams, K. A., Lacerda, F., Stevens, K. N., and Lindblom, B. (1992); Pinker, S. (1994); Werker, J. F. (1989).
4. Pinker, S. (1994); Gopnik, M., Dalalakis, J., Fukuda, E., Fukuda, S., and Kehayia, E. (1996).
5. Byatt, A. S. (1996), p. 214.
6. Pollack, R. (1994).
7. Bacon, Francis, in Byatt, A. S. (1996), p. 359.
8. Mehler, J., Jusczyck, P., Lambertz, G., Halstead, N., Bertonincini, J., and Amiel-Tison, C. (1990).
9. Fernald, A. (1992).
10. Fernald, A. (1992).
11. Fernald, A. (1992); Pinker, S. (1994).
12. Gardner, Beatrix (personal communication).
13. Hayes, C. (1951); Gardner, R. A., and Gardner, B. T. (1978).
14. Gardner, Beatrix (personal communication).
15. Terrace, H. S. (1979); Premack, D., and Premack, A. J. (1983).
16. Miles, H. L. (1983); Linden, E. (1986).
17. Rumbaugh, D., ed. (1977); Savage-Rumbaugh, E. S. (1986).
18. Savage-Rumbaugh, S., and Lewin, R. (1994), pp. 135–136.
19. Savage-Rumbaugh, S., and Lewin, R. (1994); Bass, K. (1995).

20. Savage-Rumbaugh, S., and Lewin, R. (1994); Bass, K. (1995).
21. Byatt, A. S. (1996), p. 191.
22. Langer, S. (1957).

15. ARE BABIES HUMAN?

1. Portmann, A. (1941); Martin, R. D. (1983); Martin, R. D. (1990).
2. Trevathan, W. (1987).
3. Trevathan, W. (1987); Johanson, D. C., Edgar, B., and Brill, D. (1995); Rosenberg and Trevathan (1995).
4. Trevathan, W. (1987).
5. Kaiser, I. H., and Halberg, F. (1962).
6. Malek, J. (1952); Charles, E. (1953).
7. Jolly, A. (1972); Jolly, A. (1973).
8. Van Wagenen, G. (1992); Alford, P. L., Nash, L. T., Fritz, J., and Bowen, J. A. (1992); Pickering, T. G., James, G. D., Boddie, C., Harshfield, G. A., and Laragh, J. H. (1988).
9. Trevathan, W. (1987).
10. Blauvelt, Helen (personal communication).
11. Meltzoff, A. N., and Moore, M. K. (1977); DeCasper, A. J., and Fifer, W. P. (1980).
12. Fildes, V. (1986); Trevathan, W. (1987).
13. Small, M. F. (1998).
14. Small, M. F. (1998).
15. Small, M. F. (1998).
16. UNICEF (1995); UNICEF (1997).
17. Fildes, V. (1986).
18. Fildes, V. (1988).
19. Condon, W. S., and Sander, L. W. (1974).
20. Charles-Dominique, P. (1977).
21. McKenna, J. J. (1995); Small, M. F. (1998).
22. Seth, Vikram (1986). Copyright 1986 by Vikram Seth. Reprinted by permission of Random House, Inc., and Faber & Faber Ltd.
23. Andrew, R. (1962a).
24. Lorenz, K. Z. (1952); Bowlby, J. (1969); Jolly, A. (1985).
25. Wing, L. (1972); Leslie, A. M. (1991).
26. Sinclair, Jim, in Hobson, P. R. (1993).
27. Grandin, T. (1998).
28. Sacks, O. (1995); Grandin, T. (1998).

29. Pinker, S. (1997).
30. Piaget, J. (1954).
31. Jolly, A. (1985).
32. Piaget, J. (1954).
33. Wilkie, P. (1997), p. 64.
34. Pinker, S. (1994).
35. Kagan, J. (1971).
36. Freud, S. (1910/1965).
37. Bogin, B. (1988); Bogin, B. (1994).
38. Pereira, M. E., and Fairbanks, L., eds. (1993).
39. Frisch, R. E., and MacArthur, J. W. (1974); Frisch, R. E. (1976); Chehab, F. F., Mounzih, K., Lu, R., and Lim, M. E. (1997).
40. Tanner, J. M. (1962); Short, R. V. (1976).
41. Mead, M. (1928); Freeman, D. (1983); Shore, B. (1996).
42. Stevie Smith, "The Conventionalist," in Smith, S. (1972), p. 184. Reprinted by permission of New Directions Publishing Co.
43. Jolly, A. (1985); Grabruker, M. (1988).
44. Doyle, J. A. (1985).
45. Morelli, G. A. (1997).

16. FEET FIRST, BRAINS LATER

1. Landau, M. (1991).
2. Landau, M. (1991).
3. Lee, M. O. (1994); Wilson, E. O. (1998).
4. Darwin, C. (1871).
5. Oakley, K. P., and Hoskins, C. R. (1950).
6. Dart, R. A. (1959), pp. 5–6.
7. Dart, R. A. (1959); Reader, J. (1981); Tattersall, I. (1995).
8. Holloway, R. L., and Shapiro, J. S. (1992); Falk, D., Hildebolt, C., and Vannier, M. (1994).
9. Gore, R. (1997).
10. Reader, J. (1981).
11. Tattersall, I. (1995).
12. Reader, J. (1981).
13. Tuttle, R., Webb, D., Weidl, E., and Baksh, M. (1990); Tuttle, R. H. (1990); Feibel, C. S., Agnew, N., Latimer, B., Demas, M., Marshall, F., Waane, S. A. C., and Schmid, P. (1995).
14. Johanson, D. C., and Edey, M. (1981); White, T. D., Suwa, G., and Asfaw, B. (1994); Leakey, M., Feibel, C. S., McDougal, I., and Walker, A. (1995).

15. Tattersall, I. (1995); Hager, L. D. (1997b); Zihlman, A. (1997).

16. Milner, R. (1995); Tattersall, I. (1998a).

17. Zihlman, A. (1997), pp. 108–109.

18. Johanson, D. C., and Edey, M. (1981); White, T. D., Suwa, G., and Asfaw, B. (1994); Leakey, M., Feibel, C. S., McDougal, I., and Walker, A. (1995).

19. Vrba, E. S. (1988).

20. Hunt, K. D. (1998); Tuttle, R. H., Halgrimsson, B., and Stein, T. (1998); Goma, Lamack (personal communication).

21. Tattersall, I. (1993); Falk, D. (1997).

22. Darwin, C. (1871); Dart, R. A. (1959); Jolly, Arthur (personal communication).

23. Washburn, S. L., and Lancaster, C. S. (1968); Tattersall, I. (1995); Stanford, C. B. (1999).

24. Darwin, C. (1871).

25. Isaac, G. (1978); Parker, S. T., and Gibson, K. R. (1979); Lovejoy, C. O. (1981); Zihlman, A. L. (1981); McGrew, W. C., and Feistner, A. (1992).

26. Zihlman, A. L. (1981).

27. Lovejoy, C. O. (1981).

28. Falk, D. (1997).

29. Tattersall, I. (1995).

30. Wrangham, R., and Peterson, D. (1996); Sponheimer, M., and Lee-Thorp, J. A. (1999); Hunt (1998).

31. Hrdy, S. B. (1981); Stanford, C. B. (1995a); Diamond, J. (1997).

32. Diamond, J. (1992); Diamond, J. (1997).

33. Tattersall, I. (1995); Aiello, L. (1996).

34. Gibbons, A. (1996); Wynn, T. G. (1996).

35. Foley, R. (1995); Aiello, L. (1996); Foley, R. A. (1996); Gore, R. (1997); Gibbons, A. (1998), p. 97; Foley, R. (1998).

36. Tattersall, I. (1995).

37. Campbell, B. G. (1996).

38. Balter, M. (1996).

39. Cann, R., Stoneking, M., and Wilson, A. C. (1987).

40. Eyre-Walker, A., Smith, N. H., and Maynard Smith, J. (1999).

41. Cann, R., Stoneking, M., and Wilson, A. C. (1987).

42. Wolpoff, M. H. (1992).

43. Dorit, R. L., Akashi, H., and Gilbert, W. (1995); Hammer, M. F. (1995).

44. Wills, C. (1993); Dorit, R. L., Akashi, H., and Gilbert, W. (1995); Gibbons, A. (1995); Hammer, M. F. (1995).

45. Tishkoff, S. A., Dietzsch, E., Speed, W., Pakstis, A. J., Kidd, J. R., Cheung, K., Bonné-Tamir, B., Santachiara-Benerecetti, A. S., Moral, P., Krings, M., Pääbo, S., Watson, E., Risch, N., Jenkins, T., and Kidd, K. K. (1996).
46. Gibbons, A. (1997); Seielstad, M. T., Minch, E., and Cavalli-Sforza, L. (1998).
47. Tattersall, I. (1995).
48. Diamond, J. (1992).
49. White, R. (1993a); White, R. (1993b).
50. White, R. (1993b); White, R. (1993a).
51. Krings, M., Stone, A., Schmitz, R. W., Kraintitzid, H., Stoneking, M., and Pääbo, S. (1997); Wolpoff, M. (1998); Mountain, J. L. (1998).
52. White, R. (1992).
53. Wilford, J. N. (1998).
54. Pinker, S. (1994).
55. Pinker, S. (1994).
56. Bahn, P. G. (1995/96).
57. Lock, A., and Nobbs, M. (1996); Finkel, E. (1998); O'Connell, J. F. (1998); Grayson, D. K. (1998).
58. Tattersall, I. (1998b); Gutin, J. (1995); Aiello, L. (1996); Bahn, P. G. (1995/6); Marshack, A. (1996).
59. Mellars, P. (1996).
60. Pfeiffer, J. (1982); Conkey, M. W. (1996).
61. Conkey, M. W. (1996); White, Randall (personal communication).
62. White, Randall (personal communication).
63. Conkey, M. (1991); White, R. (1995); Bisson, M., and White, R. (1996).
64. Holden, C. (1998).
65. Lewin, R. (1998).
66. Aiello, L. (1996).
67. de Laguna, G. A. (1927); Dunbar, R. (1998a).
68. Dunbar, R. I. M. (1996).
69. Falk, D. (1997).

17. DEMANDING CONTROL

1. Sahlins, M. (1972).
2. Sahlins, M. (1972).
3. Pfeiffer, J. E. (1969).
4. Freuchen, P. (1935).
5. Shostack, M. (1981), pp. 341–342.

6. MacDonald, G. F. (1996).

7. Benedict, R. (1934).

8. Tamil epic, "The Story of the Anklet," in Morris, B. (1996), p. 72.

9. Morris, B. (1996).

10. Borgognini Tarli, S. M., and Repetto, E. (1996), Cohen, M. N. (1977), p. 286.

11. Didinga tribe, Uganda, "Song for a Dance of Young Girls," trans Driberg, J. J., in Busby, M., ed. (1992), p. 2.

12. Apolebieji, Odeniyi (Yoruba, Nigeria), "The Baboon," from Babalola, S. A. (1966), in Okpewho, I., ed. (1985), pp. 100–101.

13. Geertz, C. (1963); Bloch, M. (1971).

14. Boserup, E. (1965); Barrett, J. C. (1994).

15. Tuchman, B. W. (1978).

16. Boserup, E. (1965).

17. Boserup, E. (1965).

18. Decary, R. (1969); Bloch, M. (1971); Jolly, A. (1980).

19. Geertz, C. (1963).

20. Bloch, M. (1971).

21. Geertz, C. (1963).

22. Kapilar, "Ainkurunuru 203," in Ramanunjan, A. K. (1985), p. 10. Reprinted by permission of Columbia University Press via Copyright Clearance Center, Inc.

23. Fogel, R. W. (1997).

24. Silver, L. M. (1997).

25. Silver, L. M. (1997).

26. Silver, L. M. (1997).

27. Kolata, Gina, lecture, Princeton University, May 30, 1998.

28. Silver, L. M. (1997); Holden, C. (1998).

29. Silver, L. M. (1997).

18. THE GLOBAL ORGANISM

1. Phillips, Robert F. (personal communication).

2. Wilson, E. O. (1975); Wilson, E. O., ed. (1988); Wilson, E. O. (1998).

3. Wilson, E. O. (1998), p. 11.

4. Shore, B. (1996); Kuper, A. (1999).

5. Fisher, Michael (personal communication).

6. Hobbes, T. (1651) p. 1.

7. Dyson, G. (1997), p. 214.

8. Schoettle, E. C. B., and Grant, K. (1998); Butler, D. (1999).

9. Dyson, G. (1997).

10. Maddison, A. (1995); UNDP (1997).

11. Schoettle, E. C. B., and Grant, K. (1998).

12. UNDP (1997).

13. Orwell, G. (1949).

14. Wells, H. G. (1895); Silver, L. M. (1997).

15. Ward, B., and Dubos, R. (1972).

16. Singer, Hans, interview for the UN History Project (1997).

17. Singer, Hans, interview for the UN History Project (1997).

18. Ward, B., and Dubos, R. (1972); UNICEF (1991).

19. Shurkin, J. (1979).

20. Preston, R. (1998).

21. Kightlinger, Lon, M.D. (personal communication); UNICEF (1995); UNICEF (1997).

22. UNICEF (1997).

23. UNICEF (1995); UNICEF (1997).

24. Schoettle, E. C. B., and Grant, K. (1998).

25. UNDP (1998).

26. UNICEF (1998).

27. Sivard, R. L. (1996); UNICEF (1997); UNICEF (1991); UNICEF (1998).

28. UNICEF (1998).

29. Heyser, N., Kapoor, S., and Sandler, J., eds. (1995).

30. WEDO (1998).

31. Gossett, Hattie, "World View," in Gossett, H. (1988). Copyright 1988 by Hattie Gossett.

32. Turner, B. L. I., Clark, W. C., Kates, R. W., Richards, J., Mathews, J. T., and Meyer, W. B., eds. (1990).

33. Reuters (1998); Sen, A. (1981).

34. UNDP (1998).

35. Pollan, M. (1998).

36. Wilson, E. O. (1984).

37. Mitchell, Elma, "Late Fall," in Rumens, C., ed. (1985), pp. 30–31.

REFERENCES

Adams, M. B. 1980. Sergei Chetverikov, the Kol'tsov Institute, and the Evolutionary Synethsis. In Mayr and Provine, 1980: 242–278.

Aiello, L. 1996. Terrestriality, bipedalism, and the origin of language. *Proc. British Academy* 88: 269–290.

Alford, P. L., Nash, L. T., Fritz, J., and Bowen, J. A. 1992. Effects of management practices on the timing of captive chimpanzee births. *Zoo Biology* 11: 253–260.

Allman, J. M. 1999. *Evolving Brains.* New York: Scientific American.

Alterman, L. 1995. Toxins and toothcombs: potential allospecific chemical defenses in *Nycticebus and Perodicticus*. In Alterman, L., Doyle, G. A., and Izard, M. K., 1995: 413–425.

Alterman, L., Doyle, G. A., and Izard, M. K., eds. 1995. *Creatures of the Dark: The Nocturnal Prosimians.* New York: Plenum.

Altmann, J. 1974. Observational study of behavior: sampling methods. *Behaviour* 49: 227–267.

———. 1980. *Baboon Mothers and Infants.* Cambridge, MA: Harvard University Press.

———. 1990. Primate males go where the females are. *Anim. Behav.* 39: 193–195.

Altmann, J., Hausfater, G., and Altmann, S. A. 1988. Determinants of reproductive success in savanna baboons, *Papio cynocephalus*. In *Reproductive Success*, ed. T. H. Clutton-Brock. Chicago: University of Chicago Press, 403–418.

Anderson, J. B., and Kohn, L. M. 1998. Genotyping, gene genealogies, and genomics bring fungal population genetics above ground. *TREE* 13: 444–449.

References

Andersson, M. 1982. Female choice selects for extreme tail length in a widow bird. *Nature* 299: 818–820.

Andrade, M. C. B. 1996. Sexual selection for male sacrifice in the Australian Redback spider. *Science* 271: 70–72.

Andrew, R. 1962a. The situations that evoke vocalizations in primates. *Ann. N.Y. Acad. Sci.* 102: 296–315.

———. 1962b. Evolution of intelligence and vocal mimicking. *Science* 137: 585–589.

Angier, N. 1997a. New debate over surgery on genitals. *New York Times,* May 3, C1, C6.

———. 1997b. Sexual identity not pliable after all, report says. *New York Times,* March 14, A1, A18.

Ankel-Simons, F., and Cummins, J. M. 1996. Misconceptions about mitochondria and mammalian fertilization: implications for theories on human evolution. *Proc. Nat Acad. Sci., USA* 93: 13859–13863.

Aristotle (350 BC/1961). *Physica.* Lincoln: University of Nebraska Press.

Asquith, P. 1991. Primate research groups in Japan: orientations and East-West differences. In *The Monkeys of Arashiyama: Thirty-five Years of Research in Japan and the West,* ed. L. Fedigan and P. Asquith. New York: SUNY, 81–98.

———. 1996. *Japanese Constructs of Primate Societies.* Changing Images of Primate Societies: The Role of Theory, Method, Gender. Teresopolis, Brazil.

Babalola, S. A. 1966. *The Current and Form of Yoruba Ijala.* Oxford: Oxford University Press.

Bahn, P. G. 1995/96. New developments in Pleistocene art. *Evol. Anthrop.* 4: 204–215.

Baker, R. R., and Bellis, M. A. 1993. Human sperm competition: ejaculate adjustment by males and the function of masturbation. *Anim. Behav.* 46: 861–885.

———. 1995. *Human Sperm Competition: Copulation, Masturbation, and Infidelity.* London: Chapman and Hall.

Balter, M. 1996. Cave structure boosts Neanderthal image. *Science* 271: 449.

Barash, D. P. 1978. Rape among mallards. *Science* 201: 208.

Barkow, J. H., Cosmides, L., and Tooby, J. 1992. *The Adapted Mind.* New York: Oxford University Press.

Barrett, J. C. 1994. *Fragments from Antiquity: An Archaeology of Social Life in Britain 2900–1200 BC.* Oxford: Blackwell.

Barton, R. A. 1996. Neocortex size and behavioral ecology in primates. *Proc. Roy. Soc., Series B* 254: 63–68.

References

Barton, R. A., and Dunbar, R. I. M. 1997. Evolution of the social brain. In Whiten, A., and Byrne, R. W., 1997: 240–263.

Bass, K., producer. 1995. *The Monkey in the Mirror*. Bristol: BBC Natural History Unit.

———. 1997. *Bonobos of Wamba*. Bristol: BBC Natural History Unit.

Bateman, A. J. 1948. Intrasexual selection in *Drosophila*. *Heredity* 2: 349–368.

Bateson, G. 1955. A new theory of play and fantasy. *Psychiatric Research Reports* 2: 39–51.

Bateson, P. 1980. Optimal outbreeding and the development of sexual preferences in Japanese quail. *Z. Tierpsychol.* 53: 231–244.

Bearder, S. K., and Martin, R. D. 1979. The social organization of a nocturnal primate revealed by radio tracking. In *A Handbook of Biotelemetry and Radio-Tracking*, ed. C. J. Amlaner, Jr., and D. W. MacDonald. Oxford: Pergamon, 633–648.

Beer, G. 1983. *Darwin's Plots: Evolutionary Narrative in Darwin, George Eliot, and Nineteenth-Century Fiction*. London: Routledge, Kegan Paul.

Behringer, R. R., Cafe, R. L., Froelick, G. J., Palmiter, R. D., and Brinster, R. L. 1990. Abnormal sexual development in transgenic mice chronically expressing Müllerian inhibiting substance. *Nature* 345: 167–170.

Bell, G. 1982. *The Masterpiece of Nature*. London: Croom Helm.

———. 1988. *Sex and Death in Protozoa: History of an Obsession*. Cambridge: Cambridge University Press.

Benedict, R. 1934. *Patterns of Culture*. New York: Houghton Mifflin.

Berard, J. D., Nürnberg, P., Epplen, J. T., and Schmidtke, J. 1993. Male rank, reproductive behavior, and reproductive success in free-ranging rhesus macaques. *Primates* 34: 481–489.

Bercovitch, F. B. 1995. Female cooperation, consortship maintenance, and male mating success in savanna baboons. *Anim. Behav.* 50(1): 137–149.

Berman, J. C. 1999. Bad hair days in the paleolithic: modern (re)constructions of the cave man. *Am. Anthropol.* 101(2).

Berta, P., Hawkins, J. R., Sinclair, A. H., Taylor, A., Griffiths, B. L., Goodfellow, R. N., and Fellous, M. 1990. Genetic evidence equating *SRY* and the testis-determining factor. *Nature* 348: 448–452.

Biology and Gender Study Group. 1989. The importance of feminist critique for contemporary cell biology. In *Feminism and Science*, ed. N. Tuana. Bloomington, IN: Indiana University Press, 172–187.

Birkhead, T., and Møller, A. 1993. Female control of paternity. *TREE* 8: 100–104.

References

Bishop, M. G 1942. *Spilt Milk.* New York: G. P. Putnam's Sons.

———. 1965. *A Survey of French Literature.* New York: Harcourt, Brace and World.

Bisson, M., and White, R. 1996. L'imagerie féminine du paléolithique: étude des figurines de Grimaldi. *Culture* 16: 5–47.

Bloch, M. 1971. *Placing the Dead: Tombs, Ancestral Villages, and Kinship Organization in Madagascar.* London: Berkeley Square Press.

Blurton-Jones, N. G., Hawkes, K., and O'Connell, J. 1989. Measuring and modeling costs of children in two foraging societies. In *Comparative Socioecology: The Behavioral Ecology of Humans and Other Mammals,* ed. V. Standen and R. Foley. London: Blackwell Scientific, 367–390.

Boesch, C. 1994. Cooperative hunting in wild chimpanzees. *Anim. Behav.* 48: 653–667.

Boesch, C., and Boesch, H. 1989. Hunting behavior of wild chimpanzees in the Taï National Park. *Amer. J. Phys. Anthrop.* 78: 547–574.

Boesch, D. 1991. Teaching among wild chimpanzees. *Anim. Behav.* 41: 530–533.

Bogin, B. 1988. *Patterns of Human Growth.* New York: Cambridge University Press.

———. 1994. Adolescence in evolutionary perspective. *Acta Pediatrica Supple.* 406: 29–35.

Bonner, J. T. 1988. *The Evolution of Complexity.* Princeton: Princeton University Press.

———. 1998. The origins of multicellularity. *Integrative Biology* 1: 1–10.

Borgognini Tarli, S. M., and Repetto, E. 1996. Sex differences in human populations: change through time. In Morbeck, M. E., Galloway, A., and Zihlman, A., 1996: 198–209.

Boserup, E. 1965. *The Conditions of Agricultural Growth.* Chicago: Aldine.

Bowlby, J. 1969. *Attachment and Loss.* London: Hogarth Press.

Boyd, R., and Richerson, P. J. 1985. *Culture and the Evolutionary Process.* Chicago: University of Chicago Press.

———. 1996. Why culture is common, but cultural evolution is rare. *Proc. British Academy* 88: 77–93.

Browne, R. A. 1993. Sex and the single brine shrimp. *Natural History* 5/93: 35–39.

Bruner, J. S., Jolly, A., and Sylva, K., eds. 1976. *Play: Its Role in Development and Evolution.* New York: Basic Books.

Burghardt, G. M. 1991. Cognitive ethology and critical anthropomorphism: a snake with two heads and hog-nosed snakes that play dead. In *Cognitive Ethology: The Minds of Other Animals,* ed. C. A. Ristau. Hillsdale, NJ: Lawrence Erlbaum, 53–90.

Burton, F. 1972. The integration of biology and behavior in the socialization of *Macaca sylvana* of Gibraltar. In *Primate Socialization,* ed. F. Poirier. New York: Random House, 29–62.

Busby, M., ed. 1992. *Daughters of Africa.* New York: Penguin.

Bush, S. 1978. Nettie M. Stevens and the discovery of sex determination by chromosomes. *Isis* 69: 163.

Buss, D. M. 1994a. *The Evolution of Desire: Strategies of Human Mating.* New York: Basic Books.

———. 1994b. The strategies of human mating. *Amer. Scientist* 82: 238–249.

Butler, D. 1999. Internet may help bridge the gap. *Nature* 397: 10–11.

Byatt, A. S. 1996. *Babel Tower.* New York: Random House.

Byrne, R. W. 1993a. The complex leaf-gathering skills of mountain gorillas (*Gorilla g. beringei*): variability and standardization. *Amer. J. Primatol.* 31: 241–261.

———. 1993b. Do larger brains mean greater intelligence? *Behav., and Brain Sciences* 16: 696–697.

———. 1995. *The Thinking Ape: Evolutionary Origins of Intelligence.* Oxford: Oxford University Press.

———. 1996. The misunderstood ape: cognitive skills of the gorilla. In *Reaching into Thought: The Minds of the Great Apes,* ed. A. E. Russon, K. A. Bard, and S. T. Parker. Cambridge: Cambridge University Press, 111–130.

———. 1997. The technical intelligence hypothesis: an additional evolutionary stimulus to intelligence? In Whiten, A., and Byrne, R. W., 1997: 289–311.

Byrne, R. W., and Whiten, A. 1985. Tactical deception of familiar individuals in baboons. *Anim. Behav.* 33: 669–673.

———, eds. 1988. *Machiavellian Intelligence: Social Expertise and the Evolution of Intelligence in Monkeys, Apes, and Humans.* Oxford: Oxford University Press.

———. 1990. Tactical deception in primates: the 1990 data-base. *Primate Report* 27: 1–101.

———. 1997. Machiavellian intelligence. In Whiten, A., and Byrne, R. W., 1997: 1–24.

Campbell, B. G. 1996. An outline of human phylogeny. In Lock, A., and Peters, C. R., 1996: 31–52.

Cann, R., Stoneking, M., and Wilson, A. C. 1987. Mitochondrial DNA and human evolution. *Nature* 325: 32–36.

References

Carlson, A. 1988. Chromosome count. *Ms.* October: 40–44.

———. 1991. When is a woman not a woman? *Women's Sports and Fitness,* March: 24–29.

———. 1994. Sex/gender verification in international sports: the need to reexamine policy—and our notions of femininity, physical equality, and fair play. Women's Sports Federation.

Carlson, A., Ehrhardt, A. A., and Ferris, E. 1993. Gender verification in competitive sports. *Sports Medicine* 16: 305–315.

Caro, T. M., and Hauser, M. D. 1992. Is there teaching in non-human animals? *Quart. Rev. Biol.* 67: 151–174.

Caro, T. M., Sellen, D. W., Parish, A., Frank, R., Brown, D. M., Voland, E., and Borgerhoff Mulder, M. 1995. Termination of reproduction in nonhuman and human female primates. *Int. J. Primatol.* 16(2): 205–220.

Cartmill, M. 1974. Rethinking primate origins. *Science* 184: 436–443.

———. 1991. Book review of *Primate Visions: Gender, Race, and Nature in the World of Modern Science* by Donna Haraway. *Int. J. Primatol* 12: 67–75.

Chagnon, N. A. 1988. Life histories, blood revenge, and warfare in a tribal population. *Science* 239: 985–992.

Chance, M. R. A., and Mead, A. P. 1953. Social behavior and primate evolution. *Symposia of the Soc. of Exptl. Behaviour and Evolution* 7: 395–439.

Chapais, B., Gauthier, C., Prud'homme, J., and Vasey, P. 1997. Relatedness threshold for nepotism in Japanese macaques. *Anim. Behav.* 53: 1089–1101.

Chapman, T., and Partridge, L. 1996. Sexual conflict as fuel for evolution. *Nature* 381: 189–190.

Charles, E. 1953. The hour of birth. *Brit. J. Prev. Soc. Med.* 7: 43–59.

Charles-Dominique, P. 1977. *Ecology and Behavior of the Nocturnal Primates.* London: Duckworth.

———. 1995. Food distribution and reproductive constraints in the evolution of social structure: nocturnal primates and other mammals. In Alterman, L., Doyle, G. A., and Izard, M. K., 1995: 425–438.

Chehab, F. F., Mounzih, K., Lu, R., and Lim, M. E. 1997. Early onset of reproductive function in normal female mice treated with leptin. *Science* 275: 88–90.

Cheney, D. L. 1981. Intergroup encounters among free-ranging vervet monkeys. *Folia primatol.* 35: 124–146.

Cheney, D. L., and Seyfarth, R. M. 1990. *How Monkeys See the World: Inside the Mind of Another Species.* Chicago: University of Chicago Press.

467

References

Cheney, D. L., Seyfarth, R. M., and Palombit, R. 1996. Function and mechanisms underlying baboon "contact" barks. *Anim. Behav.* 52(3): 507–518.

Cloud, P. 1978. *Cosmos, Earth and Man: A Short History of the Universe.* New Haven: Yale University Press.

Clutton-Brock, T. H., Albon, S. D., and Guinness, F. E. 1984. Maternal dominance, breeding success, and birth sex ratios in red deer. *Nature* 308: 358–360.

Clutton-Brock, T. H., Guinness, F. E., and Albon, S. D. 1982. *Red Deer: Behavior and Ecology of Two Sexes.* Chicago: University of Chicago Press.

Clutton-Brock, T. H., and Harvey, P. H. 1980. Primates, brains, and ecology. *J. Zool., Lond.* 207: 151–169.

Cohen, M. N. 1977. *The Food Crisis in Prehistory: Overpopulation and the Origins of Agriculture.* New Haven: Yale University Press.

Concar, D. 1996. Sisters are doing it for themselves. *New Scientist,* August 17, pp. 32–36.

Condon, W. S., and Sander, L. W. 1974. Neonate movement is synchronized with adult speech: interactional participation and language acquisition. *Science* 183: 99–101.

Conkey, M. W. 1991. Contexts of action, contexts for power: material culture and gender in the Magdalenian. In *Engendering Archaeology: Women in Prehistory,* ed. M. Conkey and J. Gero. Oxford: Basil Blackwell, 57–92.

———. A history of the interpretation of European "Paleolithic art": magic, mythogram, and metaphors for modernity. In Lock, A., and Peters, C. R., 1996: 288–349.

Coolidge, H. J. 1933. *Pan paniscus:* pygmy chimpanzee from south of the Congo River. *Amer. J. Phys. Anthrop.* 18: 1–57.

Cords, M. 1997. Friendships, alliances, reciprocity, and repair. In Whiten, A., and Byrne, R. W., 1997: 24–49.

Cornell University. 1997–1998. *Courses of Study.* Ithaca, New York: Cornell University.

Cosmides, L., and Tooby, J. 1992. Cognitive adaptations for social exchange. In Barkow, J. H., Cosmides, L., and Tooby, J., 1992: 163–229.

Crews, D. 1994. Constraints to parthenogenesis. In Short, R. V., and Balaban, E., 1994: 23–52.

Crick, F. 1994. *The Astonishing Hypothesis.* London: Simon and Schuster.

Crook, J. H. 1965. The adaptive significance of avian social organizations. *Symp. Zool. Soc. Lond.* 18: 181–218.

———. 1988. The experiential context of intellect. In Byrne, R. W., and Whiten, A., 1988: 347–362.

References

Crook, J. H., and Gartlan, J. S. 1966. On the evolution of primate societies. *Nature* 210: 1200–1203.

Crow, J. F. 1999. The odds of losing at genetic roulette. *Nature* 397: 293–294.

Dalton, K. 1964. *The Premenstrual Syndrome.* Springfield, IL: Charles C. Thomas.

Daly, M., and Wilson, M. 1983. *Sex, Evolution and Behavior.* Boston: Willard Grant Press.

———. 1988. Evolutionary social psychology and family homicide. *Science* 242: 462–464.

Dart, R. A. 1959. *Adventures with the Missing Link.* New York: Harper and Brothers.

Darwin, C. 1859. *On the Origin of Species by Means of Natural Selection, or the Preservation of Favoured Races in the Struggle for Life.* London: John Murray.

———. 1871. *The Descent of Man and Selection in Relation to Sex.* New York: The Modern Library, Random House.

———. 1872. *The Expression of the Emotions in Man and Animals.* London: Murray.

Dasser, V. 1988. Mapping social concepts in monkeys. In Byrne, R. W., and Whiten, A., 1988: 85–93.

Dawkins, R. 1976. *The Selfish Gene.* Oxford: Oxford University Press.

———. 1982. *The Extended Phenotype.* Oxford: W. H. Freeman.

———. 1986. *The Blind Watchmaker.* New York: Norton.

de Duve, C. 1996. The birth of complex cells. *Scientific American,* April: 50–57.

de Laguna, G. A. 1927. *Speech: Its function and Development.* Bloomington: Indiana University Press.

de Waal, F. 1982. *Chimpanzee Politics: Power and Sex among Apes.* London: Jonathan Cape.

———. 1989. *Peacemaking among the Primates.* Cambridge, MA: Harvard University Press.

———. 1995. Bonobo sex and society. *Scientific American* 272(3): 58–64.

———. 1996. *Good Natured.* Cambridge: MA., Harvard University Press.

———. 1997. The Chimpanzee's service economy: food for grooming. *Evolution and Human Behav.* 18: 375–386.

de Waal, F., and Lanting, F. 1997. *Bonobo: The Forgotten Ape.* Berkeley: University of California Press.

Decary, R. 1969. *Souvenirs et Croquis de la Terre Malgache.* Paris: Editions Maritimes et d'Outre-Mer.

References

DeCasper, A. J., and Fifer, W. P. 1980. Of human bonding: newborns prefer their mothers' voices. *Science* 208: 1174–1176.

Dennett, D. 1978. *Brainstorms: Philosophical Essays on the Mind and Psychology.* Cambridge, MA: Bradford Books/MIT Press.

———. 1991. *Consciousness Explained.* Harmondsworth, UK: Penguin Books.

———. 1995. *Darwin's Dangerous Idea: Evolution and the Meanings of Life.* New York: Simon and Schuster.

Descartes, R. 1637/1993. *Discourse on Method.* Indianapolis: Hackett Publishing Co.

Desmond, A., and Moore, J. 1991. *Darwin.* New York: Warner Books.

Diamond, J. 1992. *The Third Chimpanzee.* New York: HarperCollins.

———. 1996. Competition for brain space. *Nature* 382: 756–757.

———. 1997. *Why Is Sex Fun? The Evolution of Human Sexuality.* London: Weidenfeld and Nicolson.

Diamond, M. 1997. Changing sex is hard to do. *Science* 275: 1745.

Diamond, M., and Sigmundson, M. D. 1997. Sex reassignment at birth. *Arch. Pediatr. Adolesc. Med.* 151: 298–303.

Dickmann, M. 1981. Paternal confidence and dowry competition: a biocultural analysis of purdah. In *Natural Selection and Social Behavior,* ed. R. D. Alexander and D. W. Tinkle. New York: Chiron Press, 417–438.

Dittus, W. P. J. 1998. Birth sex ratios in toque macaques and other mammals: integrating the effects of maternal condition and competition. *Behav. Ecol. Sociobiol.* 44: 149–160.

Doran, D., and McNeilage, A. 1998. Gorilla ecology and behavior. *Evolutionary Anthropol.* 6: 120–131.

Dorit, R. L., Akashi, H., and Gilbert, W. 1995. Absence of polymorphism at the ZFY locus on the human Y chromosome. *Science* 268: 1183–1185.

Doyle, J. A. 1985. *Sex and Gender.* Dubuque, Iowa: W. C. Brown.

Drury, R. 1729. *Madagascar, or Robert Drury's Journal during Fifteen Years' Captivity on That Island.* London: W. Meadow.

Dunbar, R. 1992. Neocortex size as a constraint on group size in primates. *J. Hum. Evol.* 22: 469–493.

———. 1993. Social organization of the gelada. In *Theropithecus: The Rise and Fall of a Primate Genus,* ed. N. Jablonski. Cambridge: Cambridge University Press, 425–439.

———. 1995a. The mating system of the Callitrichid Primates: I. Conditions for the co-evolution of pair-bonding and twinning. *Anim. Behav.* 50: 1057–1070.

———. 1995b. Neocortex size and group size in primates—a test of the hypothesis. *J. Hum. Evol.* 28: 287–296.

———. 1996. Determinants of group size in primates: a general model. *Proc. British Academy* 88: 33–58.

———. 1998a. *Grooming, Gossip, and the Evolution of Language*. Cambridge, MA: Harvard University Press.

———. 1998b. The social brain hypothesis. *Evol. Anthropol.* 6: 178–190.

Dyson, G. 1997. *Darwin among the Machines: The Evolution of Global Intelligence*. Reading, MA: Addison-Wesley.

Eibl-Eibesfeldt, I. 1972. Similarities and differences between cultures in expressive movements. In *Non-Verbal Communication*, ed. R. A. Hinde. Cambridge: Cambridge University Press, 297–314.

———. 1973. The expressive movement of the deaf-and-blind born. In *Social Communication and Movement*, ed. M. von Cranach and I. Vine. London: Academic Press.

Eisenberg, J. F. 1966. The social organization of mammals. *Handbuch Zool.* 8: 1–92.

———. 1981. *The Mammalian Radiations*. Chicago: University of Chicago Press.

Eldredge, N., and Gould, S. J. 1982. Punctuated equilibria: an alternative to phyletic gradualism. In *Models in Paleobiology*, ed. T. M. Schopf. San Francisco: Freeman, Cooper and Co., 82–115.

Erikson, C. J., and Zenone, P. G. 1976. Courtship differences in male ring doves: avoidance of cuckoldry? *Science* 192: 1353–1354.

Everdell, W. R. 1997. *The First Moderns: Profiles in the Origins of Twentieth-Century Thought*. Chicago: University of Chicago Press.

Eyre-Walker, A., and Keightley, P. D. 1999. High genomic deleterious mutation rates in hominids. *Nature* 397: 344–347.

Eyre-Walker, A., Smith, N. H., and Maynard Smith, J. 1999. How clonal are mitochondria? *Proc. Roy. Soc. B* 266: 477–483.

Falk, D. 1997. Brain evolution in females: an answer to Mr. Lovejoy. In Hager, L. D., 1997a: 114–136.

Falk, D., Hildebolt, C., and Vannier, M. 1994. Relationship of squamosal suture to asterion on external skull surfaces versus endocasts of pongids: implications for Hadar Early Hominid AL 162–28. *Amer. J. Phys. Anthrop* 93: 435–440.

Fausto-Sterling, A. 1992. *Myths of Gender: Biological Theories about Male and Female*. New York: Basic Books.

References

Fausto-Sterling, A., and Balaban, E. 1993. Genetics and male sexual orientation. *Science* 261: 1257–1258.

Fay, R. E., Turner, C. F., Klassen, A. D., and Gagnon, J. H. 1989. Prevalence and patterns of same-gender sexual contact among men. *Science* 243: 338–348.

Fedigan, L. M. 1997. Is primatology a feminist science? In Hager, L. D., 1997a: 56–75.

Fedigan, L. M., and Rose, L. M. 1995. Interbirth interval variation in three sympatric species of neotropical monkey. *Amer. J. Primatol.* 37: 9–24.

Fedigan, L., and Strum, S., eds. In press. *Gender and History in Primatology.* Chicago: Chicago University Press.

Feibel, C. S., Agnew, N., Latimer, B., Demas, M., Marshall, F., Waane, S. A. C., and Schmid, P. 1995. The Laetoli hominid footprints: a preliminary report on conservation and scientific restudy. *Evol. Anthropol.* 4: 149–154.

Feldman, M., and Laland, K. 1996. Gene-culture coevolutionary theory. *Tree* 11: 453–477.

Fenchel, T., and Finlay, B. T. 1994. The evolution of life without oxygen. *Amer. Scientist* 82: 22–29.

Fernald, A. 1992. Human maternal vocalizations to infants as biologically relevant signals: an evolutionary perspective. In Barkow, J. H., Cosmides, L., and Tooby, J., 1992: 391–428.

Ferveur, J.-F., Störtkuhl, K. F., Stocker, R. F., and Greenspan, R. J. 1995. Genetic feminization of brain structures and changed sexual orientation in male *Drosophila. Science* 267: 902–905.

Fildes, V. 1986. *Breasts, Bottles and Babies.* Edinburgh: Edinburgh University Press.

———. 1988. *Wet Nursing.* London: Basil Blackwell.

Finkel, E. 1998. Aboriginal groups warm to studies of early Australians. *Science* 280: 1342–1343.

Fisher, R. A. 1930. *The Genetical Theory of Natural Selection.* Oxford: Clarendon Press.

Fleagle, J. G. 1988. *Primate Adaptation and Evolution.* San Diego: Academic Press.

Fogel, R. W. 1997. The global struggle to escape from chronic malnutrition since 1700. *WFP/UNU Seminar, May 31, 1998:* 15–29.

Foley, R. 1995. *Humans before Humanity.* Oxford: Blackwell.

———. 1996. An evolutionary and chronological framework for human social behaviour. *Proc. British Academy* 88: 95–117.

———. 1998. Genes, evolution and diverstiy: yet another look at modern human origins. *Evol. Anthrop.* 6: 191–193.

References

Foley, R. A., and Lee, P. C. 1989. Finite social space, evolutionary pathways, and reconstructing hominid behavior. *Science* 243: 901–906.

Forbes, L. S. 1997. The evolutionary biology of spontaneous abortion in humans. *TREE* 12: 446–450.

Forsyth, A. 1986. *A Natural History of Sex.* New York: Scribners.

Fossey, D. 1983. *Gorillas in the Mist.* Boston: Houghton Mifflin.

Fox Keller, E. 1994. *Refiguring Life: Metaphors of Twentieth Century Biology.* New York: Columbia University Press.

Fraser, G. M. 1994. *Flashman and the Angel of the Lord.* London: HarperCollins.

Freeman, D. 1983. *Margaret Mead and Samoa: The Making and Unmaking of an Anthropological Myth.* Cambridge, MA: Harvard University Press.

Freuchen, P. 1935. *Arctic Adventure: My Life in the Frozen North.* New York: Farrar.

Freud, S. 1910/1965. Infantile Sexuality, from Three Contributions to Sexual Theory. In *The Child,* ed. W. Kessen. New York: John Wiley and Sons, 247–267.

Frisch, R. E. 1976. Critical metabolic mass and the age at menarche. *Ann. Hum. Biol.* 3: 489.

Frisch, R. E., and MacArthur, J. W. 1974. Menstrual cycles: fatness as a determinant of minimum weight for height necessary for their maintenance or onset. *Science* 185: 949–951.

Furuichi, T. 1987. Sexual swelling, receptivity and grouping of wild pygmy chimpanzees at Wamba, Zaire. *Primates* 28: 309–318.

———. 1992. The prolonged estrus of females and factors influencing mating in a wild group of bonobos (*Pan paniscus*) in Wamba, Zaire. In *Topics in Primatology,* Vol 2: *Behavior, Ecology, and Conservation,* ed. N. Itoiga, Y. Sugiyama, G. P. Sackett, and K. R. Thompson. Tokyo: University of Tokyo Press, 179–190.

Gadgil, M., and Bossert, W. 1970. Life history consequences of natural selection. *Am. Nat.* 104: 1–24.

Gagneux, P., Boesch, C., and Woodruff, D. S. 1999. Female reproductive strategies, paternity and community structure in wild West African Chimpanzees. *Anim. Behav.* 57: 9–12.

Gahr, M. 1994. Brain structure: causes and consequences of brain sex. In Short, R. V., and Balaban, E., 1994: 273–302.

Galdikas, B. 1981. Orang-utan reproduction in the wild. In *Reproductive Biology of the Great Apes,* ed. C. E. Graham. New York: Academic Press, 281–300.

———. 1988. Orangutan diet, range and activity at Tanjung Puting. *Int. J. Primatol.* 9: 1–37.

————. 1995a. *Reflections of Eden*. Boston: Little, Brown.

————. 1995b. Social and reproductive behavior of wild adolescent female orang-utans. In *The Neglected Ape*, ed. R. D. Nadler, B. M. F. Galdikas, L. K. Sheeran, and N. Rosen. New York: Plenum Press, 163–182.

Galef, B. G. 1992. The question of animal culture. *Hum. Nat.* 3: 157–178.

Gallup, G. G. J. 1970. Chimpanzees: self-recognition. *Science* 167: 86–87.

————. 1982. Self-awareness and the emergence of mind in primates. *Amer. J. Primatol.* 2: 237–248.

Gallup, G. G. J., Povinelli, D. J., Suarez, S. D., Anderson, J. R., Lethmate, J., and Menzel, E. W., Jr. 1995. Further reflections on self-recognition in primates. *Anim. Behav.* 50(6): 1525–1532.

Garber, P. A. 1997. One for all and breeding for one: cooperation and competition as a tamarin reproductive strategy. *Evol. Anthrop.* 5: 187–198.

Garber, P. A., Encarnacion, F., Moya, L., and Pruetz, J. D. 1993. Demographic and reproductive patterns in mustached tamarin monkeys (*Saguinus mystaxI*): implications for reconstructing platyrrhine mating systems. *Amer. J. Primatol.* 29: 235–254.

Gardner, R. A., and Gardner, B. T. 1969. Teaching sign language to a chimpanzee. *Science* 165: 664–672.

————. 1978. Comparative psychology and language acquisition. *Ann. N.Y. Acad. Sci.* 309: 37–76.

Gaulin, S. J. C., and Boster, J. S. 1990. Dowry as female competition. *Amer. Anthropologist* 92: 994–1005.

Geertz, C. 1963. *Agricultural Involution: The Process of Ecological Change in Indonesia*. Berkeley: University of California Press.

Gibbons, A. 1995. The mystery of humanity's missing mutations. *Science* 267: 35–36.

————. 1996. *Homo erectus* in Java: a 250,000-year anachronism. *Science* 274: 1841–1842.

————. 1997. Y Chromosome shows that Adam was an African. *Science* 278: 804–805.

————. 1998. Solving the brain's energy crisis. *Science* 280: 1345–1347.

Gibson, R. M., and Langen, T. A. 1996. How do animals choose their mates? *TREE* 11: 468–470.

Gibson, W. 1996. *Idoru*. New York: G. P. Putnam.

Gigerenzer, G. 1997. The modularity of social intelligence. In Whiten, A., and Byrne, R. W., 1997: 264–289.

Goldizen, A. W. 1990. A comparative perspective on the evolution of tamarin and marmoset mating systems. *Int. J. Primatol.* 11: 63–84.

References

Goldschmidt, K. 1995. *The Organism.* New York: Zone Books.

Goodall, J. 1963. My life among the wild chimpanzees. *Nat. Geog.* 124(2): 272–308.

———. 1964. *Miss Goodall and the Wild Chimpanzees.* Washington, DC: National Geographic Society.

———. 1968. The behavior of free-living chimpanzees in the Gombe Stream Reserve. *Anim. Behav. Monographs* 1(3): 165–311.

———. 1986. *The Chimpanzees of Gombe: Patterns of Behavior.* Cambridge, MA: Harvard University Press.

Goodwin, B. 1997. *The Emergence of Order.* International Congress on Artificial Life, Brighton, UK.

Gopnik, M., Dalalakis, J., Fukuda, E., Fukuda, S., and Kehayia, E. 1996. Genetic language impairment: unruly grammars. *Proc. British Academy* 88: 223–251.

Gore, R. 1997. The first steps. *Nat. Geographic* 191: 72–99.

Gossett, H. 1988. *Presenting . . . Sister No Blues.* Ithaca, NY: Firebrand Books.

Gould, J. L., and Gould, C. G. 1997. *Sexual Selection: Mate Choice and Courtship in Nature.* New York: Scientific American Library, W. H. Freeman and Co.

Gould, S. J. 1977. *Ever Since Darwin.* London: Burnett, 21–27.

———. 1987. *An Urchin in the Storm.* New York: Norton.

———. 1989. *Wonderful Life: The Burgess Shale and the Nature of History.* New York: Norton.

———. 1997a. Darwinian fundamentalism. *New York Review of Books* June 12, pp. 34–37.

———. 1997b. Evolution: the pleasures of pluralism. *New York Review of Books,* June 26, pp. 47–52.

Gould, S. J., and Eldredge, N. 1993. Punctuated equilibrium comes of age. *Nature* 336: 223–227.

Gould, S. J., and Lewontin, R. C. 1979. The spandrels of San Marco and the Panglossian paradigm: a critique of the adaptationist programme. *Proc. Roy. Soc. Lond. Series B* 205: 281–288.

Gouzoules, S., Gouzoules, H., and Marler, P. 1984. Rhesus monkey (*Macaca mulatta*) screams: representative signaling in the recruitment of agonistic aid. *Anim. Behav.* 32: 182–193.

Grabruker, M. 1988. *There's a Good Girl.* London: Women's Press.

Graham, C. E., ed. 1981. *Reproductive Biology of the Great Apes\.* New York: Academic Press.

Grandin, T. 1998. An Inside View of Autism. Center for the Study of Autism, www.autism.org/temple/inside/html.

Grant, B. R., and Grant, P. R. 1989. *Evolutionary Dynamics of a Natural Population: The Large Cactus Finch of the Galapagos.* Princeton: Princeton University Press.

Grant, P. R. 1986. *Ecology and Evolution of Darwin's Finches.* Princeton: Princeton University Press.

Grayson, D. K. 1998. Confirming antiquity in the Americas. Review of T. D. Dillehay, *Monte Verde: A Late Pleistocene Site in Chile* (Washington, DC: Smithsonian Press, 1997). *Science* 282: 1425–1426.

Greenwood, P. J., and Adams, J. 1987. *The Ecology of Sex.* London: Edward Arnold.

Griffin, D. R. 1976. *The Question of Animal Awareness.* New York: The Rockefeller University Press.

———. 1984. *Animal Thinking.* Cambridge, MA: Harvard University Press.

Gross, J., ed. 1995. *The Oxford Book of Comic Verse.* Oxford: Oxford University Press.

Guinee, L., and Payne, K. 1988. Rhyme-like repetitions in songs of humpback whales. *Ethology* 79: 295–306.

Gutin, J. 1995. Do Kenya tools root the birth of modern thought in Africa? *Science* 270: 1118–1119.

Hager, L. D. 1997a. *Women in Human Evolution.* London: Routledge.

———. 1997b. Sex and gender in paleoanthropology. In Hager, L. D., 1997a: 1–28.

Haig, D. 1993. Genetic conflicts in human pregnancy. *Quart. Rev. Biol* 68: 495–532.

Haig, D., and Trivers, R. 1995. Genomic imprinting: causes and consequences. In *Genomic Imprinting: Causes and Consequences,* ed. R. Ohlsson, K. Hall, and M. Ritzen. Cambridge: Cambridge University Press, 17–28.

Hamer, D. H., Hu, S., Magnuson, V. L., Hu, N., and Pattatucci, A. M. 1993. A linkage between DNA markers on the X chromosome and male sexual orientation. *Science* 291: 321–327.

Hamilton, W. D. 1964. The genetical theory of social behavior, I, II. *J. Theoretical Biology* 7: 1–52.

———. 1996. *The Narrow Roads of Gene Land: The Collected Papers of W. D. Hamilton.* Oxford: W. H. Freeman/Spektrum.

Hamilton, W. D., and Lenton, T. M. 1998. Spora and Gaia: how microbes fly with their clouds. *Ethology, Ecology and Evolution* 10: 1–16.

References

Hamilton, W. D., and Zuk, M. 1982. Heritable true fitness and bright birds: a role for parasites? *Science* 218: 384–387.

Hammer, M. F. 1995. A recent common ancestry of human Y chromosomes. *Nature* 378: 376–378.

Haqq, C. M., King, C.-Y., Ukiyama, E., Falsafi, S., Haqq, T. N., Donahoe, R. K., and Weiss, M. A. 1994. Molecular basis of mammalian sexual determination: activation of Müllerian inhibiting substance gene expression by SRY. *Science* 266: 1494–1499.

Haraway, D. 1989. *Primate Visions: Gender, Race and Nature in the World of Modern Science.* New York: Routledge.

———. 1991. *Simians, Cyborgs, and Women: The Reinvention of Nature.* New York: Routledge.

———. 1997. *Modest-Witness@Second-Millenium.FemaleMan-Meets-Onco-Mouse: Feminism and Technoscience.* New York: Routledge.

Harcourt, A. H. 1988. Evolution and family homicide (letter). *Science* 243: 462–463.

Harcourt, A. H., Harvey, P. H., Larson, S. G., and Short, R. 1981. Testis weight, body weight, and breeding system in primates. *Nature* 293: 55–57.

Harcourt, A. H., Purvis, A., and Liles, L. 1995. Sperm competition: mating system, not breeding season, affects testes size of primates. *Functional Ecology* 9(3): 468–476.

Hartung, J. 1976. On natural selection and the inheritance of wealth. *Curr. Anthropology* 17: 612–613.

Hatziolos, M., and Caldwell, R. L. 1983. Role reversal in courtship in the stomatopod *Psuedosquilla ciliata* (crustacea). *Anim. Behav.* 31: 1077–1087.

Hausfater, G., and Hrdy, S. B., eds. 1984. *Infanticide: Comparative and Evolutionary Perspectives.* Hawthorne, NY: Aldine.

Hayaki, H. 1985. Social play of juvenile and adolescent chimpanzees in the Mahale Mountains National Park, Tanzania. *Primates* 26: 343–360.

Hayes, C. 1951. *The Ape in Our House.* New York: Harper and Brothers.

Hayes, J. M. 1996. The earliest memories of life on earth. *Nature* 384: 21–22.

Hayes, K., and Hayes, C. 1952. Imitation in a home-raised chimpanzee. *J. Comp. Physiol. Psychol.* 45: 450–459.

Heinrich, B. 1989. *Ravens in Winter.* New York: Summit Books.

Helmstaedt, K., and Freeman, A. 1998. Blowing the whistle on drugs. *Toronto Globe and Mail,* October 31, pp. A26–A24.

References

Herrnstein, R. J., Loveland, D. H., and Cable, C. 1976. Natural concepts in pigeons. *J. Exp. Psychol.: Animal Behavior Processes* 2: 285–302.

Heurtebize, G. 1986. *Histoire des Afomarolahy (Extrème-sud de Madagascar.* Paris: CNRS.

Heyser, N., Kapoor, S., and Sandler, J., eds. 1995. *A Commitment to the World's Women.* New York: UNIFEM.

Hobbes, T. 1651. *Leviathan; or the Matter, Forme, and Power of a Commonwealth Ecclesiasticall and Civill.* London: Andrew Crooke.

Hobson, P. R. 1993. *Autism and the Development of the Mind.* E. Hove. Sussex, UK: Lawrence Erlbaum Associates.

Hoffmann, E. T. A. 1816/1971. The Sandman. In *A Romantic Storybook,* ed. M. G. Bishop. Ithaca, NY: Cornell University Press, 91–120.

Hofstadter, D. R. 1979. *Gödel, Escher, Bach: An Eternal Golden Braid.* New York: Basic Books.

Holden, C. 1992. Twin study links genes to homosexuality. *Science:* 33.

———. 1995. More on genes and homosexuality. *Science* 268: 1571.

———. 1998. The first gene marker for IQ? *Science* 280: 681.

———. 1999. Neanderthals left speechless? *Science* 283: 1111.

Holloway, R. L., and Shapiro, J. S. 1992. Relationship of squamosal suture to asterion in Pongids *(Pan):* relevance to early hominid brain evolution. *Amer. J. Phys. Anthrop.* 89: 275–282.

Hood, L. C., and Jolly, A. 1995. Troop fission in female *Lemur catta* at Berenty Reserve, Madagascar. *Int. J. Primatol.* 16: 997–1016.

Houde, A. E. 1997. *Sex, Color, and Mate Choice in Guppies.* Princeton: Princeton University Press.

Hrdy, S. B. 1977. *The Langurs of Abu.* Cambridge, MA: Harvard University Press.

———. 1981. *The Woman That Never Evolved.* Cambridge, MA: Harvard University Press.

———. 1993. Sex and the mating game. In *Reinventing the Future,* ed. T. A. Bass. Reading, MA: Addison-Wesley: 7–25.

Hubbard, R. 1979. Have only men evolved? In *Discovering Reality,* ed. S. Harding and M. B. Hintikka. New York: Schenkman, 45–69.

Humphrey, N. 1976. The social function of intellect. In *Growing Points in Ethology,* ed. P. P. G. Bateson and R. A. Hinde. Cambridge: Cambridge University Press, 307–317.

Hunt, K. D. 1998. Ecological morphology of *Australopithecus afarensis:* travelling terrestrially, eating arboreally. *Primate Locomotion: Recent Advances,* ed. E. Strasser, J. Fleagle, A. Rosenberger, and H. McHenry. New York: Plenum Press, 397–418.

Hunt, L. 1998. Send in the clouds. *New Scientist.* May 30, pp. 28–33.

Hunter, R. H. F. 1995. *Sex Determination, Differentiation, and Intersexuality in Placental Mammals.* Cambridge: Cambridge University Press.

Hurst, L. D., and Hamilton, W. D. 1992. Cytoplasmic fusion and the nature of the sexes. *Proc. Royal Soc. London: Series B* 247: 189.

Hurst, L. D., and McVean, G. T. 1996. . . ., and scandalous symbionts. *Nature* 381: 650–651.

Hutchinson, G. E. 1981. Random adaptation and innovation in human evolution. *Amer. Scientist* 69: 161–165.

Huxley, A. 1967. Fifth philosopher's song. In *Collected Poems.* London: Chatto & Windus.

Huxley, J. S. 1914. The courtship habits of the great crested grebe (*Podiceps cristatus*) with an addition to the theory of sexual selection. *Proc. Zool. Soc. Lond.* 35: 491–562.

Huxley, T. H. 1861. On the brain of *Ateles paniscus. Proc. Zool. Soc. Lond.* 1861: 249–260.

———. 1863. *Man's Place in Nature.* London: Macmillan.

Hyman, S. E. 1998. Neurobiology: a new image for fear and emotion. *Nature* 393: 417–418.

Imanishi, K. 1965. The origin of the human family: a primatological approach. In *Japanese Monkeys: A Collection of Translations,* ed. K. Imanishi and S. A. Altmann. Edmonton, Alberta: S. A. Altmann, 113–140.

Inoue, N., and Matsuzawa, T. In press. Chimpanzees' learning of nut-cracking in Bossue, Guinea.

Isaac, G. 1978. The food-sharing behavior of proto-human hominids. *Sci. American* 238(4): 90–108.

Itani, J., Tokuda, K., Furuya, Y., Kano, K., and Shin, Y. 1963. The social construction of natural troops of Japanese Monkeys in Takasakiyama. *Primates* 4(3): 1–42.

Jakubowski, M., and Terkel, J. 1982. Infanticide and caretaking in non-lactating *Mus musculus:* influence of genotype, family group and sex. *Anim. Behav.* 30: 1029–1035.

James, W. 1892/1920. *Psychology: Briefer Course.* New York: Henry Holt.

Jane Goodall Institute. 1998. JGI Newsletter, Jane Goodall Institute, Box 14890, Silver Spring, MD 20911–4890.

Johanson, D. C., and Edey, M. 1981. *Lucy: The Beginnings of Humankind.* New York: Simon and Schuster.

References

Johanson, D. C., Edgar, B., and Brill, D. 1995. *From Lucy to Language.* New York: Simon and Schuster.

Jolly, A. 1964. Prosimians' manipulation of simple object problems. *Anim. Behav.* 12: 560–570.

———. 1966. Lemur social behavior and primate intelligence. *Science* 153: 501–506.

———. 1972. Hour of birth in primates and man. *Folia primatol.* 18: 108–121.

———. 1973. Primate birth hour. *Int'l. Zoo Yearbk.* 13: 391–397.

———. 1980. *A World Like Our Own: Man and Nature in Madagascar.* New Haven: Yale University Press.

———. 1985. *The Evolution of Primate Behavior,* 2nd ed. New York: MacMillan.

———. 1991. Conscious chimpanzees? In *Cognitive Ethology: The Minds of Other Animals,* ed. C. A. Ristau. Hillsdale, NJ: Lawrence Erlbaum, 231–252.

———. 1996. Primate communication, lies, and ideas. In Lock, A., and Peters, C. R., 1996: 167–177.

———. 1997. Social intelligence and sexual reproduction: evolutionary strategies. In Morbeck, M. E., Galloway, A., and Zihlman, A., 1996: 262–269.

———. 1998. Pair-bonding, female aggression, and the evolution of lemur societies. *Folia primatol.* 69(Suppl. 1): 1–13.

Jolly, A., and Jolly, M. 1990. A view from the other end of the telescope. Review of Haraway, *Primate Visions. New Scientist,* April 21, p. 58.

Jones, I. L., and Hunter, F. M. 1993. Mutual selection in a monogamous seabird. *Nature* 362: 238–239.

Jones, J. S. 1981. Models of speciation—the evidence from *Drosophila. Nature* 289: 743.

Kabnick, K. S., and Peattie, D. A. 1991. *Giardia:* a missing link between prokaryotes and eukaryotes. *Amer. Scientist* 79: 34–43.

Kagan, J. 1971. *Change and Continuity in Infancy.* New York: John Wiley & Sons.

Kaiser, I. H., and Halberg, F. 1962. Circadian periodic aspects of birth. *Ann. N.Y. Acad. Sci.* 98: 1056–1067.

Kano, T. 1992. *The Last Ape: Pygmy Chimpanzee Behavior and Ecology.* Stanford, CA: Stanford University Press.

Kappeler, P. M. 1993. Reconciliation and post-conflict behavior in ringtailed *(Lemur catta)* and redfronted *(Eulemur fulvus rufus)* lemurs. *Anim. Behav.* 45: 901–915.

Kauffman, S. 1992. *The Origins of Order.* Oxford: Oxford University Press.

References

Kawai, M. 1958. On the system of social ranks in a natural troop of Japanese monkeys, 1: basic rank and dependent rank. *Primates* 1: 111–130.

Kellogg, W. N., and Kellogg, L. A. 1993. *The Ape and the Child: A Study of Environmental Influences upon Early Behavior.* New York: McGraw-Hill.

Kilner, R., and Johnstone, R. A. 1997. Begging the question: are offspring solicitation behaviors signs of need? *TREE* 12: 11–15.

King, M.-C. 1993. Sexual orientation and the X. *Nature* 364: 288–289.

Kinsey, C., Pomeroy, W. B., and Martin, C. E. 1949. *Sexual Behavior in the Human Male.* Philadelphia: Saunders.

Kitcher, P. 1985. *Vaulting Ambition.* Cambridge, MA: MIT Press.

Knight, R. D., Freeland, S. J., and Landweber, L. F. 1999. Selection, history, and chemistry: the three faces of the genetic code. *Trends in Biochem. Sci.* 24: 241–247.

Koestler, A. 1968. *The Ghost in the Machine.* New York: Macmillan.

Komdeur, J., Daan, S., Tinbergen, J., and Mateman, C. 1997. Extreme adaptive modification of sex ratio of the Seychelles warber's eggs. *Nature* 385: 522–525.

Kondrashov, A. S. 1988. Deleterious mutations and the evolution of sexual reproduction. *Nature* 336: 435–440.

Koopman, P., Gubbay, J., Vivian, N., Goodfellow, P., and Lovell-Badge, R. 1991. Male development of chromosomally female mice transgenic for *Sry. Nature* 351: 117–121.

Korach, K. S. 1994. Insights from the study of animals lacking a functional estrogen receptor. *Science* 266: 1524–1527.

Krings, M., Stone, A., Schmitz, R. W., Krainitzid, H., Stoneking, M., and Pääbo, S. 1997. Neandertal DNA sequences and the origin of modern humans. *Cell* 90: 1–20.

Kuhl, P. K., Williams, K. A., Lacerda, F., Stevens, K. N., and Lindblom, B. 1992. Linguistic experience alters phonetic perception by 6 months of age. *Science* 255: 606–608.

Kummer, H. 1967. Tripartite relations in hamadryas baboons. In *Social Communication among Primates*, ed. S. A. Altmann. Chicago: University of Chicago Press, 63–73.

———. 1995. *In Quest of the Sacred Baboon: A Scientist's Journey.* Princeton: Princeton University Press.

Kummer, H., Daston, L., Gegerenzer, G., and Silk, J. 1997. The social intelligence hypothesis. In *Human by Nature: Between Biology and the Social Sciences*, ed. P. Weingart, P. Richerson, S. D. Mitchell, and S. Maasen. Hillsdale, NJ: Lawrence Erlbaum, 157–179.

References

Kuper, A. 1999. *Culture: The Anthropologists' Account.* Cambridge, MA: Harvard University Press.

Landau, M. 1991. *Narratives of Human Evolution.* New Haven: Yale University Press.

Langer, S. 1957. *Philosophy in a New Key.* Cambridge, MA: Harvard University Press.

Lattimore, R. 1960. *Greek Lyrics.* Chicago: Phoenix Books, University of Chicago Press.

Lawick-Goodall, J. v. 1971. *In the Shadow of Man.* London: Collins.

Leakey, M., Feibel, C. S., McDougal, I., and Walker, A. 1995. New four-million-year-old hominid species from Kanapoi and Allia Bay, Kenya. *Nature* 376: 565–571.

Lee, M. O. 1994. *Wagner's Ring: Turning the Sky Round.* New York: Limelight Editions.

Leigh, E. G. 1971. *Adaptation and Diversity: Natural History and the Mathematics of Evolution.* San Francisco: Freeman, Cooper.

Leslie, A. M. 1991. The theory of mind impairment in autism: evidence for a modular mechanism in development? In *Natural Theories of Mind,* ed. A. Whiten. Oxford: Basil Blackwell, 63–78.

Lessing, D. 1994. *Under My Skin: Volume One of My Autobiography, to 1949.* London: HarperCollins.

Lewin, R. 1992. *Complexity: Life at the Edge of Chaos.* New York: Macmillan.

———. 1998. *Principles of Human Evolution.* Malden, MA: Blackwell Science.

Liberman, A. M., and Mattingly, I. 1989. A specialization for speech perception. *Science* 243: 489–494.

Linden, E. 1986. *Silent Partners: The Legacy of the Ape Language Experiments.* New York: Times Books, Random House.

Lively, C. M. 1987. Red Queen hypothesis supported by parasitism in sexual and clonal fish. *Nature* 344: 864–866.

Lock, A., and Nobbs, M. 1996. Australian aboriginal art. In Lock, A., and Peters, C. R., 1996: 351–357.

Lock, A., and Peters, C. R. 1996. *Handbook of Human Symbolic Evolution.* Oxford: Clarendon Press

Lorenz, K. 1935/1970. Companions as factors in the bird's environment. In *Konrad Lorenz: Studies in Animal and Human Behavior,* vol. 1, ed. R. Martin. London: Methuen, 101–258.

———. 1952. *King Solomon's Ring: New Light on Animal Ways.* London: Methuen.

———. 1966. *On Aggression.* London: Methuen and Co, Ltd.

References

———. 1975. *Evolution and Modification of Behavior.* Chicago: Chicago University Press.

Lovejoy, C. O. 1981. The Origin of Man. *Science* 211: 341–348.

Lovelock, J. E. 1979. *Gaia: A New Look at Life on Earth.* Oxford: Oxford University Press.

Low, B. S., Alexander, R. D., and Noonan, K. M. 1987. Human hips, breasts and buttocks: is fat deceptive? *Ethology and Sociobiology* 8: 249–258.

MacArthur, R. H., and Wilson, E. O. 1967. *The Theory of Island Biogeography.* Princeton: Princeton University Press.

MacDonald, G. F. 1996. *Haida Art.* Seattle: University of Washington Press.

Machiavelli, N. 1513/1961. *The Prince.* Harmnondsworth: Penguin Books.

MacKinnon, J. 1974. The behavior and ecology of wild orangutans. *Anim. Behav.* 22: 3–74.

———. 1979. Reproductive behavior in wild orangutan populations. In *The Great Apes,* ed. D. A. Hamburg and E. R. McCown. Menlo Park, CA: Benjamin Cummings, 257–274.

Maddison, A. 1995. *Monitoring the World Economy 1820–1922.* Paris: Organization for Economic Cooperation and Development.

Malek, J. 1952. The manifestation of biological rhythms in delivery. *Gyneaecologia* 133: 365–372.

Malthus, T. R. 1826. *An Essay on the Principle of Population.* London: Murray.

Margulis, L. 1970. *The Origin of Eukaryotic Cells.* New Haven: Yale University Press.

———. 1981. *Symbiosis in Cell Evolution.* San Francisco: W. H. Freeman and Co.

Margulis, L., and Sagan, D. 1986. *Origins of Sex: Three Billion Years of Genetic Recombination.* New Haven: Yale University Press.

Marsh, C. W. 1979. Comparative aspects of social organization in the Tana river red colobus *(Colobus badius rufomitratus).* Z. Tierpsychol. 51: 337–362.

Marshack, A. 1996. A middle Paleolithic symbolic composition from the Golan Heights: the earliest known depictive image. *Curr. Anthropol.* 37: 356–365.

Martin, E. 1991. The egg and the sperm: how science has constructed a romance based on stereotypical male-female roles. *Signs: J. of Women in Culture and Society* 16: 485–501.

Martin, R. D. 1983. Human brain evolution in ecological context. 52nd James Arthur Lecture on the Evolution of the Human Brain. New York: American Museum of Natural History.

———. 1990. *Primate Origins and Evolution: A Phylogenetic Reconstruction.* Princeton:, Princeton University Press.

References

Martineau, H. 1832. *Illustrations of Political Economy. No.6: Weal and Woe in Garveloch.* Boston: Leonard C. Bowles.

Marx, J. 1995. Snaring the genes that divide the sexes for mammals. *Science* 269: 1824–1825.

Maynard Smith, J. 1976. Group selection. *Quart. Rev. Biol.* 51: 277–283.

———. 1978. *The Evolution of Sex.* Cambridge: Cambridge University Press.

———. 1982. *Evolution and the Theory of Games.* Cambridge: Cambridge University Press.

———. 1989. Weismann and modern biology. *Oxford Studies in Evolutionary Biology* 5: 1–12.

Maynard Smith, J., and Szathmáry, E. 1995. *The Major Transitions in Evolution.* Oxford: W. H. Freeman.

Mayr, E. 1991. *One Long Argument: Charles Darwin and the Genesis of Modern Evolutionary Thought.* Cambridge, MA: Harvard Univeristy Press.

Mayr, E., and Provine, W. B., eds. 1980. *The Evolutionary Synthesis.* Cambridge, MA: Harvard University Press.

McGrew, W. C. 1981. The female chimpanzee as a human evolutionary prototype. In *Woman the Gatherer,* ed. F. Dahlberg. New Haven: Yale University Press, 35–72.

———. 1992. *Chimpanzee Material Culture: Implications for Human Evolution.* Cambridge: Cambridge University Press.

McGrew, W. C., and Feistner, A. 1992. Two nonhuman primate models for the evolution of human food sharing: chimpanzees and callitrichids. In Barkow, J. H., Cosmides, L., and Tooby, J., 1992: 229–243.

McGrew, W. C., Ham, R. M., White, L. J. T., Tutin, C. E. G., and Fernandez, M. 1997. Why don't chimpanzees in Gabon crack nuts? *Int. J. Primatol.* 18: 353–374.

McKenna, J. J. 1995. The potential benefits of infant-parent co-sleeping in relation to SIDS prevention: overview and critique of epidemiological bed-sharing studies. In *Sudden Infant Death Syndrome: New Trends in the Nineties,* ed. T. O. Rognum. Oslo: Scandinavian University Press, 256–265.

Mead, M. 1928. *Coming of Age in Samoa.* New York: W. Morrow.

Mealey, L., and Mackey, W. 1990. Variation in offspring sex ratio in women of differing social status. *Ethology and Sociobiology* 11: 83–95.

Mehler, J., Jusczyk, P., Lambertz, G., Halstead, N., Bertonincini, J., and Amiel-Tison, C. 1990. A precursor of language acquisition in young infants. *Cognition* 29: 143–178.

Mellars, P. 1996. The emergence of modern populations in Europe: a social and cognitive 'revolution?' *Proc. British Acad.* 88: 179–203.

References

Meltzoff, A. N., and Moore, M. K. 1977. Imitation of facial and manual gestures by human neonates. *Nature* 198: 75–78.

Menzel, C. R. 1997. Primates knowledge of their natural habitat: as indicated in foraging. In Whiten, A., and Byrne, R. W., 1997: 207–240.

Miles, H. L. 1983. Apes and language: the search for communicative competence. In *Language in Primates,* ed. J. de Luce and H. T. Wilder. New York: Springer-Verlag, 43–62.

Miles, H. L., Mitchell, R. W., and Harper, S. 1996. Simon says: the development of imitation in an enculturated orangutan. In *Reaching into Thought: The Minds of the Great Apes,* ed. A. E. Russon, K. A. Bard, and S. T. Parker. Cambridge: Cambridge University Press, 278–299.

Miller, G. F. 1997. Protean primates: the evolution of adaptive unpredictability in competition and courtship. In Whiten, A., and Byrne, R. W., 1997: 312–340.

Miller, S. K. 1993. To catch a killer gene. *New Scientist.* April 5, pp. 37–41.

Milner, R. 1995. Portraits of prehistory. *Natural History* 12/95: 44–47.

Milton, K. 1985. Mating patterns of woolly spider monkeys: implications for female choice. *Behav. Ecol. Sociobiol.* 17: 53–59.

———. 1988. Foraging behavior and the evolution of primate intelligence. In Byrne, R. W., and Whiten, A., 1988: 285–305.

Mohnot, S. M. 1971. Some aspects of social changes and infant-killing in the hanuman langur, *Presbytis entellus* (Primates, Cercopithecidae) in western India. *Mammalia* 35: 175–178.

Mojzsis, S. J., Arrhenius, G., McKeegan, K. D., Harrison, T. M., Nutman, A. P., and Friend, C. R. L. 1996. Evidence for life on earth before 3,800 million years ago. *Nature* 384: 55–59.

Møller, A. P. 1992. Female swallow preference for symmetrical male sexual ornaments. *Nature* 357: 238–240.

Money, J., and Ehrhardt, A. 1972. *Man and Woman, Boy and Girl: The Differentiation of Gender from Conception to Maturity.* Baltimore: Johns Hopkins Press.

Montgomery, S. 1991. *Walking with the Great Apes.* Boston: Houghton Mifflin.

Morbeck, M. E., Galloway, A., and Zihlman, A., eds. 1997. *The Evolving Female.* Princeton: Princeton University Press.

Morelli, G. A. 1997. Growing up female in a forager community and a farmer community. In Morbeck, M. E., Galloway, A., and Zihlman, A., 1996: 209–219.

Mori, A. 1979. Analysis of population changes by measurement of body weight in the Koshima Troop of Japanese monkeys. *Primates* 20: 371–398.

Morris, B. 1996. *Ecology and Anarchism.* Malvern Wells: Images Publishing.

Morris, D. 1962. *The Biology of Art.* London: Methuen.

———. 1967. *The Naked Ape.* London: Jonathan Cape.

———. 1957/1970. "Typical intensity" and its relationship to the problem of ritualization. In *Patterns of Reproductive Behavior: Collected Papers,* ed. D. Morris. London: Jonathan Cape, 230–243.

———. 1977. *Manwatching.* London: Jonathan Cape.

Morris, J. 1974. *Conundrum.* New York: Harcourt Brace.

Morris, J. S., Öhman, A., and Dolan, R. J. 1998. Conscious and unconscious emotional learning in the human amygdala. *Nature:* 467–470.

Mountain, J. L. 1998. Molecular evolution and modern human origins. *Evol. Anthropol.* 7: 21–38.

Moynihan, M. 1970. Control, suppression, decay, disappearance, and replacement of displays. *J. Theor. Biol.* 29: 85–112.

Muller, H. J. 1964. The relation of recombination to mutational advance. *Mutation Research* 1: 2–9.

Nagel, T. 1974. What is it like to be a bat? *Philosophical Rev.* 83: 435–450.

Nakamichi, M., and Koyama, N. 1997. Social relationships among ring-tailed lemurs *(Lemur catta)* in two free-ranging troop at Berenty Reserve, Madagascar. *Int. J. Primatol.* 18: 73–93.

Napier, J. R. 1961. Prehensility and opposability in the hands of primates. *Symp. Zool. Soc. Lond.* 5: 115–132.

Nelson, H. 1993. *Gorilla Blessing.* Moravia, NY: Falling Tree Press.

———. 1997. *Bone Music.* Troy, ME: Nightshade Press.

Nichols, G. 1984. *The Fat Black Woman's Poems.* London: Virago Press.

Nilsson, L. A., Rabakonandrianina, E., and Pettersson, B. 1992. Exact tracking of pollen transfer and mating in plants. *Nature* 360: 666–668.

Nishida, T. 1983. Alpha status and agonistic alliance in wild chimpanzees *(Pan troglodytes schweinfurthii).* *Primates* 24: 318–336.

Nishida, T., Hiraiwa-Hasegawa, M., Hasegawa, T., and Takanata, Y. 1985. Group extinction and female transfer in wild chimpanzees of the Mahale Mountains National Park. *Z. Tierpsychol.* 67: 284–301.

Nishida, T., and Hosaka, K. 1996. Coalition strategies among adult male chimpanzees of the Mahale Mountains, Tanzania. In *Great Ape Societies,* ed. W. C. McGrew, L. F. Marchant, and T. Nishida. Cambridge: Cambridge University Press, 114–134.

Nishida, T., Hosaka, K., Nakamura, M., and Hamai, M. 1995. A within-group gang attack on a young adult male chimpanzee: ostracism of an ill-mannered member? *Primates* 36(2): 207–212.

References

Normile, D. 1998. Habitat seen playing larger role in shaping behavior. *Science* 279: 1454–1455.

O'Connell, H. E., Hutson, J. M., Anderson, C. R., and Plenter, R. J. 1998. Anatomical relationship between the urethra and the clitoris. *J. Urology* 159: 1892–1897.

O'Connell, J. F. 1998. When did humans first arrive in Greater Australia and why is it important to know? *Evolutionary Anthropol.* 6: 132–146.

Oakley, K. P. 1961. *Man the Tool-Maker.* Chicago: University of Chicago.

Oakley, K. P., and Hoskins, C. R. 1950. New evidence on the antiquity of Piltdown man. *Nature* 165: 379–382.

Okpewho, I., ed. 1985. *The Heritage of African Poetry.* Burnt Mill, Harlow, Essex, UK: Longmans.

Orwell, G. 1949. *1984, a Novel.* London: Secker and Warburg.

Packer, C., Tatar, M., and Collins, A. 1998. Reproductive cessation in female mammals. *Nature* 392: 807–811.

Paley, W. 1802. *Natural Theology, or Evidences of the Existence and Attributes of the Deity, Collected from the Appearances of Nature.* Oxford: J. Vincent.

Parker, S. T., and Gibson, K. R. 1979. A developmental model for the evolution of language and intelligence in the early hominids. *Behav. and Brain Sci.* 2: 367–408.

Parmigiani, S., and vom Saal, F., eds. 1994. *Infanticide and Parental Care.* London: Harwood Academic Publishers.

Patterson, F., and Linden, E. 1981. *The Education of Koko.* New York: Holt, Rinehart and Winston.

Patterson, F. G. P., and Cohn, R. H. 1994. Self-recognition and self-awareness in lowland gorillas. In *Self-Awareness in Animals and Humans: Developmental Perspectives,* ed. S. T. Parker, R. W. Mitchell, and M. L. Boccia. New York: Cambridge University Press, 273–290.

Payne, K., and Payne, R. 1983. Large-scale changes over 19 years in songs of humpback whales in Bermuda. *Z. fur Tierpsychol.* 68: 89–114.

Payne, K., Tyack, P., and Payne, R. 1983. Progressive changes in the songs of humpback whales *(Megaptera novaeangeliae)*: a detailed analysis of two seasons in Hawaii. In *Communication and Behavior of Whales,* ed. R. Payne. Washington, DC: American Association for the Advancement of Science Symposia, 9–57.

Pear, R. 1997. States pass laws to regulate uses of genetic testing. *New York Times,* pp. A1, A9.

Pereira, M. E. 1993. Agonistic interaction, dominance relations, and ontogenetic trajectories in ringtailed lemurs. In *Juvenile Primates,* ed.

M. E. Pereira and L. Fairbanks. New York: Oxford University Press, 285–309.

———. 1995. Development and social dominance among group-living primates. *Amer. J. Primatol.* 37: 143–176.

Pereira, M. E., and Fairbanks, L., eds. 1993. *Juvenile Primates.* New York: Oxford University Press.

Pereira, M. E., and Weiss, M. L. 1991. Female mate choice, male migration, and the threat of infanticide in ringtailed lemurs. *Behav. Ecol. Sociobiol.* 28: 141–152.

Perrigo, G., and vom Saal, F. 1994. Behavioral cycles and the neural timing of infanticide and parental behavior in male house mice. In *Infanticide and Parental Care,* ed. S. Parmigiani and F. vom Saal. London: Harwood Academic, 365–369.

Pfeiffer, J. 1982. *The Creative Explosion.* New York: Harper and Row.

———. 1969. *The Emergence of Man.* New York: Harper and Row.

Piaget, J. 1954. *The Construction of Reality in the Child.* New York: Basic Books.

Pickering, T. G., James, G. D., Boddie, C., Harshfield, G. A., and Laragh, J. H. 1988. How common is white-coat hypertension? *JAMA* 259(2): 225–228.

Pinker, S. 1994. *The Language Instinct.* New York: William Morrow and Co.

———. 1997. *How the Mind Works.* New York: W. W. Norton.

Plomin, R. 1990. *Nature and Nurture: An Introduction to Human Behavioral Genetics.* Pacific Grove, CA: Brooks/Cole.

Plomin, R., Owen, M. J., and McGuffin, P. 1994. The genetic basis of complex human behaviors. *Science* 264: 1733–1739.

Pollack, R. 1994. *Signs of Life: The Language and Meanings of DNA.* Boston: Houghton Mifflin Co.

Pollan, M. 1998. Playing God in the garden. *New York Times Sunday Magazine,* October 25, 1998, pp. 44–51, 62–63, 82, 92–93.

Pollock, J. I. 1979. Female dominance in *Indri indri. Folia primatol.* 31: 143–164.

Pond, C. 1997. The biological origins of adipose tissue in humans. In Morbeck, M. E., Galloway, A., and Zihlman, A., 1996: 147–162.

Pool, R. 1993. Evidence for a homosexuality gene. *Science* 261: 291–292.

Portmann, A. 1941. Die Tragezeiten der Primaten und die Dauer der Schwangerschaft beim Menschen: Ein Prolem der vergleichenden Biologie. *Rev. suisse Zool.* 48: 511–518.

Povinelli, D. J., and Eddy, T. J. 1996. Chimpanzees: joint visual attention. *Psychological Science* 7: 129–135.

Povinelli, D. J., Nelson, K. E., and Boysen, S. T. 1990. Inferences about guessing and knowing by chimpanzees *(Pan troglodytes). J. Comp. Psychol.* 104: 203–210.

———. 1992. Comprehension of role reversal in chimpanzees: evidence of empathy? *Anim. Behav.* 43: 633–640.

Povinelli, D. J., Parks, K. A., and Novack, M. A. 1992. Role reversal by rhesus monkeys; but no evidence of empathy. *Anim. Behav.* 44: 269–281.

Povinelli, D. J., Rulf, A. R., Landau, K. R., and Beirschwale, D. T. 1993. Self-recognition in chimpanzees *(Pan troglodytes):* distribution, ontogeny, and patterns of emergence. *J. Comp. Psychol.* 107: 347–372.

Premack, D. 1988. Does the chimpanzee have a theory of mind? revisited. In Byrne, R. W., and Whiten, A., 1988: 160–179.

Premack, D., and Premack, A. J. 1983. *The Mind of an Ape.* New York: Norton.

Premack, D., and Woodruff, G. 1978. Does the chimpanzee have a theory of mind? *Behavioral and Brain Sciences* 1: 515–526.

Preston, R. 1998. The bioweaponeers. *The New Yorker.* March 9, pp. 52–65.

Profet, M. 1993. Menstruation as a defense against pathogens transported by sperm. *Quart. Rev. Biol.* 68: 335–386.

Pusey, A. E., and Packer, C. 1994. Infanticide in lions: consequences and counterstrategies. In *Infanticide and Parental Care,* ed. S. Parmigiani and F. vom Saal. London: Harwood Academic, 277–300.

Ramanunjan, A. K. 1985. *Poems of Love and War, from the Eight Anthologies and Ten Long Poems of Classical Tamil.* New York: Columbia University Press.

Raps, S., and White, F. J. 1996. Female social dominance in semi-free ranging ruffed lemurs *(Varecia variegata. Folia primatol.* 65: 163–168.

Reader, J. 1981. *Missing Links.* London: Collins.

Reichard, U. 1995. Extra-pair copulations in monogamous wild white-handed gibbons (Hylobates lar. *Z. Saugetierkunde* 60(3): 186–188.

Reidl, R. 1978. *Order in Living Organisms.* Chichester: John Wiley.

Reuters. 1998. China unveils plan to fight summer floods. *International Herald Tribune.*

Reynolds, J. D., and Harvey, P. H. 1994. Sexual selection and the evolution of sex differences. In Short, R. V., and Balaban, E., 1994: 53–70.

Rice, W. 1996. Sexually antagonistic male adaptation triggered by experimental arrest of female evolution. *Nature* 381: 232–234.

Richard, A. F., Goldstein, S. J., and Dewar, R. E. 1988. Weed macaques: the evolutionary implications of macaque feeding ecology. *Int. J. Primatol.* 10: 569–594.

Ridley, M. 1994a. *The Red Queen: Sex and the Evolution of Human Nature.* Harmondsworth, UK: Penguin Books.

————. 1994b. Why Presidents have more sons. *New Scientist*. December 3, pp. 28–31.

Robinson, J. G. 1986. Seasonal variation in the use of time and space by the wedge-capped capuchin monkey, *Cebus olivaceous:* Implications for foraging theory. *Smithsonian Contributions to Zoology* 431: 1–60.

Rodman, P. 1979. Individual activity patterns and the solitary nature of orangutans. In *The Great Apes,* ed. D. A. Hamburg and E. R. McCown. Menlo Park, CA: Benjamin/Cummings, 235–256.

Rogers, E. M., Abernethy, K. A., Fontaine, B., Wickings, J. E., White, L. J. T., and Tutin, C. E. G. 1996. Ten days in the life of a Mandrill horde in the Lope Reserve, Gabon. *Amer. J. Primatol.* 40(4): 297–314.

Rosenberg, K., and Trevathan, W. 1995. Bipedalism and human birth: the obstetrical dilemma revisited. *Evol. Anthropol.* 4:161–168.

Rozema, J., van de Staaij, J., Björn, L. O., and Caldwell, M. 1997. UV-B as an environmental factor in plant life: stress and regulation. *TREE* 12: 22–28.

Rubenstein, D. 1994. The ecology of female social behavior in horses, zebras, and asses. In *Animal Societies: Individuals, Interactions, and Organization,* ed. P. J. Jarman and A. Rossiter. Kyoto: Kyoto University Press, 13–28.

Rumbaugh, D., ed. 1977. *Language Learning by a Chimpanzee: The Lana Project.* New York: Academic Press.

Rumens, C., ed. 1985. *Making for the Open: The Chatto Book of Post-Feminist Poetry 1964–1984.* London: Chatto and Windus.

Russon, A. E. 1997. Exploiting the expertise of others. In Whiten, A., and Byrne, R. W., 1997: 174–206.

Russon, A. E., and Galdikas, B. M. F. 1993. Imitation in free-ranging rehabilitant orangutans (*Pongo pygmaeus*). *J. Comp. Psychol.* 107: 147–161.

————. 1995. Constraints on great apes' imitation: model and action selectivity in rehabilitant orangutan (*Pongo pygmaeus*) imitation. *J. Comp. Psychol.* 109: 5–17.

Ryan, M. J. 1991. Sexual selection and communication in frogs. *TREE* 6: 351–355.

Ryan, M. J., Fox, J. H., Wilczynski, W., and Rand, A. S. 1990. Sexual selection for sensory exploitation in the frog *Physalaemus pustulosus*. *Nature* 343: 66–67.

Sacks, O. 1995. *An Anthropologist on Mars: Seven Paradoxical Tales.* New York: Alfred A. Knopf.

Sahlins, M. 1972. *Stone Age Economics.* Chicago: Aldine-Atherton.

Said, E. 1978. *Orientalism.* New York: Pantheon.

References

Sapolsky, R. M. 1980. Styles of dominance and their endocrine correlates among wild olive baboons (*Papio anubis*). *Amer. J. Primatol.* 18: 1–13.

Sapolsky, R. M. 1990. Stress in the wild. *Scientific American*, January, pp. 116–123.

Savage-Rumbaugh, E. S. 1986. *Ape Language: From Conditioned Response to Symbol*. New York: Columbia University Press.

Savage-Rumbaugh, S., and Lewin, R. 1994. *Kanzi: The Ape at the Brink of the Human Mind*. New York: John Wiley.

Schaller, G. 1972. *The Serengeti Lion: A Study of Predator-Prey Relations*. Chicago: University of Chicago Press.

Scheibinger, L. 1999. *Has Feminism Changed Science? If So How, If Not Why Not?* Cambridge, MA: Harvard University Press.

Schiller, P. H. 1951. Figural preferences in the drawings of a chimpanzee. *J. Comp. Physiol. Psychol.* 44: 101–111.

Schmid, J., and Kappeler, P. M. 1994. Sympatric mouse lemurs (*Microcebus* spp.) in Western Madagascar. *Folia primat.* 63: 162–170.

Schoettle, E. C. B., and Grant, K. 1998. Globalization: a discussion paper. New York: The Rockefeller Foundation, 1–89.

Scott, M. P. 1990. Brood guarding and the evolution of male parental care in burying beetles. *Behav. Ecol. Sociobiol* 26: 31–39.

Seielstad, M. T., Minch, E., and Cavalli-Sforza, L. 1998. Genetic evidence for a higher female migration rate in humans. *Nature Genetics* 20: 278–280.

Sen, A. 1981. *Poverty and Famines: An Essay on Entitlement and Deprivation*. Oxford: Clarendon Press.

Seth, V. 1986. "The Golden Gate." In *The Golden Gate*. London: Faber & Faber Ltd.

Sewell, E. 1951. *The Structure of Poetry*. London: Routledge and Kegan Paul.

Shakespeare, W. 1599. *Henry V.*

Shore, B. 1996. *Culture in Mind: Culture, Cognition, and the Problem of Meaning*. Oxford: Oxford University Press.

Short, R. V. 1976. The evolution of human reproduction. *Proc. Roy. Soc. Lond. Series B* 195: 3–24.

———. 1994. Why sex? In *The Differences between the Sexes*. In Short, R. V., and Balaban, E., 1994: 3–23.

Short, R. V., and Balaban, E., eds. 1994. *The Differences between the Sexes*. Cambridge: Cambridge University Press.

Shostack, M. 1981. *Nisa: The Life and Words of a !Kung Woman*. Cambridge, MA: Harvard University Press.

Shurkin, J. 1979. *The Invisible Fire.* New York: G. P. Putnam's Sons.

Sillén-Tullberg, B., and Møller, A. P. 1993. The relationship between concealed ovulation and mating systems in Anthropoid primates: a phylogenetic analysis. *Amer. Nat.* 141: 1–25.

Silver, L. M. 1997. *Remaking Eden.* New York: Avon Books.

Simon, H. A. 1962. The architecture of complexity. *Proc. Amer. Philos. Soc.* 106: 470–473.

———. 1982. *Models of Bounded Rationality.* Cambridge, MA: MIT Press.

Simons, E. 1993. The fossil history of primates. In *The Cambridge Encyclopedia of Human Evolution,* ed. S. Jones, R. Martin, and D. Pilbeam. Cambridge: Cambridge University Press, 199–208.

Simpson, J. L., Ljungqvist, A., de la Chappelle, A., Ferguson-Smith, M. A., Genel, M., Carlson, A., Ehrhardt, A. A., and Ferris, E. 1993. Gender verification in competitive sports. *Sports Medicine* 16: 305–315.

Sinervo, B., and Lively, C. M. 1996. The rock-paper-scissors game and the evolution of alternative male strategies. *Nature* 380: 240–243.

Sivard, R. L. 1996. *World Military and Social Expenditures.* Washington, DC: World Priorities.

Small, M., ed. 1984. *Female Primates: Studies by Women Primatologists.* New York: Alan R. Liss.

———. 1998. *Our Babies, Ourselves: How Biology and Culture Shape the Way We Parent.* New York: Anchor Books.

Smith, P. K. 1988. The cognitive demand of children's interaction with peers. In Whiten, A., and Byrne, R. W., 1997: 309–326.

Smith, S. 1972. *The Collected Poems of Stevie Smith.* London: New Directions Publishing Co.

Smulders, T. V. 1997. How much memory do tits need? *TREE* 12: 417–418.

Smuts, B. 1992. Male aggression against women: an evolutionary perspective. *Human Nature* 3: 1–44.

———. 1995. The evolutionary origins of patriarchy. *Human Nature* 6(1): 1–32.

Smuts, B. B., and Smuts, R. W. 1992. Male aggression and sexual coercion of females in nonhuman primates and other mammals: evidence and theoretical implications. In *Advances in the Study of Behavior,* vol. 22, ed. P. J. B. Slater, J. S. Rosenblatt, M. Milinski and C. T. Snowdon. London: Academic Press, 1–61.

Snyder, R. G. 1961. The sex ratio of offspring of pilots of high performance military aircraft. *Human Biol.* 33: 1–10.

Sober, E., and Wilson, D. S. 1998. *Unto Others: The Evolution and Psychology of Unselfish Behavior.* Cambridge, MA: Harvard University Press.

Sponheimer, M., and Lee-Thorp, J. A. 1999. Isotopic evidence for the diet of an early hominid, *Australopithecus africanus*. *Science* 283: 368–370.

Stanford, C. B. 1995a. Chimpanzee hunting behavior and human evolution. *Amer. Scientist* 83: 256–261.

———. 1995b. The influence of chimpanzee predation on group size and anti-predator behavior in red colobus monkeys. *Anim. Behav.* 49: 577–587.

Stanford, C. B. 1999. *The Hunting Apes: Meat Eating and the Origins of Human Behavior*. Princeton: Princeton University Press.

Stearns, S. C. 1987. The evolution of sex and the difference it makes. In *The Evolution of Sex and Its Consequences*. S. C. Stearns. Basel: Birkhauser, 15–31.

———. 1992. *The Evolution of Life Histories*. New York: Oxford University Press.

Sterling, E. J. 1994. Evidence for nonseasonal reproduction in wild aye-ayes (*Daubentonia madagascariensis*). *Folia primatol* 62: 46–53.

Strassman, B. I. 1996a. Energy economy in the evolution of menstruation. *Evolutionary Anthropol.* 5: 157–64.

———. 1996b. The evolution of endometrial cycles and menstruation. *Quart. Rev. Biol.* 71: 181–220.

Strier, K. 1994. Brotherhoods among apelines: kinship, affiliation, and competition. *Behaviour* 130: 151–167.

Sugiyama, Y. 1965. On the social change of hanuman langurs in their natural condition. *Primates* 6: 381–418.

———. 1967. Social organization of Hanuman langurs. In *Social Communication among Primates*, ed. S. A. Altmann. Chicago: University of Chicago Press, 221–236.

———. 1995a. Drinking tools of wild chimpanzees at Bossou. *Amer. J. Primatol.* 37(3): 263–270.

———. 1995b. Tool-use for catching ants by chimpanzees at Bossou and Mont Nimba, West Africa. *Primates* 36(2): 193–206.

Sugiyama, Y., and Koman, J. 1979. Tool-using and tool-making behavior in wild chimpanzees at Bossou, Guinea. *Primates* 20: 513–524.

Sugiyama, Y., and Ohsawa, H. 1982. Population dynamics of Japanese monkeys with special reference to the effect of artificial feeding. *Folia primatol.* 39: 238–263.

Sussman, R. W., Cheverud, J. M., and Bartlett, T. Q. 1995. Infant killing as an evolutionary strategy: reality or myth? *Evol. Anthropol.* 3: 149–151.

Swain, A., Narvaez, V., Burgoyne, P., Camerino, G., and Lovell-Badge, R. 1988.

References

Dax1 antagonizes *Sry* action in mammalian sex determination. *Nature* 391: 761–777.

Swift, J. 1782. *Travels into Several Remote Nations of the World, by Lemuel Gulliver.* London: Harrison.

Takahata, H., and Takahata, Y. 1989. Inter-unit group transfer of an immature male of the common chimpanzee and his social interactions in the non-natal group. *African Study Monog.* 9: 209–220.

Takahata, Y., Ihobe, H., and Idani, G. 1996. Comparing copulations of chimpanzees and bonobos: do females exhibit proceptivity or receptivity? In *Great Ape Societies,* ed. W. C. McGrew, L. F. Marchant, and T. Nishida. Cambridge: Cambridge University Press, 146–155.

Takasaki, H. In press. Traditions of the Kyoto School of Field Primatology in Japan. *Changing Images of Primate Societies: The Role of Theory, Method and Gender,* ed. L. Fedigan and S. Strum. Chicago: Chicago University Press.

Tanner, J. M. 1962. *Growth at Adolescence.* Oxford: Blackwell.

Tattersall, I. 1993. *The Human Odyssey: Four Million Years of Human Evolution.* New York: Prentice Hall.

———. 1995. *The Fossil Trail: How We Know What We Think We Know about Human Evolution.* New York: Oxford University Press.

———. 1998a. The Laetoli diorama. *Scientific American.* September 1998: 53.

———. 1998b. The origin of the human capacity. *68th James Arthur Lecture on the Evolution of the Human Brain.* New York: American Museum of Natural History.

Temerlin, M. K. 1975. *Lucy: Growing up Human.* London: Souvenir Press.

Tennyson, A. L. 1850. *In Memoriam.* London: Edward Moxon.

Terrace, H. S. 1979. *Nim.* New York: Knopf.

Tinbergen, N. 1951. *The Study of Instinct.* Oxford: Oxford University Press.

Tishkoff, S. A., Dietzsch, E., Speed, W., Pakstis, A. J., Kidd, J. R., Cheung, K., Bonné-Tamir, B., Santachiara-Benerecetti, A. S., Moral, P., Krings, M., Pääbo, S., Watson, E., Risch, N., Jenkins, T., and Kidd, K. K. 1996. Global patterns of linkage disequilibrium at the CD4 locus and modern human origins. *Science* 271: 1380–1387.

Tomasello, M., Savage-Rumbaugh, E. S., and Kruger, A. C. 1993. Imitative learning of actions on objects by children, chimpanzees, and enculturated chimpanzees. *Child Devel.* 64: 688–705.

Tooby, J., and Cosmides, L. 1992. The psychological foundations of culture. In Barkow, J. H., Cosmides, L., and Tooby, J., 1992: 19–136.

References

Trevathan, W. 1987. *Human Birth: An Evolutionary Perspective.* New York: Aldine de Gruyter.

Trivers, R. L. 1971. The evolution of reciprocal altruism. *Quart. Rev. Biol.* 46: 35–57.

———. 1972. Parental investment and sexual selection. In *Sexual Selection and the Descent of Man,* ed. B. Campbell. Chicago: Aldine, 136–179.

———. 1974. Parent-offspring conflict. *Amer. Zool.* 14: 249–264.

———. 1985. *Social Evolution.* Menlo Park, CA: Benjamin Cummings.

———. 1997. Genetic basis of intrapsychic conflict. In *Uniting Psychology and Biology,* ed. N. Segal, G. Weisfeld, and C. Weisfeld. Washington, DC: American Psychological Association, 385–395.

Trivers, R. L., and Willard, D. E. 1973. Natural selection of parental ability to vary the sex ratio of offspring. *Science* 179: 90–92.

Tuchman, B. W. 1978. *A Distant Mirror: The Calamitous Fourteenth Century.* New York: Knopf.

Turner, B. L. I., Clark, W. C., Kates, R. W., Richards, J., Mathews, J. T., and Meyer, W. B., eds. 1990. *The Earth as Transformed by Human Action.* Cambridge: Cambridge University Press.

Tutin, C. E. G. 1979. Mating patterns and reproductive strategies in a community of wild chimpanzees. *Behav. Ecol., and Sociobiol.* 6: 29–38.

Tuttle, R. H. 1990. The pitted pattern of Laetoli feet. *Natural History* 1990(3): 60–65.

———. 1998. Global primatology in a new millennium. *Int. J. Primatol.* 19: 1–12.

Tuttle, R. H., Halgrimsson, B., and Stein, T. 1998. Heel, squat, stand, stride: function and evolution of hominid feet. *Primate Locomotion: Recent Advances,* ed. E. Strasser, J. Fleagle, A. Rosenberger, and H. McHenry. New York: Plenum Press.

Tuttle, R., Webb, D., Weidl, E., and Baksh, M. 1990. Further progress on the Laetoli trails. *J. Archeol. Sci.* 17: 347–362.

UNDP. 1997. *Human Development Report 1997.* New York: Oxford University Press.

———. 1998. *Human Development Report 1998.* New York: Oxford University Press.

UNICEF. 1991. *State of the World's Children 1991.* New York: Oxford University Press.

———. 1995. *State of the World's Children 1995.* New York: Oxford University Press.

———. 1997. *State of the World's Children 1997*. New York: Oxford University Press.

———. 1998. *State of the World's Children 1998*. New York: Oxford University Press.

United Nations. 1995. *The World's Women 1995: Trends and Statistics*. New York: United Nations.

van Schaik, C. 1995. *The Evolution of Cathemerality and Its Social Consequences*. Biology and Conservation of the Prosimians, Chester Zoo, UK.

———. 1996. Social evolution in primates: the role of ecological factors and male behavior. *Proc. British Academy* 88: 1–9.

van Schaik, C. P., and Dunbar, R. I. M. 1990. The evolution of monogamy in large primates: a new hypothesis and some crucial tests. *Behaviour* 115: 30–52.

van Schaik, C. P., and Hrdy, S. B. 1991. Intensity of local resource competition shapes the relationship between maternal rank and sex ratios at birth in cercopithecine primates. *Am. Nat.* 138: 1555–1562.

van Schaik, C. P., and Kappeler, P. M. 1993. Life history, activity period, and lemur social systems. In *Lemur Social Systems and their Ecological Basis*, ed. P. M. Kappeler and J. U. Ganzhorn. New York: Plenum, 241–260.

———. 1996. The social systems of gregarious lemurs: lack of convergence with Anthropoids due to evolutionary disequilibrium? *Ethology* 102: 915–941.

Van Valen, L. 1973. A new evolutionary law. *Evolutionary Theory* 1: 1–30.

Van Wagenen, G. 1992. Vital statistics from a breeding colony. Reproductive and pregnancy outcome in *Macaca mulatta*. *J. Med. Primatol.* 1: 3–28.

Vander Wall, S. B. 1982. An experimental analysis of cache recovery in Clark's nutcracker. *Anim. Behav.* 30: 84–94.

Vasey, P. L. 1995. Homosexual behavior in primates: a review of evidence and theory. *International Journal of Primatology* 16(2): 173–204.

———. 1998. Female choice and intersexual competition for female sexual partners in Japanese macaques. *Behaviour* 135: 597–597.

Vick, L. G., and Pereira, M. E. 1989. Episodic targeted aggression and the histories of Lemur social groups. *Behav. Ecol. Sociobiol.* 25: 3–12.

Vines, G. 1992. Last Olympics for the sex test? *New Scientist*, July 4, pp. 39–42.

Vrana, P. B., Guan, X.-J., Ingram, R. S., and Tilghman, S. M. 1998. Genomic imprinting is disrupted in interspecific *Peromyscus* hybrids. *Nature Genetics* 20: 362–365.

Vrba, E. S. 1988. Late Pleistocene climatic events and hominid evolution. In

Evolutionary History of the "Robust" Australopithecines, ed. F. E. Grines. New York: Aldine de Gruyter: 405–426.

Waddell, P. J., and Penny, D. 1996. Evolutionary trees of apes and humans from DNA sequences. In Lock, A., and Peters, C. R., 1996: 53–73.

Waley, A., trans. 1919. *One Hundred Seventy Chinese Poems.* New York: Alfred A. Knopf.

Wallis, J. 1995. Seasonal influence on reproduction in chimpanzees of Gombe National Park. *Int. J. Primatol.* 16(3): 435–452.

Wang, H., Paesen, G. C., Nuttall, P. A., and Barbour, A. G. 1998. Male ticks help their mates to feed. *Nature* 391: 763–764.

Ward, B., and Dubos, R. 1972. *Only One Earth: The Care and Maintenance of a Small Planet.* Harmondsworth: Penguin Books.

Warner, S. T. 1967/1989. *T. H. White.* Oxford: Oxford University Press.

Waser, P. M. 1978. Postreproductive survival and behavior in a free-ranging female mangabey. *Folia primatol.* 29: 142–160.

Washburn, S. L., and Lancaster, C. S. 1968. The evolution of hunting. In *Man the Hunter,* ed. R. B. Lee and I. DeVore. Chicago: Aldine, 293–303.

Wasser, S. K., and Norton, G. 1993. Baboons adjust secondary sex ratio in response to predictors of sex-specific offspring survival. *Behav. Ecol. Sociobiol.* 32: 273–281.

Wasserthal, L. T. 1998. Deep flowers for long tongues. *TREE* 13: 460–461.

Watts, D. P. 1985. Relationships between group size and composition and feeding competition in mountain gorilla groups. *Anim. Behav.* 33: 72–85.

———. 1989. Infanticide in mountain gorillas: new cases and a reconsideration of the evidence. *Ethology* 81: 1–18.

———. 1990. Ecology of gorillas and its relation to female transfer in mountain gorillas. *Int. J. Primatol.* 11: 21–46.

———. 1998. Coalitionary mate guarding by male chimpanzees at Ngogo, Kibale Park, Uganda. *Behav. Ecol. Sociobiol.* 44: 43–55.

Webster, G., and Goodwin, B. 1996. *Form and Transformation: Generative and Relational Principles in Biology.* Cambridge: Cambridge University Press.

Wedekind, C., and Furi, S. 1997. Body odour preferences in men and women: do they aim for specific MHC combinations or simply heterozygosity? *Proc. Roy. Soc. Lond. B* 264: 1471–1479.

Wedekind, C., Seebeck, T., Bettens, F., and Paepke, A. J. 1995. MHC-dependent mate preferences in humans. *Proc. Roy. Soc. Lond. B* 260: 245–249.

WEDO (Women's Environment and Development Organization). 1998. *Map-*

ping Progress: Assessing Implementaion of the Beijing Platform 1998. New York: WEDO.

Weismann, A. 1891. *Essays upon Heredity and Kindred Biological Problems.* Oxford: Oxford University Press.

Wells, H. G. 1895. *The Time Machine, An Invention.* London: W. Heinemann.

Werker, J. F. 1989. Becoming a native listener. *Amer. Sci.* 77: 54–59.

Westergaard, G. C., and Suomi, S. J. 1997. Modification of clay forms by tufted capuchins *(Cebus apella). Intl. J. Primatol.* 18: 455–468.

White, R. 1992. Beyond art: toward an understanding of the origins of material representation in Europe. *Ann. Rev. Anthropol.* 21: 537–564.

———. 1993a. The dawn of adornment. *Natural History* 5/93: 62–66.

———. 1993b. Technological and social dimensions of "Aurignacian-Age" body ornaments across Europe. In *Before Lascaux: The Complex Record of the Upper Paleolithic,* ed. H. Knecht, A. Pike-Tay, and R. White. Boca Raton: CRC, 277–299.

———. 1995. Les images feminines paleolithiques: Un coup d'oeil sur quelques perspectives americaines. *La Dame de Brassempouy, Actes du Colloque de Brassempouy (juillet 1994).* Liege, ERAUL. 74: 285–298.

White, T. D., Suwa, G., and Asfaw, B. 1994. *Australopithecus ramidus,* a new species of early hominid from Aramis, Ethiopia. *Nature* 371: 306–312.

White, T. H. 1938/1971. *The Sword in the Stone.* London: Collins.

Whiten, A. 1997. The Machiavellian mindreader. In Whiten, A., and Byrne, R. W., 1997: 144–173.

Whiten, A., and Byrne, R. W., eds. 1997. *Machiavellian Intelligence II: Extensions and Evaluations.* Cambridge: Cambridge University Press.

Whitman, W. 1855/1986. *Leaves of Grass.* New York: Penguin Books USA.

Wilford, J. N. 1998. Neanderthal or cretin? a debate over iodine. *New York Times,* F1, F4.

Wilkie, P. 1997. *Voyager.* Calstock, Cornwall: Peterloo Poets.

Williams, G. C. 1966. *Adaptation and Natural Selection.* Princeton: Princeton University Press.

———. 1975. *Sex and Evolution.* Princeton: Princeton University Press.

Williamson, S. 1998. The truth about women. *New Scientist,* August, pp. 34–35.

Wills, C. 1993. *The Runaway Brain.* New York: Basic Books.

Wilson, E. O. 1975. *Sociobiology.* Cambridge, MA: Harvard University Press.

———. 1984. *Biophilia.* Cambridge, MA: Harvard University Press.

———, ed. 1988. *Biodiversity.* Washington, DC, National Academy Press.

———. 1994. *Naturalist.* Washington, DC: Island Press.

———. 1998. *Consilience: The Unity of Knowledge.* New York: Alfred A. Knopf.

References

Wilson, J. D. 1994. Translating gonadal sex into phenotypic sex. In Short, R. V., and Balaban, E., 1994: 203–212.

Wimmer, H., and Perner, J. 1983. Beliefs about beliefs: representation and constraining function of wrong beliefs in young children's understanding of deception. *Cognition* 13: 103–128.

Wing, L. 1972. *Autistic Children.* Secaucus, NJ: The Citadel Press.

Wolfe, L. D., Gray, J. P., Robinson, J. G., Lieberman, L. S., and Peters, E. H. 1982. Models of human evolution. *Science* 217: 302.

Wolpoff, M. H. 1992. Multiregional evolution: the fossil alternative to Eden. In *The Human Revolution: Behavioral and Biological Perspectives on the Origin of Modern Humans,* ed. P. Mellars and C. Stringer. Princeton: Princeton University Press.

———. 1998. Concocting a divisive theory. *Evol. Anthropol.* 7: 1–3.

Wrangham, R. W. 1980. An ecological model of female-bonded groups. *Behaviour* 75: 262–300.

Wrangham, R. W., McGrew, W. C., de Waal, F. B. M., and Heltne, P. G. 1994. *Chimpanzee Cultures.* Cambridge, MA: Harvard University Press.

Wrangham, R., and Peterson, D. 1996. *Demonic Males.* Boston: Houghton Mifflin.

Wright, P. C. 1990. Patterns of paternal care in primates. *Int. J. Primatol.* 11: 89–102.

Wynn, T. G. 1996. The evolution of tools and symbolic behavior. In Lock, A., and Peters, C. R., 1996: 263–285.

Wynne-Edwards, V. C. 1962. *Animal Dispersion in Relation to Social Behavior.* Edinburgh: Oliver and Boyd.

Zahavi, A. 1975. Mate selection—a selection for a handicap. *J. of Theoretical Biology* 53: 205–214.

Zhao, Q.-K. 1997. Intergroup interactions in Tibetan macaques at Mt. Emei, China. *Amer. J. Phys. Anthrop.* 104: 459–470.

Zihlman, A. 1997. The paleolithic glass ceiling: women in human evolution. In Hager, L. D., 1997a: 91–113.

———. 1981. Women as shapers of the human adaptation. In *Woman the Gatherer,* ed. F. Dahlberg. New Haven: Yale University Press, 75–120.

Zihlman, A., Cronin, J. E., Cramer, D. L., and Sarich, V. M. 1978. Pygmy chimpanzee as a possible prototype for the common ancestor of humans, chimpanzees, and gorillas. *Nature* 275: 744–746.

INDEX

Index

Chemotron, 31, 32

Cheney, D. L., 289

Childbirth, 3, 4, 48, 92, 262, 263, 322, 324–327, 329, 331, 344–350

Childcare, 190, 191, 192, 196, 197, 325, 362, 382; cooperative, 350, 363, 367

Children, 116, 118, 122, 140, 196, 402; number of, 69, 118–120, 274, 426; parental investment in, 69, 70–71, 111, 116; reproductive value of, 101–102; abandonment of, 116, 117, 216, 305; as allies, 116, 346; sex ratio of, 120–123, 121, 122–123, 305; dependency of, 190, 192; first year, 333–340; first to fifth years, 340–343; sexual interest of, 342, 343–344; fifth year to puberty, 343–345; duration of childhood, 350, 366, 367, 368; protection of, 362; health and diseases, 421, 422–423; death rates, 426. *See also* Offspring

Chimpanzee Politics (De Waal), 207

Chimpanzee(s), 4, 9, 55, 145, 159, 161, 302, 406, 410, 430, 432; sexual swelling/display, 2, 174, 184, 193; intelligence, 3, 77, 207, 302; relationships, 10, 176, 212; deceptive behavior, 11, 67, 208–209, 286; competition, 67, 75–77, 180, 182; alpha males, 75–77, 78, 174, 175; warfare, 75–77, 173–174, 176, 178, 194, 290, 360, 364; male-bonded groups, 76, 172–175, 178, 182, 193, 364–365, 382; consciousness, 77; female migration, 77, 116, 177, 192, 193, 373; learning capacity, 77, 299; mating and mate choice, 78–79, 81, 174–175; testes, 90; estrous females, 93, 185, 253; as human ancestor, 109, 139, 367–368; individualism, 147; food supply and sharing, 149, 176, 202, 204–205, 302, 361; predators, 149; common, 156, 212, 364; infanticide, 178; societies, 178–179, 180, 193, 264; sperm competition, 182; sexual activity, 184; births and children, 190–191, 323, 326–327; reproduction cycles, 195; tools and toolmaking, 200–201, 202, 218, 302; hunting, 203–205, 365; male politics, 207–208, 209; cooperation, 212, 298; imagination, 288–289, 290–292; play and toys, 290–292, 300; signing and language, 294, 298, 313–315; intent, 298–299; brain, 323, 367; lineage, 359; size ratio, 364; growth rate, 366; grooming, 382. *See also* Bonobo(s)

Chlamydia, 50

Chloroplasts, 40, 44

Cholera, 23, 67

Chromosome(s), 17, 28, 49, 251, 270; X and Y, 33, 58, 133, 234–235, 239–240, 241–242, 248, 270–271, 371–372; in cell division, 36, 42, 49, 234; hybrid, 36; ring, 39; anomalies, 48, 59–60; reassortment, 57; number, 59; genes and, 63, 236; extra/added, 417

Claustration, 304–306

Clones and cloning, 2, 4, 17, 19, 35, 38, 44, 49–50, 64, 370, 390, 401–402, 431; as gene copying, 50–52; of identical offspring, 53–54; advantages of, 54, 56; information storage in, 58

Color and brilliance, 157, 158; for courtship, 83–85, 87, 89; emotions and, 265; innate behavior and, 266

Coming of Age in Samoa (Mead), 345

Communication, 210, 408–409, 431–432; global, 411–413, 414

Communities, 22, 28, 92, 124, 229; cellular, 38–42; evolution and, 103; tribal, 123–125. *See also* Cell(s): as community of one; Societies/social groups

Competition, 3, 5, 21–22, 54, 256, 402, 409, 410; food, 9, 194, 363; direct, 21–22; environmental, 22; individual, 22, 23, 108; sexual, 43–44, 50, 71, 176, 256; reproduction rates and, 53; sibling, 54; species, 54–55; scramble, 74–75, 256; contest, 75–78; evolution of, 99; peer, 107; between-group, 108; genetic, 110; intergroup, 124–125, 176; female, 166, 171, 172–173, 186, 194; male, 171, 176, 182, 200; territorial, 194; global, 419. *See also* specific animals and organisms

Complexity, 12, 16, 22, 24, 26, 225, 229, 230–233, 256, 270, 356, 407, 431; theory, 28–29, 284, 418; environmental, 205–206, 429; genetic, 267–274, 277; brain, 281–282, 356

Computers, 23, 220, 282, 316, 411–413, 432

Index

Dobson, Jerome, 375

Dogs, 268; wolves, 204–205

Dominance rank, 123, 213, 320; male-female, 1, 165–166, 169–171, 172, 184, 186, 193, 359

Down syndrome, 49, 59, 269

Dowries, 120, 304, 305

Dreams, 282, 283, 287–288

Dunbar, Robin, 216, 381–382

Dyson, George, 411

Economics, 6, 22, 367, 390, 393, 413, 419–421, 429, 431

Education, 425, 426

Egg(s), 34, 35, 45–48, 64, 235, 401; size, 34, 43–44, 52, 70; DNA/genes in, 35, 44, 229; division of, 42, 49; fertilized, 43, 48, 51, 58, 64, 83, 104, 226, 234, 267, 347; nourishment for, 43–44, 49, 69, 71, 229; one-celled, 45; /sperm bond, 46; number of, 48, 69, 71; fertile, 57; implantation of, 59; anomalies in, 59–60; female investment in, 69–70; production, 69–70; competition, 89–91; release, 93; survival rate, 226; nucleus, 229; cells, 236; sex ratio of, 305

Einstein, Albert, 145

Emergence, 4, 227–230, 407

Emotions, 26, 140, 231, 260, 265, 282, 343; bonding with, 65, 179; during menstruation and menopause, 251, 252–253, 254; language and, 312, 321; puberty and, 345

Energy, 30, 41, 64, 69, 120, 252, 414, 432, 434; sources, 30, 31, 39–41, 43, 44, 45, 58, 367, 414; metabolic, 64, 367; for mating and sex, 75–76, 82–84, 86, 150, 174; for gaining resources, 150, 203; for food finding, 202–203

Engels, Friedrich, 168

Environment, 9, 126, 131, 132, 389, 419, 421; control of, 5, 390–398, 403–404, 420–421, 427–429, 432; mental, 11, 34, 227; adaptation to, 19, 21, 29, 30–31, 32, 37, 53; genes and, 19, 256, 260, 277; threats of, 22, 57, 120, 285–286; complexity of, 53–54, 205–206, 207; changing, 54, 61, 66, 87, 207, 278; niches, 54, 61,

66, 128, 130, 352, 371; fixed sites, 106; development and, 226–227; sex determination and, 234, 243, 244, 246, 254–255

Enzymes, 36

Estradiol, 238, 244

Estrogen, 238, 241–242, 243, 252, 253, 254

Estrus, 78, 80, 92, 93, 365; concealed, 92, 93, 94, 184–186; simulated, 114; females, 162, 172, 173, 174, 204; mating during, 184. *See also* Ovulation; *specific animals*

Eugenics, 53, 60–61, 125

Eve, *see* Foremother, human

Eukaryotes, 39

Everdell, William, 230

Evolution, 3, 4, 14–15, 22, 54, 63, 110, 126, 256, 359, 406, 408, 410, 413; transitions in, 2, 4, 28, 431; human, 6, 185, 364, 371–372; mechanism of, 12, 25, 225; chemical, 32; equilibrium, 44; reproduction rates and, 55; to commonest life form, 56; innovative, 63; alternative strategies, 87–88; play in the system (by-products), 127, 128, 129, 169; continuity of, 129, 152, 155; role of chance in, 130–131; at population level, 131; past, 132; future, 147, 152; within and between species, 264; of ritualization, 265; of social signals, 266–267; hero and heroine stories, 351, 352, 360–363; history of human status, 351–352; narrative structure of, 351–352; speed of, 372; principles of, 405, 406; biological, 409; information transfer in, 413; purpose and, 431–434. *See also* Natural selection; *specific animals and organisms*

Extinction, 5, 11, 12, 16, 21, 22, 40, 409

Eyes, 24–25, 281, 288, 334

Falk, Dean, 355

Family, 5, 66, 117, 124, 180; cooperation, 100, 151; gene sharing in, 104; extended, 124, 177, 189, 191; supportive, 136; size and structure, 155, 183, 189, 363; power shifts in, 207

Famine, 13, 16, 21, 428

Fashion, 97–98, 98

Fat, body, 92–93, 94, 344

Index

Klein bottle, 280, 407
Knowledge, 197, 352, 354, 385. *See also* Information
Koestler, Arthur, 29, 229
Kohler, Wolfgang, 292
Komdeur, Jan, 120
Koshima Island, Japan, 170
K selection, 119
Kummer, Hans, 165, 187–188, 213
!Kung San tribe, 191–192, 205, 326, 344, 362, 386
Kwakiutl people, 387
Kyoto School, 147

Labor, 377, 389, 393–396, 402; division of, 32, 204, 361, 413, 415, 416
Lactase, 277
Lactation, 69, 70, 114, 164, 182, 183, 187, 252, 330, 331
Lactoferrin, 329
Lactose, 276–277
Lamarck, Chevalier de, 12, 13, 19–20
Land, 391–392, 393, 396, 403, 427, 428
Language, 2, 4, 62, 286, 294, 339, 379–383, 402, 416; learning/acquisition, 2, 311–313, 341, 354, 379; speech (phonemes), 10, 309–310, 311, 354, 366, 432; complexity, 282–283; as communication, 297; as discrete system, 309, 311, 312; grammar and vocabulary, 310, 312, 315, 317, 341, 375, 379; word meanings (morphemes), 310, 319–321; evolution, 320, 380, 381; capacity/capability, 375, 376, 380–381, 432; symbolic, 375, 376, 380; dialect, 377; sexual, 382; social, 382. *See also* Apes: signing and language
Langurs of Abu, The (Hrdy), 114
Lascaux, France, cave art, 378–379
Laughter, 263, 264
Leakey, Louis, 145, 201, 367
Leakey, Maeve, 367
Leakey, Mary, 356, 367
Leakey, Richard, 367
Learning, 64–67, 278, 299, 312, 343, 344; capacity, 62, 379; from others, 62, 65, 67, 311, 347; flexible, 66, 260, 262, 278; instinct and, 258, 261, 262, 264, 346–347; visual copying, 264, 276; genes and, 276–

277; channeled, 278; trial and error, 278, 299; of conscious actions, 285; by imitation, 299–301, 302, 303; social, 302, 317. *See also* Teaching
Lemurs, 116, 156, 211, 212–217, 406; intelligence, 3, 54; sifaka, 26, 27, 154, 156; pygmy mouse, 129, 156; extinct, 156, 157; indri, 156, 163; sloth, 156; aye-ayes, 156–157, 162, 202; infanticide, 157, 184; slow loris, 157; monogamy, 162; deceptive behavior, 208; societies, 212–213, 220; targeted aggression, 213–214. *See also* Prosimians
Lemurs, ring-tailed, 1; scent-marking, 1, 2, 8, 289; violence and warfare, 75, 213–214; sex competition, 75–78, 140; infanticide, 140, 216–217; female-bonded groups, 168; mate choice, 171–172; food supply and diet, 206; societies, 212–213, 214; bipedalism of, 360
Leptin, 93, 344
Lessing, Doris, 71
Lévi-Strauss, Claude, 177
Lewanika, Inonge M., 433
Lewontin, Richard, 125
Lies, 335–336
Life histories, 119
Life span, 195–197, 218, 398, 399; sex and, 53, 54, 55–56, 119
Linneaus, Carolus, 70
Lorenz, Konrad, 65, 75, 276, 333
Love, 2, 3, 4, 23, 106, 110, 143, 177–178, 186, 189, 258, 347, 368, 396, 402, 434
Lovejoy, Owen, 361, 362–363, 364
Lovelock, James, 17
Lucy (African australopithecine), 2, 5, 175, 323, 324, 357–360, 360
Lysenko, Trofim Denisovich, 20–21

Macaques: Japanese, 116, 168, 170, 195, 215, 302; toque, 122, 158; Barbary, 158, 289–290; crab-eating, 158; pigtailed, 158; rhesus, 158, 168, 211, 213; Tibetan, 158; group structure, 169; mate choice, 171; sexual behavior, 171, 211–212; societies, 211–212, 213; stumptailed, 211–212; violence in, 213; infant carrying, 289–290; social status, 289–290; cooperation, 298